D0947703

BARRON'S NEW YORK STATE

GRADE 8 SCIENCE TEST

THIRD EDITION

Edward J. Denecke, Jr., M.A.
Science Teacher
Formerly, William H. Carr JHS 194Q
Whitestone, New York

Acknowledgments

I wish to thank the following sources for their kind permission to use many of the illustrations that appear in this book:

Figure 1.6: The Visual Display of Quantitative Information (E.R. Tufte; Graphics Press).
Figure 3.3: Trippensee Planetarium Co. Photo courtesy of Science First.
Figure 3.5: NOAA GeoSpatial Data Information and Climate Services.
Figure 6.1: United States Department of Agriculture.
Figures 6.3, 9.9(c): Smith, Robert Leo, and Thomas M. Smith, *Elements of Ecology*. (Reading, MA: Addison Wesley Longman, Inc., 2000).
Figures 7.10(b), 7.13, 9.7: Jantzen, Paul G., and Judith L. Michel, *Macmillan Life Science*. (New York: Macmillan Publishing Co., 1986).
Figure 8.2: "Figure 17.10" illustrated by Paula Di Santo Bensadoun, from *Biological Science, Second Edition* by William T. Keeton. Copyright © 1972, 1967 by W.W. Norton & Company, Inc. Used by permission of W.W. Norton & Company, Inc.
Figure 8.3: From the experiments of Dr. H.B.D. Kettlewell, Oxford University.
Figure 8.4: Zumberge, James H., *Elements of Geology*. Copyright © 1958, John Wiley & Sons, Inc. Reprinted by permission of John Wiley & Sons, Inc.
Figure 10.1(a): Photo by William K. Hartmann.
Figure 10.1(b): Chapman, Robert D., *Discovering Astronomy*. (New York: W. H. Freeman, 1978).
Figure 10.5(a): Moché, Dinah L., *Astronomy: A Self-Teaching Guide*, 4th Ed. Copyright © 1993, John Wiley & Sons, Inc. Reprinted by permission of John Wiley & Sons, Inc.
Figures 10.5(b), 10.20: From Seeds. *Horizons: Exploring the Universe*, 2nd Ed. © 1987, Brooks/Cole, a part of Cengage Learning, Inc. Reproduced by permission. www.cengage.com/permissions
Figures 10.14 (b), 12.19: Earth Science Curriculum Project, *Investigating the Earth. Part I: Teacher's Guide*. Sponsored by the American Geological Institute and the National Science Foundation. Copyright © 1967, American Geological Institute.
Figure 11.2: Bridges, E. M., *World Geomorphology*. (New York: Cambridge University Press, 1990). Reprinted with the permission of Cambridge University Press.
Figures 11.4, 11.5, 11.8, 11.9, 11.19, 12.7, 12.9, 14.2(b), Table 10.1: The State Education Department, *Earth Science Reference Tables—2010 Edition* (Albany, New York: The University of the State of New York).
Figure 11.10: Thomas McGuire.
Figure 11.16: Courtesy of the New York State Museum, Albany, NY: *Educational Leaflet #28*.
Figure 12.17: From *Earth Science on File*. Copyright © 2004 by Facts on File, Inc., an imprint of Infobase Publishing.
Figure 13.5: The State Education Department, *Chemistry Reference Tables—2010 Edition* (Albany, New York: The University of the State of New York).
Figure 14.4: Beiser, Arthur, *Physics*, 2nd Ed. © 1978, p. 365. Reprinted by permission of Pearson Education, Inc., Upper Saddle River, NJ.
Figure 15.5(b): http://www.physicsclassroom.com/Class/newtlaws/u211d.cfm#balanced

All inquiries should be addressed to:
Barron's Educational Series, Inc.
250 Wireless Boulevard
Hauppauge, New York 11788
www.barronseduc.com

ISBN: 978-0-7641-4621-3
International Standard Serial No.: 1932-3093

Printed in the United States of America
9 8 7 6 5 4 3 2 1

Contents

Preface

TO THE STUDENT

This book has been written to help you understand and review intermediate-level (grades 5, 6, 7, and 8) science in preparation for the New York State Grade 8 Intermediate-Level Science Test. Note that although you will take this test at the end of grade 8, it covers four years of study in science. Preparing for such a comprehensive test may seem like an immense task. However, if you approach it in a logical, step-by-step fashion, you can accomplish it.

The *New York State Grade 8 Science Test* includes a general review of concepts in life, earth, and physical sciences. It also includes process-oriented skills along with practice exams with answers and explanations.

The text presents the major ideas of each topic in a clear, easy-to-understand manner. The numerous illustrations are designed to help you understand the concepts presented. The answer sections include explanations for all multiple-choice, constructed-response, and extended constructed-response questions. To help you prepare and to become familiar with these types of questions, general test-taking advice and model questions with answers are provided throughout the book.

The contents of this book generally follow the New York State *Learning Standards in Mathematics, Science, and Technology*. They are divided into three units: Analysis, Inquiry, and Design; The Living Environment; and The Physical Setting.

Analysis, inquiry, and design are the abilities needed to *do* science. They include mathematical skills, such as constructing and interpreting graphs; science inquiry skills, such as stating a hypothesis or designing a controlled experiment; and engineering design skills, such as planning and constructing a model of a technological solution to a problem.

This test will assess your ability to explain, analyze, and interpret scientific processes and phenomena more than your ability to recall specific facts. Therefore, the focus of this book is on understanding important relationships, processes, and mechanisms and on applying concepts and process skills, rather than on the exhaustive coverage of content. We hope that students will find this approach an effective way to review the four years of study leading up to this rigorous test.

WHAT THE EXAM IS ABOUT

You are completing the 8th grade, and you are looking forward to entering high school. However, before the end of your intermediate-school experience, your science teacher tells you that every 8th grader in New York State must take the Grade 8 Intermediate-Level Science Test.

The test, like the 4th grade science test that you took in elementary school, will contain three types of questions. Some are multiple-choice questions. Others, constructed-response questions, require you to supply an answer rather than choose one from a list. The third type consists of several laboratory performance tasks that you must perform. Do not panic. If you have been doing well on your classroom tests all year and you devote some time to review and preparation, you will have no trouble doing well on this test.

Why do I say this? The New York State Education Department has given your teachers helpful guidelines. Throughout your elementary and intermediate school years, you have been taught specific skills and concepts relating to science. These skills and concepts are called standards. In the 1990s, the New York State Education Department published seven learning standards for mathematics, science, and technology. Five of these major standards are emphasized in science.

*STANDARD 1: ANALYSIS, INQUIRY, AND DESIGN

Students will use mathematical analysis, scientific inquiry, and engineering design, as appropriate, to pose questions, seek answers, and develop solutions.

*STANDARD 2: INFORMATION SYSTEMS

Students will access, generate, process, and transfer information using appropriate technologies.

*STANDARD 4: SCIENCE

Students will understand and apply scientific concepts, principles, and theories pertaining to the physical setting and living environment and recognize the historical development of ideas in science.

*STANDARD 6: INTERCONNECTEDNESS: COMMON THEMES

Students will understand the relationships and common themes that connect mathematics, science, and technology and apply the themes to these and other areas of learning.

*STANDARD 7: INTERDISCIPLINARY PROBLEM SOLVING

Students will apply the knowledge and thinking skills of mathematics, science, and technology to address real-life problems and make informed decisions.

Intermediate-level (grades 5–8) science courses are designed to help you master Standards 1, 2, 4, 6, and 7. Standard 4 is divided into two main parts: The Living Environment, which deals with life science concepts, and The Physical Setting, which deals with concepts in the physical sciences—earth science, chemistry, and physics. Standards 1, 2, 6, and 7 describe process skills common to mathematics, science, and technology.

So even before you review, you should know a great deal of the material that will be on the test. For example, you already know that individual organisms and species change over time. You have been taught the major human organ systems and their functions. You have been shown how the motions of objects in the solar system explain phenomena such as the day, the year, phases of the moon, and eclipses. You understand that energy exists in many forms, and that when these forms change, energy is conserved. You also understand how the motion of particles helps to explain the phases of matter as well as the characteristics of solids, liquids, and gases.

THE NATURE OF THE GRADE 8 INTERMEDIATE-LEVEL SCIENCE TEST

Let us take a look at the specifics of the 8th grade test. The test is administered in two sessions: performance test given in late May or early June and written test given in June.

THE PERFORMANCE TEST (FORM A)

The Performance Test, or Form A, consists of three tasks and represents 15% of the test. Each task will assess your mastery of laboratory and process skills based on Standard 4. Laboratory work typically represents at least 15% of your classroom time. Table P.1 lists examples of general process skills, living environment skills, and physical setting skills. Each task will take about 15 minutes to complete and may involve one or more of these skills. For example, you may be asked to view a microscopic object *and* determine its size. You might be asked to determine the volume of a regular-shaped solid by using water displacement, *and* use a balance to measure its mass, *and* determine its density.

The best preparations for this part of the exam are the laboratory activities that you do as a normal part of your classroom work. For example, during your study of life science, you learned how to operate the microscope. You probably used it many times as you studied plant and animal cells and tissue. During your study of physical science, you probably used the triple-beam balance many times to measure mass during laboratory activities.

Although you may not practice the actual tasks used for this part of the exam, you can look at a test sampler with three performance task stations that are similar to those on the actual test. This sampler is available at *http://www.p12.nysed.gov/ciai/mst/pub/2interscisam.pdf* on the New York State Education Department website. In the sampler, the performance test is called Part D, but it refers to what is now Form A. These sample tasks will provide you with a good idea of what to expect on this part of the exam.

Table P.1 **Process Skills**

General Skills

1. Follow safety procedures in the classroom and laboratory.
2. Safely and accurately use the following measurement tools:

❑ metric ruler	❑ stopwatch	❑ thermometer	❑ voltmeter
❑ balance	❑ graduated cylinder	❑ spring scale	

3. Use appropriate units for measured or calculated values.
4. Recognize and analyze patterns and trends.
5. Classify objects according to an established scheme and a student-generated scheme.
6. Develop and use a dichotomous key.
7. Sequence events.
8. Identify cause-and-effect relationships.
9. Use indicators and interpret results.

Living Environment Skills

1. Manipulate a compound microscope to view microscopic objects.
2. Determine the size of a microscopic object, using a compound microscope.
3. Prepare a wet-mount slide.
4. Use appropriate staining techniques.
5. Design and use a Punnett square or a pedigree chart to predict the probability of certain traits.
6. Classify living things according to a student-generated scheme and an established scheme.
7. Interpret and/or illustrate the energy flow in a food chain, energy pyramid, or food web.
8. Identify pulse points and pulse rates.
9. Identify structure and function relationships in organisms.

Physical Setting Skills

1. If given the latitude and longitude of a location, indicate its position on a map, and determine the latitude and longitude of a given location on a map.
2. By using identification tests and a flow chart, identify mineral samples.
3. Use a diagram of the rock cycle to determine geologic processes that led to the formation of a specific rock type.
4. Plot the location of recent earthquake and volcanic activity on a map, and identify patterns of distribution.
5. Use a magnetic compass to find cardinal directions.
6. Measure the angular elevation of an object, using appropriate instruments.
7. Generate and interpret field maps including topographic and weather maps.
8. Predict the characteristics of an air mass based on the origin of the air mass.
9. Measure weather variables such as wind speed and direction, relative humidity, barometric pressure, and so on.
10. Determine the density of liquids and both regular- and irregular-shaped solids.
11. Determine the volume of a regular- and an irregular-shaped solid, using water displacement.
12. By using the periodic table, identify an element as a metal, nonmetal, or noble gas.
13. Determine the identity of an unknown element, using physical and chemical properties.
14. By using appropriate resources, separate the parts of a mixture.
15. Determine the electrical conductivity of a material, using a simple circuit.
16. Determine the speed and acceleration of a moving object.

THE WRITTEN TEST (PARTS I AND II)

The written test, administered in one 2-hour session, consists of two parts as described below. The exact number of questions in each part of the written test may vary from year to year but will remain within the given ranges.

Part I (45 points)—multiple-choice questions that test your knowledge of life, earth, and physical science concepts from Standard 4.

EXAMPLE:

SCIENCE CONCEPTS FROM STANDARD 4:

Genes are composed of DNA that makes up the chromosomes of cells. In sexual reproduction, typically half of the genes come from each parent.

PART I QUESTION:

A male chimpanzee has 48 chromosomes in each of his regular body cells. How many chromosomes would be found in each of his sperm cells?
1. 96 2. 48 3. 24 4. 12

Part II (40 points)—constructed-response questions that test your ability to use science concepts from Standard 4 and skills from Standard 1 to apply, analyze, and evaluate material. In other words, answering Part II questions requires both science knowledge and skills.

The Part II "constructed-response" questions shown below require *you* to supply an answer by applying both science knowledge and skills. Your answer may be a single word or sentence, as in the first question, or it may require a paragraph or two in which you form and defend a logical argument, as in the second question.

EXAMPLE:

SCIENCE CONCEPTS FROM STANDARD 4:

Food webs identify feeding relationships among producers, consumers, and decomposers in an ecosystem.

PROCESS SKILLS FROM STANDARDS 1, 2, 6, AND 7:

Interpret and/or illustrate the energy flow in a food chain, energy pyramid, or food web.

Form and defend a logical argument about cause-and-effect relationships.

PART II CONSTRUCTED-RESPONSE QUESTION:

Base your answers on the food web shown below.

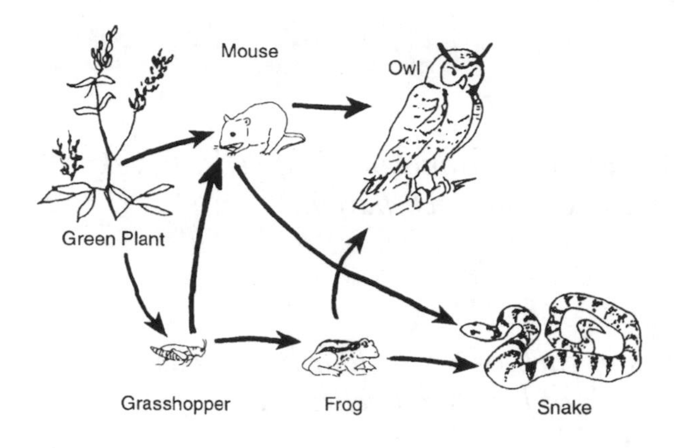

1. *Identify an herbivore in this food web.*
2. *Explain why removing the snake from this food web might result in a decrease in the grasshopper population.*

Some Part II constructed-response questions test your ability to use knowledge of science concepts and skills to address real-world situations. Such questions are called "extended" constructed-response questions because they are more complex—requiring the application of a wider range of science knowledge and/or process skills. You may, for example, need to perform numerical calculations, draw or interpret a graph, draw a diagram, or provide an extended written response to a question or problem.

EXAMPLE:

SCIENCE CONCEPT FROM STANDARD 4:

Energy can be considered to be either kinetic energy, which is the energy of motion, or potential energy, which depends on relative position.

An object's motion is the result of the combined effect of all forces acting on the object.

Friction is a force that opposes motion.

PROCESS SKILLS FROM STANDARDS 1, 2, 6, AND 7:

Refine and clarify questions so they are subject to scientific investigation.
Design a simple controlled experiment.
Independently formulate a hypothesis.
Identify independent variables, dependent variables, and constants in a simple controlled experiment.

PART II EXTENDED CONSTRUCTED-RESPONSE QUESTION:

A student plays tennis several times a week. She notices that the tennis ball seems to bounce higher on some courts than on other courts. She wonders if this has something to do with the surface of the court. Design an experiment to see if her hypothesis is correct. Include these elements in your response:

State the hypothesis.
Identify the factor to be varied.
Identify two factors that should be held constant.
Clearly describe the procedures.

To help you prepare and to become familiar with all these types of questions, general test-taking advice and model questions with answers are provided throughout the book. The last part of this book has full practice exams so that you can test yourself and identify any areas of weakness. If you read this book carefully and answer all the review questions, you will have a pretty decent understanding of what will be expected of you on this test.

Although the New York State Education Department does not permit anyone to publish past Grade 8 science tests, copies of past tests and answer keys are available online at *http://www.nysedregents.org/Grade8/Science/home.html.*

NOTE TO TEACHERS

The Answer Key and Model Examination and Answers are on pages that can be cut along dotted lines for easy removal.

Unit 1

Analysis, Inquiry, and Design

Standard 1: Students will use mathematical analysis, scientific inquiry, and engineering design, as appropriate, to pose questions, seek answers, and develop solutions.

Why do leaves change color in the fall? What is a rainbow? How did the Grand Canyon form? The search for answers to questions like these is what science is all about. Science is a great adventure, a never-ending search for understanding that people everywhere can take part in, and have for many centuries. The goal of science is to figure out how the world works. HOW we go about figuring out how the world works, though, is just as important as WHAT we figure out. Therefore, science has a dual nature.

Science is often defined as "a body of knowledge based on observation and experimentation." This definition pinpoints the dual nature of science: it is not just *what* we know but *how* we know it. Scientific knowledge is knowledge arrived at in a particular way—by observation and experimentation!

In this unit, you will learn about the process of doing science. Doing science involves posing questions, seeking answers, and developing solutions. Science has been so successful because it unites the fields of science, mathematics, and technology in the endeavor to understand how the world works. The chapters in this unit will emphasize the ways in which science, mathematics, and technology depend on and reinforce one another. In later units, you will learn about what scientists have figured out about how the world works.

Chapter 1

Mathematical Analysis

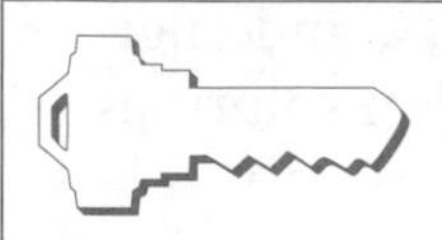

Key Idea: Abstraction and symbolic representation are used to communicate mathematically.

THE LANGUAGE OF MATHEMATICS

As you study science, you will have lots of ideas. You will have ideas about what objects are, why things happen, or what something means. Some of these ideas will involve ***abstraction***, or considering something apart from a particular object or event. Take, for example, the idea of a predator. When you think of a predator, lots of examples may pop into your mind, such as a lion, a fox, a hawk, or a snake. The idea of a predator is an abstraction. It is not tied to any one animal. Rather, it is the idea of an animal that hunts and eats other animals.

The Nobel Prize–winning physicist Richard Feynman once said, "Nature talks to us in the language of mathematics." In the language of mathematics, symbols are used to stand for ideas and the ways in which ideas are related to one another. Mathematics is a way of communicating the ideas that we have when we try to solve problems in science, technology, and everyday life.

Symbols are just things that stand for other things. Symbols can be objects, markings such as letters or numbers, or even sounds. For example, candles on a birthday cake stand for years. The number "3" can stand for how many birds you saw in a nest. The long and slow clicks "•••———•••" can stand for the letters "SOS," which, in turn, stand for the distress call "save our ship." Symbols can also stand for relationships between things—"10 > 5" stands for "10 is greater than 5." Using symbols to stand for things and how they are related to one another is called ***symbolic representation.***

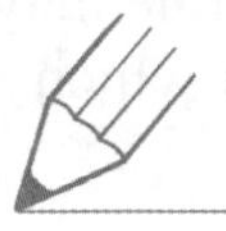

TRY THIS

In science, the language of mathematics is used to express ideas. Try writing the following science ideas mathematically using symbolic representation.

1. The number of birds in a population equals the number of females plus the number of males.

   ____

2. The density of a rock is the ratio of its mass to its volume.

   ____

3. You learn that for every 1,000 meters you rise in the atmosphere, the temperature drops by 6°C. You have the idea that if you know the temperature at the base of a mountain and its height, you can figure out the temperature at the top of the mountain.

   ____

4. You observe that one fox eats 100 rabbits each year. You have the idea that every increase of one in the fox population will cause the rabbit population to decrease by 100.

   ____

VARIABLES

When doing science, you often find that changing one thing causes changes in other things. Something that can change or be changed is called a ***variable***. For example, suppose you made a paper airplane and it did not fly very well. You might decide to change something about the airplane to make it fly better. Maybe you refolded it to change the shape of the wing or tried to throw it more gently. Then, when you flew it again, you noticed that the plane flew differently. Perhaps it flew longer, or straighter, or farther. All of these factors—wing shape, weight, flight time, flight distance, and straight travel—are variables. However, some factors remained the same, such as the size of the paper, the texture of the paper, the color of the paper, and the weight of the paper. Factors that do not change are called ***constants***.

Notice that in the example just described, you may have purposefully changed some variables, and other variables may have responded to the changes you made. If you purposefully changed the wing shape, perhaps the plane flew longer. In any investigation, a variable that is purposefully changed, or manipulated, is called an ***independent variable***. The variable that changes in response is called the ***dependent variable***. You may find it helpful to think of the independent variable as the "you changed it variable" and the dependent variable as the "it changed variable."

When you purposefully change an independent variable, the dependent variable may respond in a number of ways, or it may not respond at all. Some of the ways in which variables can be related to one another is summarized in Table 1.1.

Table 1.1 **Types of Relationships Between Variables**

Relationship	Purposeful Change to Independent Variable	How the Dependent Variable Responds
Direct	Increase	Increases
Inverse	Increase	Decreases
Cyclic	Increase	Increases and decreases in a repeating pattern
Nonrelated	Increase	Not at all or randomly

Suppose you had a contest with some of your friends to see whose paper airplane could fly the longest. Would throwing each plane just once be the fairest way to decide the winner? If you have ever flown a paper airplane, you have probably found that even if you try not to change anything, it rarely flies exactly the same way each time you throw it. This is because chance errors and unknown variables can affect the way an airplane flies. To be fair, you would probably want to do several airplane flights and average the results. Testing an idea several times is called doing ***repeated trials.*** Repeated trials reduce the effects of chance errors or unknown variables on the overall results.

Since errors and unknown variables can affect the way an airplane flies, how can you be sure that the change you purposefully made was really what affected its flight? One way would be to fly an unchanged airplane as part of each trial to see if its flight remained the same. Then, if the changed airplane flew differently but the unchanged airplane flew the same, you could be surer that the change you purposefully made affected the airplane's flight. A parallel experiment, in which no variables have been purposefully changed, is called a ***control.*** Comparing results of repeated trials with a control allows you to be surer that the change in the dependent variable is really the result of the change you purposefully made in the independent variable.

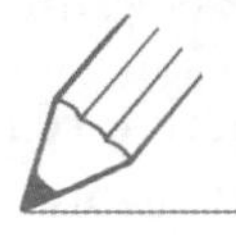

TRY THIS

In the following scenario, identify the independent variable and the dependent variable. Then list as many factors as you can think of that should be held constant.

Ben heard that some paper towels soaked up water faster than others. He decided to determine which brand soaked up water the fastest. He cut identical strips of towel from each of 5 different brands. Next, he filled five jars half-full with water and suspended a strip of a different brand in each jar so that the end of the strip was 5 cm below the water surface. Then he measured the height the water rose in each paper towel strip every ten seconds for one minute.

1. Independent variable: ______________________________
2. Dependent variable: ______________________________
3. Constants: ______________________________

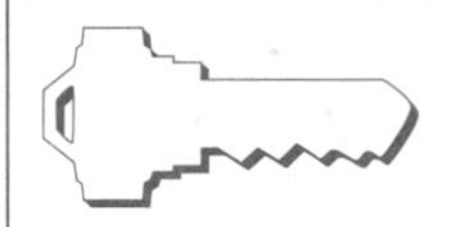

Key Idea: Deductive and inductive reasoning are used to reach mathematical conclusions.

We mentioned earlier that in science, *how* we figure out the way the world works is just as important as *what* we figure out. Scientific thinking is based on using logic to figure out things. Logical reasoning means having a valid reason for forming a conclusion. Two forms of logical reasoning are *inductive* and *deductive* reasoning.

INDUCTIVE AND DEDUCTIVE REASONING

Inductive reasoning means reasoning from specific observations and experiments to more general hypotheses and theories. In other words, this means going from the specific to the general.

Suppose you observed that fertilizing your lawn made it grow faster. If you then proposed that fertilizing plants would make them grow faster, you would be reasoning inductively. From your observations of the effect of fertilizer on one *specific* plant, the grass in your lawn, you reasoned that fertilizer would have the same effect on plants in *general.*

Deductive reasoning means reasoning from general theories to account for specific experimental results. In other words, this means going from the general to the specific.

A good example of deductive reasoning is the way in which the theory of plate tectonics has been used to find new deposits of valuable minerals and oil. By using plate tectonics, scientists have reconstructed how the continents once fit together. Knowing where ores and oil were found on one side of an ocean allowed them to find deposits on the other side of that ocean. Based on the general theory of plate tectonics, scientists determined the specific locations in which to search for ores and oil.

INTERPOLATION AND EXTRAPOLATION

Interpolation, or estimating a value between two known values, is a form of deductive reasoning. For example, suppose you measured the temperature outside to be 30°C at noon, 32°C at 1:00 P.M., and 34°C at 2:00 P.M. You might conclude that between noon and 2:00 P.M., the temperature was increasing at a rate of 2°C per hour. From the pattern of known values for the *general* time period of noon to 2:00 P.M., you might estimate values for more *specific* times between the known values, e.g., that at 12:30 P.M., the temperature was 31°C.

Extrapolation, or estimating by extending or projecting beyond known information, is a form of inductive reasoning. You start with specific information you do have, identify a pattern, and then extend it beyond what you know. If you start with the pattern of temperature change for the *specific* time in the example above and then extend it to time in *general*, you might estimate that at 11:00 A.M. the temperature was 28°C or that at 4:00 P.M. it will be 38°C.

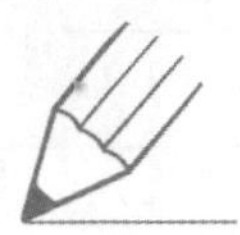

TRY THIS

In each of the following examples, make a prediction and state whether you arrived at the prediction by using interpolation or extrapolation.

1. Jennifer measured air temperature and air pressure outside her house at noon for five days. Her measurements are shown below.

	Monday	Tuesday	Wednesday	Thursday	Friday
Air Temperature (°C)	18	5	12	18	16
Air Pressure (millibars)	1,013	1,022	1,018	1,012	1,015

   1. If the air pressure at noon on Saturday is 1,012 millibars, predict the air temperature. ________
   2. Extrapolation or interpolation? ________

2. A high school gym teacher tested the school's top athletes to see how many sit-ups they could do in 2 minutes. The chart to the right shows their ages and the number of sit-ups they did.

Student Name	Age (years)	Number of Sit-ups
Bill	18	97
Dave	14	88
Joe	15	12
Juan	14.5	90
Luis	15.5	92
Paul	17	95
Victor	16	93

   1. How many sit-ups should the gym teacher expect a 16.5-year-old athlete to do in 2 minutes?

      ____

   2. Extrapolation or interpolation?

      ____

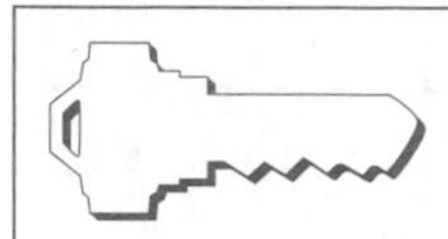

Key Idea: Critical thinking skills are used in the solution of mathematical problems.

Scientists communicate their findings through speaking and writing as well as through organized representations such as data tables, charts, graphs, and diagrams. Interpolating or extrapolating from information is easier if the information is organized in data tables, charts, and graphs. Diagrams are drawings that display information in such a way that complex relationships can be more easily understood.

CONSTRUCTING DATA TABLES AND CHARTS

Although there are no official rules for constructing data tables and charts, you should follow some conventions (generally agreed upon practices). Table 1.2 is a typical data table. In this example, a student compared the rate at which rounded quartz pebbles of different diameters sank through a 1-meter column of water.

When making data tables and charts, the independent variable is usually recorded in the left column and the dependent variable in the right column. When repeated trials are done, the column for the dependent variable may be subdivided. If observations are analyzed mathematically, the derived data may be recorded in an additional column to the right. Note that all columns are titled. The title includes the name, and units of measure, of the variable or derived data. Note, too, that the values of the independent variable are ordered. Data are usually arranged from largest to smallest or from smallest to largest. The title of the data table clearly communicates the purpose of the experiment and refers to the variables being investigated.

Table 1.2 **The Effect of Particle Size on the Settling Time of Rounded Quartz Pebbles**

Independent Variable Column	Dependent Variable Column			Mathematically Derived Data
Diameter of Rounded Quartz Pebbles (mm)	Settling Time in 1 meter of Water (sec) Trials			Average Settling Time (sec)
	1	2	3	
2.0	3.1	3.2	3.0	3.1
4.0	2.8	2.7	2.9	2.8
6.0	2.4	2.5	2.3	2.4
8.0	2.2	2.2	2.2	2.2
10.0	2.0	1.9	2.1	2.0

Table 1.3 **Checklist for Evaluating Data Tables and Charts**

Table Element	Criteria	✓
Title	Communicates experiment purpose; identifies variables being investigated.	
Independent-variable column	Vertical column, usually placed on the left.	
	Label contains name/unit of the independent variable.	
	Values of independent variable ordered.	
Dependent-variable column	Vertical column, usually placed on the right.	
	Label contains name/unit of the dependent variable.	
	Column subdivided for repeated trials.	
Derived-quantity column	Vertical column, usually placed to the right of the dependent variable column.	
	Label contains the name and unit of the derived quantity.	
	Derived quantity correctly calculated.	

On the 8th grade assessment, you may be asked to construct a data table or chart as part of a constructed-response question. Table 1.3 is a handy checklist you can use to evaluate data tables or charts you construct.

CONSTRUCTING GRAPHS

Patterns, trends, and other relationships are usually easier to recognize when observations are shown in the form of a well-constructed graph. Graphs communicate the observations made in an experiment in the form of a picture in two dimensions—horizontal and vertical. Two commonly used types of graphs are line graphs and bar graphs.

Let us go through the process of constructing a line graph for the following example. A student wanted to know if temperature affected the germination of radish seeds. She made up plastic bags, each containing 10 radish seeds, on a moist paper towel. She placed one into the freezer (0°C), one into the refrigerator (5°C), one into the basement (10°C), one into a closet (20°C), and one into the attic (30°C). After one week, she examined the bags and counted the number of seeds that had germinated. She did three trials during the month of October. Her results are summarized in Table 1.4.

Table 1.4 **The Effect of Temperature on the Germination of Radish Seeds**

Temperature (°C)	Average Germination Rate (%)
0	0
5	10
10	20
20	60
30	90

DRAWING AND LABELING AXES

By convention, scientists place the independent variable on the horizontal or *x*-axis and the dependent variable on the vertical or *y*-axis. See Figure 1.1*a*. The axes of a graph should be labeled so the reader knows what they represent. The unit of measurement is usually placed in parentheses next to or below the variable. Figure 1.1*b* shows correctly labeled graph axes for this data.

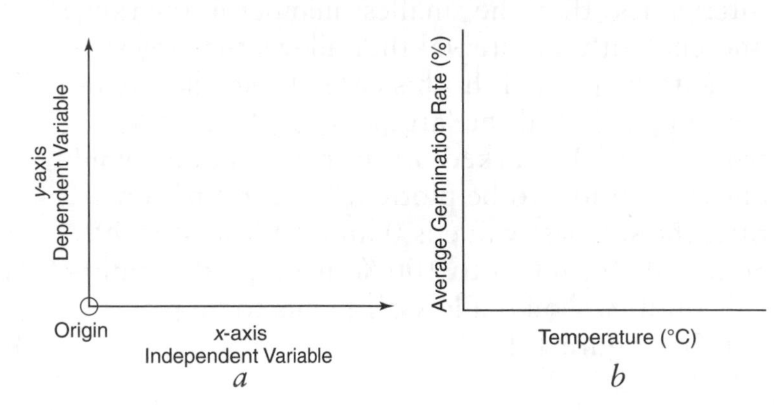

Figure 1.1
(*a*) The axes of a graph. (*b*) Correctly labeled axes for a graph of the germination rate data in Table 1.4.

WRITING DATA PAIRS

The coordinates of points on a graph are represented by data pairs. By convention, the value for the horizontal or *x*-axis is written first, followed by the value for the vertical or *y*-axis. The two values are separated by a comma and enclosed in parentheses. For example, the data pairs for temperature and average germination rate are (0,0), (5,10), (10,20), (20,50), and (30,90).

CHOOSING THE SCALES FOR AXES

The *scale* of a graph means the size of the axes and the intervals into which they are divided. Choosing the appropriate scale for a graph can be difficult. If you use too many intervals, the graph looks crowded. If you use too few, plotting points is difficult. A good rule of thumb is to find the range, or difference between the smallest value and the largest value, for a variable and then divide by 5. Next, round off the quotient to the nearest number easily counted in multiples, such as 1, 2, 5, or 10. For the radish seed experiment, the process would go something like this:

	x-axis (Temperature) Interval	*y*-axis (Germination Rate) Interval
Largest value	30	90
Smallest value	0	0
Range (difference)	30	90
Range divided by 5	6	18
Round to nearest number with easy multiples	5	20

The scale for each axis should begin with an interval less than the smallest number in the range and end with an interval that allows the largest value to be plotted. In this case, the smallest temperature is 0° and the largest is 30°. So a scale from 0° to 30° marked off in intervals of 5° would allow all points to be plotted. For germination rate, the smallest value is 0 and the largest is 90. So a scale from 0% to 100% marked off in intervals of 20% would allow all points to be plotted. See Figure 1.2.

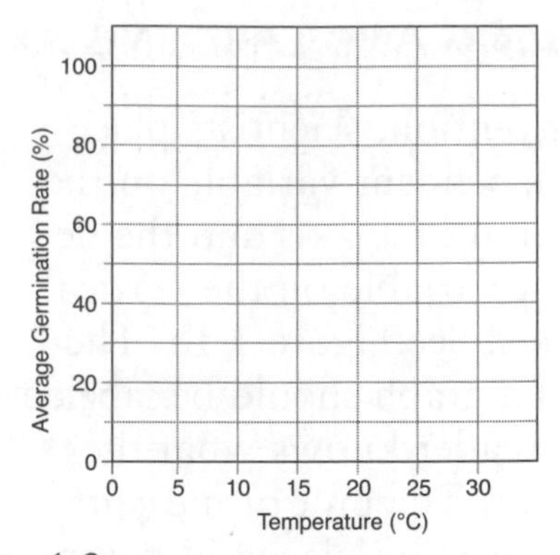

Figure 1.2
Correctly labeled and scaled graph for germination rate data.

PLOTTING DATA PAIRS

A data pair is plotted by locating the first number on the x-axis and the second number on the y-axis. For the data pair (5,10), locate 5 on the x-axis and 10 on the y-axis. By imagining a line running straight up from the 5 and another running straight across from the 10, a point can be plotted where the two lines cross. As shown in Figure 1.3, this procedure can be used to plot other data pairs, such as (20,50) and (30,90).

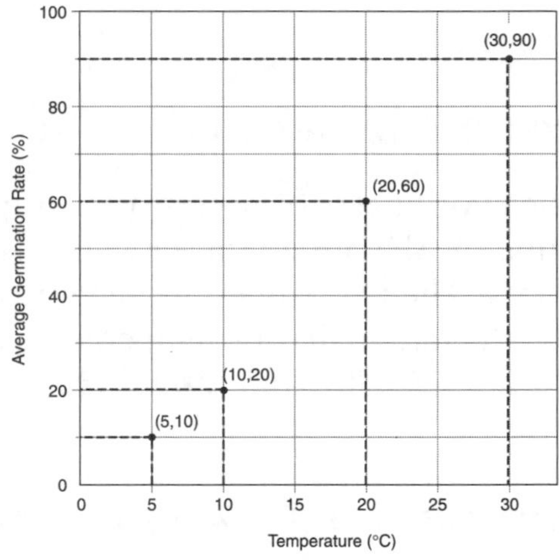

Figure 1.3 Procedure for plotting points.

ANALYZING TRENDS

A ***trend*** is a pattern of observations that are occurring in a particular direction (for example, increasing or decreasing). Since experimental data often contain errors, the data points on a graph are not always directly connected. Instead, a best-fit line is drawn to indicate the general pattern of the data. A best-fit line is one that is drawn so that an equal number of data points fall to either side of the line. See Figure 1.4.

A best-fit line shows what happens to the dependent variable when the independent variable is changed. The trend shown by the line can also be communicated by writing a sentence. For example, as the temperature of the bag of radish seeds increased, the germination rate of the radish seeds also increased.

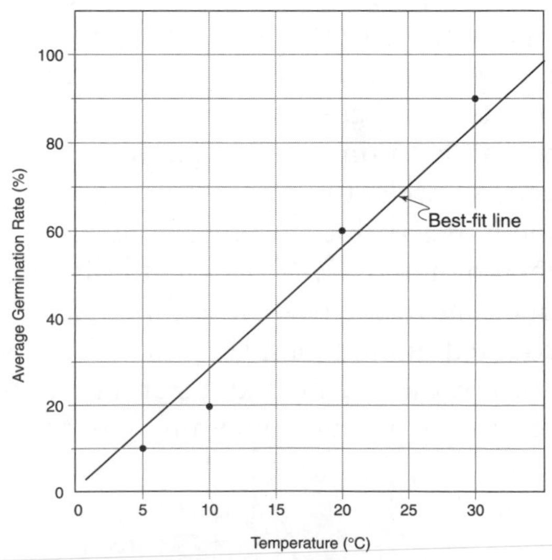

Figure 1.4 Drawing a best-fit line.

The line on a graph is a picture showing how the dependent variable responds when you purposefully change the independent variable. Table 1.5 (page 12) shows line graphs representing different relationships among variables.

BAR GRAPHS

Bar graphs are used when observations or measurements of a variable are not a continuous quantity but separate categories, such as brands, gender, kind of plant, or color. A good way to decide whether a line graph or a bar graph should be used to show data is to ask yourself whether the data can have a value between the intervals shown on the graph. If it can, a line graph is used; if it cannot, a bar graph is used.

Suppose you measured the effect of a fertilizer, measured in grams, on plant height, measured in centimeters. Both grams and centimeters can have values between any intervals shown on a graph. Although marked off in 5-cm intervals on a graph, plant height could have a value between the 5- and 10-cm interval, such as 5.5 cm or 8 cm. The amount of fertilizer could also have values between the 100-g and 200-g interval, such as 125 g or 163 g. Therefore, in this case, a line graph should be used.

Now, suppose you compared the gasoline mileage of different brands of compact cars. There are no values *between* brands of cars, such as Ford and Chrysler. They are distinctly separate categories. Therefore, a bar graph should be used.

When constructing a bar graph, axes are drawn and labeled, and data pairs are written as they are for a line graph. However, when determining the scale for the axis that has separate categories instead of continuous values, the categories are evenly distributed along the axis, leaving a space between each value. Next, the data pairs are plotted. However, when analyzing trends, the data points are not connected with a line. Instead, a bar is drawn from the category to the corresponding value of the other variable, leaving spaces between each bar. Finally, any trends can be summarized with a descriptive sentence. See Figure 1.5.

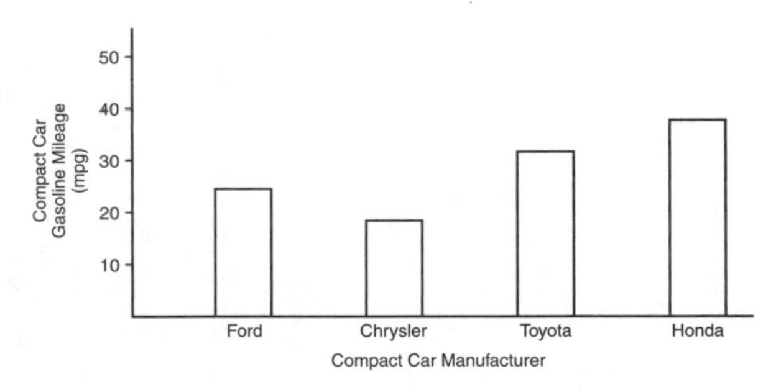

Figure 1.5
Compact cars made by foreign car companies get higher gasoline mileage than those made by American car companies.

Table 1.5 Line Graphs Showing Different Relationships Between Variables

Relationship	Line Graphs
Direct	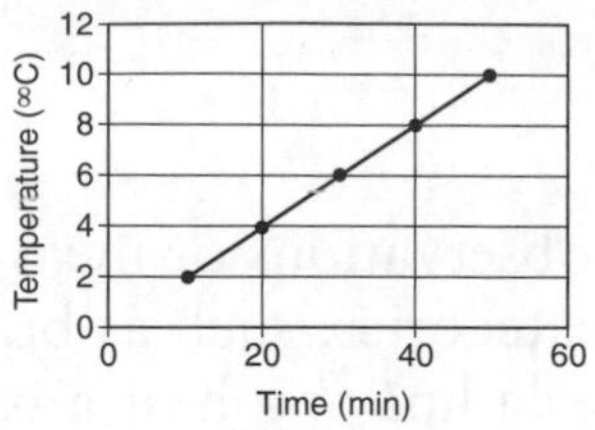 Temperature (°C) Time (min) As time *increases*, temperature *increases*.
Inverse	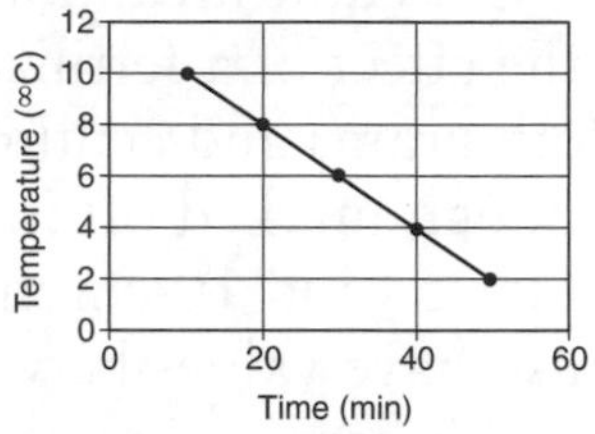 Temperature (°C) Time (min) As time *increases*, temperature *decreases*.
Cyclic	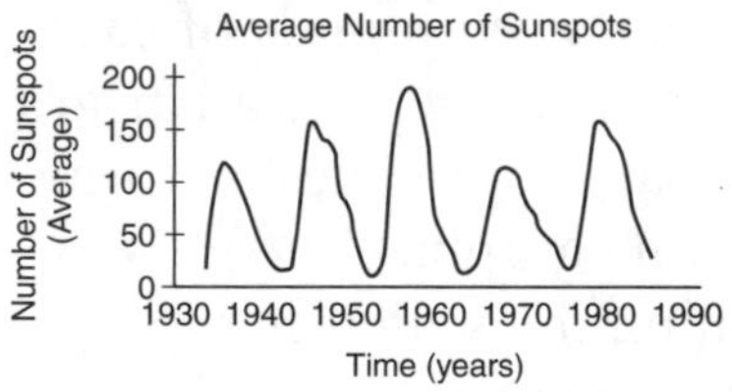 Average Number of Sunspots Number of Sunspots (Average) Time (years) As time increases, the number of sunspots increases and decreases in a repeating pattern.
Nonrelated	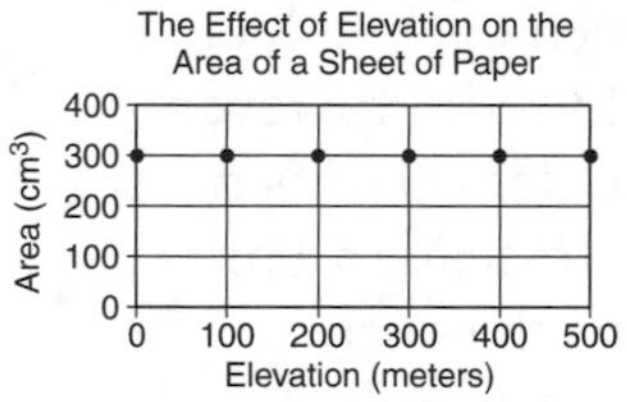 The Effect of Elevation on the Area of a Sheet of Paper Area (cm^3) Elevation (meters) As elevation *increases*, the area of a sheet of paper *does not change*.
	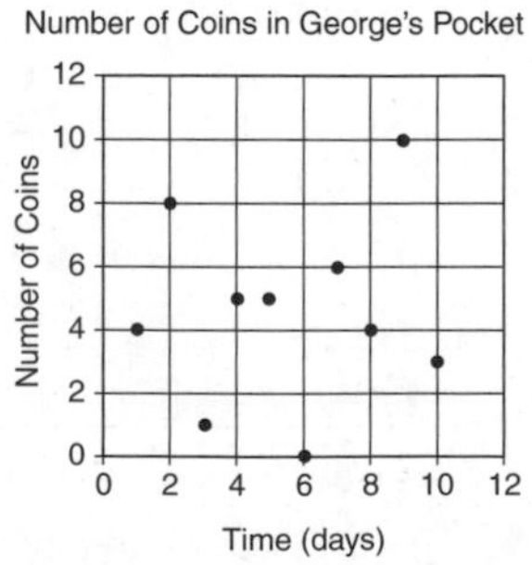 Number of Coins in George's Pocket Number of Coins Time (days) As time *increases*, there is a *random change* in the number of coins in George's pocket.

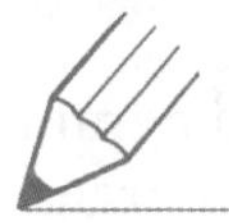

TRY THIS

For each of the following, decide whether a line graph or a bar graph should be used to represent the data. Then follow these steps to construct a graph. (1) Draw and label axes. (2) Write data pairs. (3) Determine the scales for the axes. (4) Plot the data pairs. (5) Draw a line, or bars, as appropriate. (6) Write a descriptive sentence summarizing any data trends.

Table 1.6 **The Effect of Ramp Angle on the Travel Distance of a Toy Car**

Ramp Angle (degrees)	Average Travel Distance (cm)
0	0
10	167
15	226
20	282
25	320
30	357
35	390
40	345

Table 1.7 **A Comparison of the Average Life of Different Types of AA Cells in a Portable CD Player**

Type of AA Cell	Average Life (hours)
Carbon-zinc	2
NiCad	5
NiMH	6
Alkaline	9
Lithium	11

On the 8th-grade assessment, you may be asked to construct or interpret a graph as part of a constructed-response question. Table 1.8 is a handy checklist you can use to evaluate line or bar graphs you construct.

Table 1.8 **Checklist for Evaluating Line and Bar Graphs**

Graph Element	Criteria	✓
Title	Communicates experiment purpose; identifies variables being investigated.	
Draw and label axes	*x*-axis correctly labeled with name/unit of the independent variable.	
	y-axis correctly labeled with name/unit of the dependent variable.	
Determine scales for axes	*x*-axis correctly subdivided—into a scale for a line graph, into categories for a bar graph.	
	y-axis correctly subdivided into a scale.	
Plot data points	Data pairs correctly written.	
	Data pairs correctly plotted.	
Analyze data trends	Line graph: data trend summarized with a best-fit line.	
	Bar graph: bars for data pairs drawn correctly.	
	Data trends summarized with a descriptive sentence.	

SCIENTIFIC DIAGRAMS

Scientific diagrams are pictures that display information. The information in a diagram may be in the form of images, numbers, words, or art. Scientists use diagrams to organize and display observations and to make complex ideas and relationships easier to understand. Scientific diagrams are complex and rich in meaning and require *careful* study. For example, the diagram in Figure 1.6 shows the life cycle of the Japanese beetle.

In one easy-to-understand picture, thousands of observations made over many years have been summarized. At first glance, you may notice the months of the year across the top and realize that the diagram shows where in its life cycle the beetle is during any given month. *Careful* study reveals deeper insights.

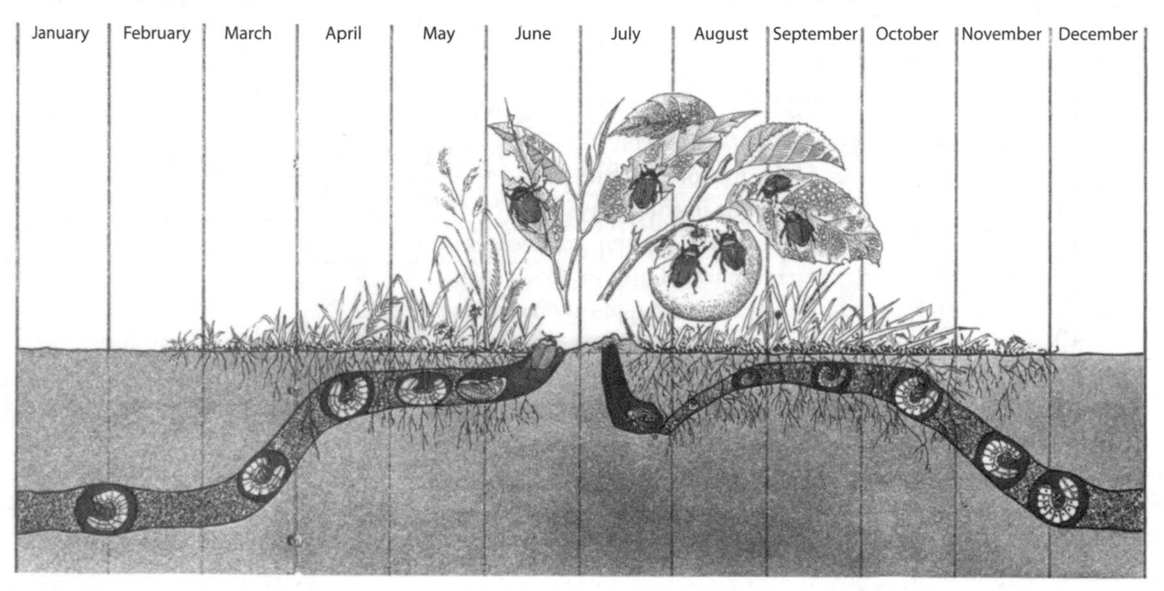

Figure 1.6
The life cycle of the Japanese beetle.

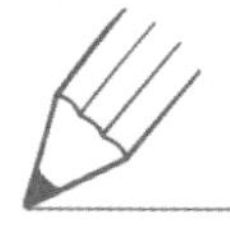

TRY THIS

Can you tell from the diagram:

- When the adult beetle emerges from the ground?
- What the adult beetle eats?
- What the grubs eat?
- When Japanese beetles mate?
- Where females lay their eggs?
- When are the eggs laid?
- What does a leaf damaged by a Japanese beetle look like?
- During which months do the grubs grow the fastest?
- How do the grubs keep from freezing in the winter?
- When do the grubs form pupae?

Practice Review Questions

1. For each experiment title listed, state whether the experiment should be graphed as a bar or a line graph:

1. The Rate at Which Water Drains Through Different Brands of Coffee Filters

__________graph

2. The Effect of Depth on Water Temperature in a Lake

__________graph

3. The Effect of Temperature on Heart Rate

__________graph

2.* The graph shows the progress made by an ant moving along a straight line.

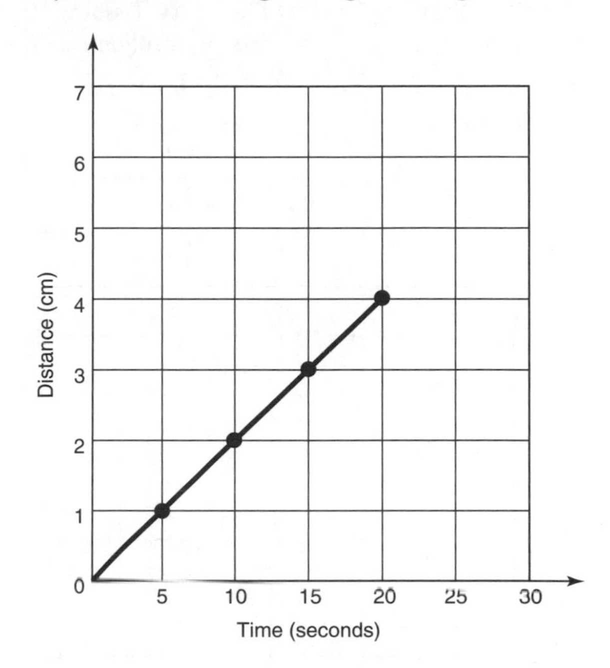

If the ant keeps moving at the same speed, how far will it have traveled at the end of 30 seconds?

1. 5 cm
2. 6 cm
3. 20 cm
4. 30 cm

GO ON

3. Construct an appropriate graph, including correctly labeled and scaled axes, for the following sets of data (a–b):

a. Toxic Chemical Residues in Humans

Year	DDT in Fatty Tissues (parts per million)
1970	8.09
1973	6.09
1976	4.68
1979	3.14

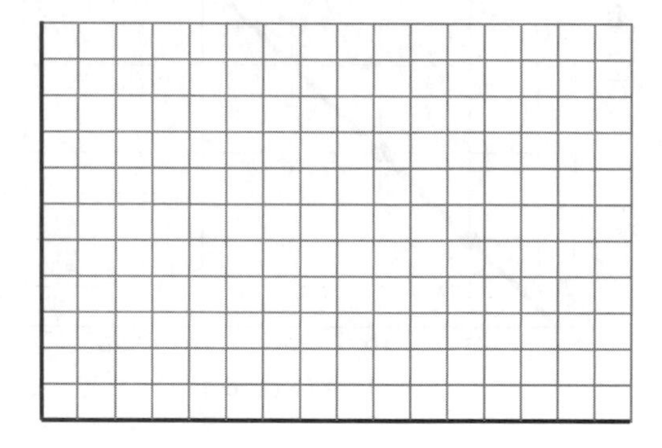

b. Comparative Fuel Efficiency in Transportation

Mode of Transportation	Average Fuel Economy (mpg)
Motorcycle	50.0
Passenger car	19.2
Bus	6.0

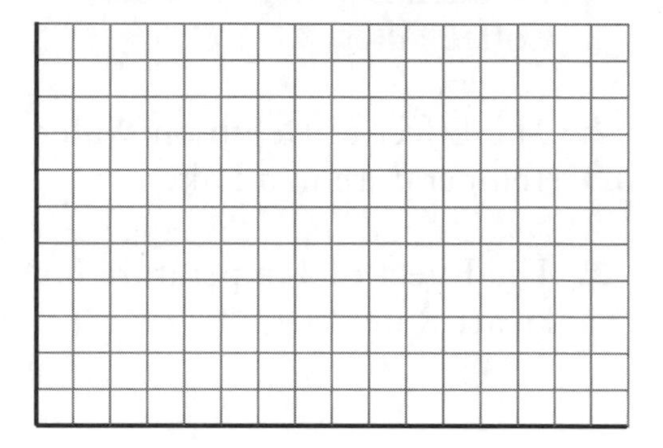

4. Describe the relationship between the variables graphed in each part of question 3.

GO ON →

5.** If the population increases by the same rate from the year 1990 to the year 2000 as in the years from 1980 to 1990, approximately what is the expected population by the year 2000?

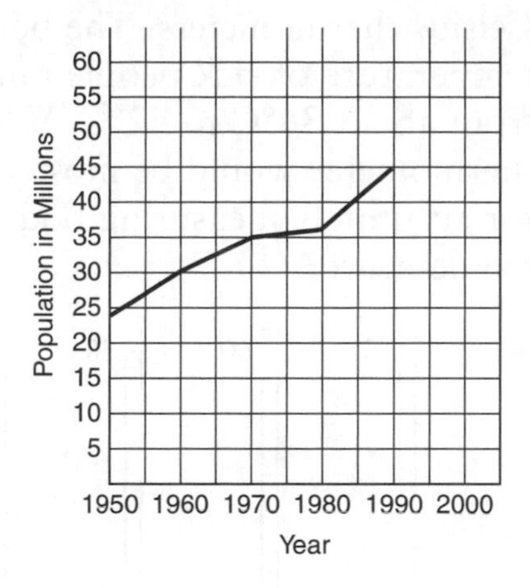

1. 47 million
2. 50 million
3. 53 million
4. 58 million

6.* The graph shows the distance traveled before coming to a stop after the brakes are applied for a typical car traveling at different speeds.

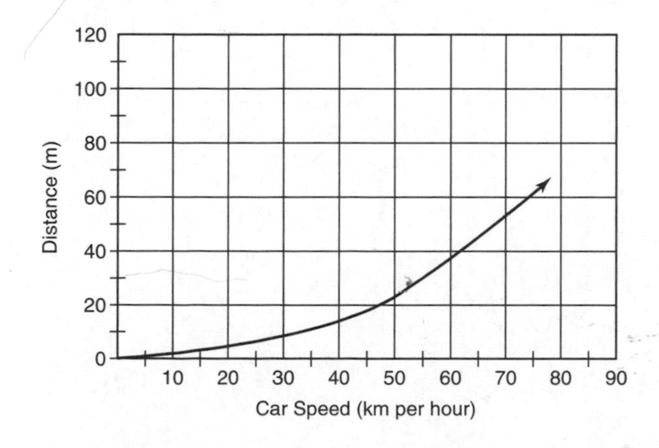

A car traveling on a highway stopped 30 m after the brakes were applied. About how fast was the car traveling?

1. 48 km per hour
2. 55 km per hour
3. 70 km per hour
4. 160 km per hour

7.** Kelly went for a drive in her car. During the drive, a cat ran in front of the car. Kelly slammed on the brakes and missed the cat.

Slightly shaken, Kelly decided to return home by a shorter route. The graph below is a record of the car's speed during the drive.

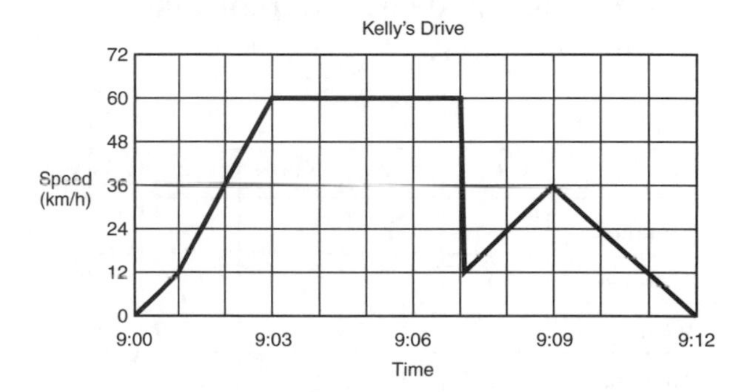

1. What was the maximum speed of the car during the drive?

____

2. What time was it when Kelly slammed on the brakes to avoid the cat?

____

GO ON ➡

8.** A TV reporter showed this graph and said,

"There's been a huge increase in the number of robberies this year."

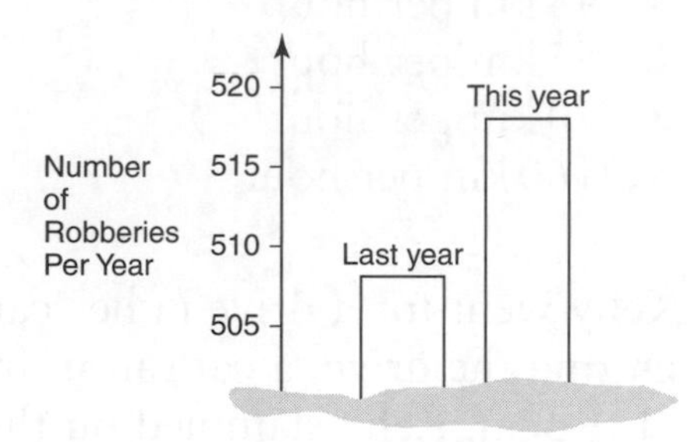

Do you consider the reporter's statement to be a reasonable interpretation of the graph? Briefly explain.

9.* This chart shows temperature readings made at different times on four days.

Temperatures

	6 A.M.	9 A.M.	NOON	3 P.M.	8 P.M.
Monday	15°	17°	20°	21°	19°
Tuesday	15°	15°	15°	10°	9°
Wednesday	8°	10°	14°	13°	15°
Thursday	8°	11°	14°	17°	20°

When was the highest temperature recorded?

1. Noon on Monday
2. 3 P.M. on Monday
3. Noon on Tuesday
4. 3 P.M. on Wednesday

10.* Draw a diagram to show how the water that falls as rain in one place may come from another place that is far away.

11.* The diagram shows five different Celsius thermometers. The body temperature of sick people ranges from about 36°C to 42°C. Which thermometer would be most suited for accurately measuring body temperature?

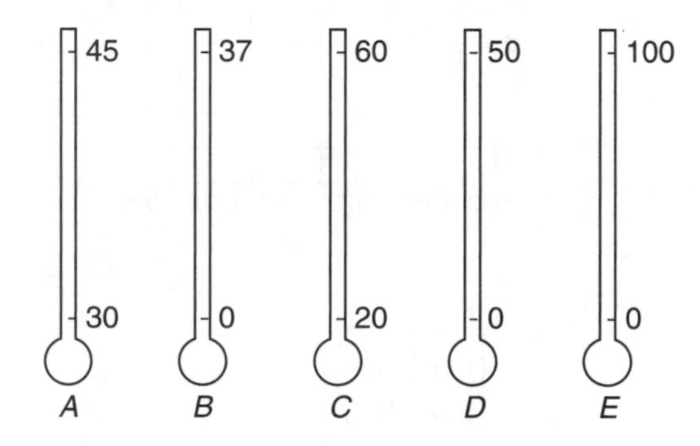

1. thermometer *A*
2. thermometer *B*
3. thermometer *C*
4. thermometer *D*
5. thermometer *E*

Chapter 2

Scientific Inquiry

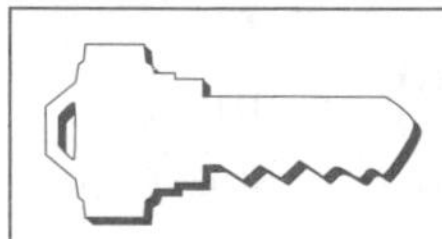

Key Idea: The central purpose of scientific inquiry is to develop explanations of natural phenomena in a continuing, creative process.

Scientific inquiry is *how* scientists go about their work—the particular ways in which scientists observe, think, experiment, and check whether their ideas and information are valid. Scientific inquiry both adds to the body of scientific knowledge and corrects it.

An important part of science is the understanding that as we gain knowledge, our view of the natural world may change again and again. In finding answers to one set of questions about the world, new questions undoubtedly arise. New instruments may be invented or new techniques developed that lead to new observations and new questions. In response, new experiments and analyses will be done. Some of the new findings may add new knowledge to science. Some may challenge existing theories. Then the theory may have to be corrected or, in rare cases, an entirely new theory invented. For example, people once thought that Earth's continents were immovable. However, when new technology allowed new observations and measurements to be made, they found that continents *do* move. A new theory was then invented—plate tectonics. Scientists understand that scientific knowledge is always open to improvement and can never be thought of as *absolutely* certain. Therefore, the quest for scientific knowledge by the continuous, creative process of scientific inquiry will likely continue as long as people are curious about the world around them.

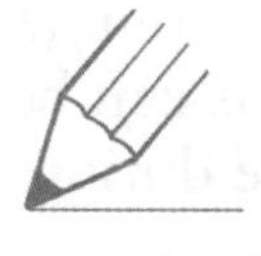

TRY THIS

Name some ideas about the natural world that have changed over time because of new observations.

THINKING LIKE A SCIENTIST

Scientists think about the world in ways that are not always shared by nonscientists. They think that by working together over time, people *can* figure out how the world works. They think that the universe is a single, vast system in which the basic rules are the same everywhere, so knowledge gained by studying one part of it can be applied to other parts (although they also know that this is not always the case). They also think that scientific knowledge is both certain and subject to change.

These ideas lead scientists to *do* science in a particular way. Let us now consider the particular ways in which scientists observe, think, experiment, and check whether their ideas and information are valid.

OBSERVING

Scientific inquiry begins with scientific observation of the natural world. Observing is simply noticing, or becoming aware of, something by using one or more of our senses. Seeing, hearing, tasting, touching, and smelling are all ways of observing. Scientific ***observation*** means taking careful notice of natural phenomena in a systematic way and making a record of what is observed. When scientists observe, they try to describe things as accurately as possible and make a record of what they observe. This record enables them to compare their observations with those of others.

Wherever possible, scientists use instruments to make measurements and record the quantities obtained. ***Instruments***, such as thermometers, rulers, microscopes, or balances, are devices that give more information about things than can be obtained by observing with our senses alone. Measurements also make observations more objective and make it easier for scientists to share information. For example, observing that a tree is "pretty tall" is not very useful—it could mean different things to different people. However, observing a tree by measuring that it is "10 meters tall" means the same thing to all scientists and would allow that observation to be compared with observations made by other people or at other times. In science, if different people make different observations of the same thing, they usually make some fresh observations instead of just arguing about the differences.

You might think that with only five senses, not too many types of observations can be made. Consider, though, that with just our eyes, we can observe color, size, shape, patterns, motion, phase of matter, behaviors, and many other characteristics. With just our sense of touch, we can observe characteristics such as size, shape, heat or cold, pressure, texture, hardness, and flexibility. Add measuring instruments and the list of possible observations gets even longer—length, mass, time, temperature, and angles to name just a few. Many quantities can also be derived from observations, such as rate, volume, density, and so on. Thus, an incredible number of different types of observations can be made with our senses. Scientists choose the specific observations they will make during an investigation based on the question they are studying.

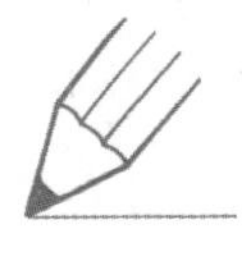

TRY THIS

1. Make a list of some observations that could be made with just our sense of hearing.
2. Describe two situations in which the use of instruments would enable you to obtain more information than would be possible with your senses alone.

COLLECTING INFORMATION

Scientists' observations are called *data*. Data may be measurements, descriptions of what things are like or what is happening somewhere, or other facts. However, data are facts without context. Data become information when they inform the researcher about the problem at hand. When information is correctly interpreted, it creates knowledge—an understanding of the problem or process being investigated.

Observations are more easily interpreted when they are made in a purposeful, systematic way. ***Investigations*** are systematic approaches to collecting observations. Three general types of investigations used in science are field studies, surveys, and experiments.

Field studies are used by scientists such as a field biologist or geologist who must make observations under conditions that cannot be perfectly controlled. A field biologist may not be able observe the natural behavior of an animal in a laboratory cage and may have to wait for the right animal to wander by in the field. However, the field scientist must have a plan in mind of what data to collect, how the data will be recorded, and how the observations will be interpreted.

Surveys involve making observations that describe things as they are, without drawing any inferences. For example, a field biologist may survey the number of different animals living in a city park. The observations are not being made to test an inference but simply to obtain information that may serve as the basis for inferences.

An ***experiment*** is a situation set up by a researcher to test an idea. All experiments are either to prove or to disprove something. Doing an experiment means creating a way in which a phenomenon can be observed so that all of the factors except those being studied are controlled or eliminated. An experiment is not just doing something to see what happens. It is a test of a proposed explanation.

ORGANIZING INFORMATION

You can make all of the possible observations and measurements of a situation and still not understand what the data you collected really mean. One way to make sense out of observations is to organize them. How observations are organized can be just as important as the observations themselves.

One way to organize data is to arrange them in some order or place them into groups. Arranging data into some order is called ***sequencing***. Placing data into groups is called ***classification***. For example, suppose you were given the following data. A student noticed that the goldfish in a backyard pond seemed to open and close their gill covers at different rates at different times. The student wondered whether temperature had anything to do with this and decided to make observations of the goldfish as the weather got warmer and colder. The student measured the water temperature and counted the number of times per minute a goldfish opened and closed its gill covers for several weeks. During week 1, the water temperature was 25°C, and the gill-opening rate was 57 times per minute. During week 2, the observations were 15°C and 25 times per minute; week 3 was 18°C and 30 times per minute; week 4 was 10°C and 15 times per minute; week 5 was 20°C and 38 times per minute; week 6 was 23°C and 60 times per minute; and week 7 was 27°C and 25 times per minute.

First, the student decided to organize the data by grouping them according to type of observation (temperature and gill-opening rate). Then the student sequenced them in the order in which they were collected. The result is shown in Table 2.1.

The data do not seem to follow any pattern. However, the same set of observations can be organized in many ways. The student next decided to order the observations by water temperature, as shown in Table 2.2.

Once observations are organized, you may be able to conclude something from the organized data that you could not have concluded on the basis of any one of the separate observations. A conclusion that logically follows from a set of observations is called an ***inference***. An inference is an interpretation of a set of observations. Based on Table 2.2, the student inferred that the gill-opening rate increases as water temperature increases.

Observations are data; inferences are not! The same set of observations can often be interpreted in a number of ways. Suppose you observed that your friend Pat was crying. You might infer that Pat had gotten hurt. Someone else might infer that Pat had failed a test. Yet another person might infer that Pat had just won the lottery and was crying tears of joy. The observation was the same in all three cases, but the interpretation of the data changed. Which inference is correct? The only way to find out is to investigate further and to obtain more data. In science, recognizing the difference between an observation and an inference is important. Not presenting inferences *as if they were* observations is particularly important, because doing so can lead to misinterpretations of data. Knowledge in science is not gained through gut feelings or common sense but through systematic research.

Table 2.1 **Observations of Water Temperature and Gill-Opening Rate**

Week	Water Temperature (°C)	Gill-Opening Rate (openings/minute)
1	25	57
2	15	25
3	18	30
4	10	15
5	20	38
6	23	60
7	27	25

Table 2.2 **Observations of Water Temperature and Gill-Opening Rate (Organized by Water Temperature)**

Week	Water Temperature (°C)	Gill-Opening Rate (openings/minute)
4	10	15
2	15	25
3	18	30
5	20	38
6	23	60
1	25	57
7	27	25

TRY THIS

Place an "O" before the statements that are observations, and an "I" before the statements that are inferences.

______ 1. The beaker contains 250 ml of water.
______ 2. The water in the beaker looks polluted.
______ 3. Thirty centimeters of snow fell overnight.
______ 4. When the snow melts, the stream will flood.
______ 5. A yellow solid formed when I mixed the two liquids.
______ 6. The bark on the aspen tree was white.
______ 7. Those clouds are dark gray and mushroom shaped.
______ 8. It will rain when those clouds reach us.
______ 9. The dog is barking.
______10. That sounded like a mean dog.

QUESTIONING

The first step in systematic research is realizing that you do not know something. When you try to discover something about the natural world, you must ask a question about the world and then create an experiment or decide what data need to be collected in order to answer that question with a "yes" or "no."

In the goldfish investigation, the student inferred that increasing water temperature increased gill-opening rate. However, when making inferences, you cannot just ignore observations that do not seem to agree with the general trend of the data. For example, the student also noted that at the highest temperature, the gill-opening rate decreased. This raised many new questions in the student's mind. "Did I make a mistake counting that day? Does this mean that gill-opening rate doesn't *always* increase with temperature? Is there some other factor that affects gill-opening rate? Why would an increase in water temperature cause the gill-opening rate to increase up to 25°C and then decrease so much if the water got just 2°C warmer? Were there drops in gill-opening rates at temperatures other than those that were observed?"

FORMULATING A HYPOTHESIS

Some of these questions may be answered simply by making more observations. Others may require the student to design a new experiment. When a scientist observes a phenomenon, such as the relationship between water temperature and gill-opening rate, he or she usually tries to create a possible explanation for it. The purpose of an experiment is to test whether this possible explanation is correct. A possible explanation for a set of observations and that can be tested by further investigation is called a ***hypothesis***.

Nonscientists often define a hypothesis as an educated guess. However, a hypothesis is more than a mere guess. It is an inference based upon observations, a possible explanation for something observed in the natural world.

In order to create a possible explanation, you need to know something about the topic. So, when formulating a hypothesis, begin by reading! Find background information about the area being investigated. Consult periodicals, journals, and the Internet. Try to discover whether or not other people have conducted experiments similar to the one you want to do and the kinds of results they obtained. Then you will be ready to formulate your own hypothesis.

A hypothesis is often stated in the form of an *if* . . . (change in independent variable), *then* . . . (change in dependent variable) statement. For example, *if* I increase water temperature, *then* a goldfish will open and close its gill covers at a faster rate.

An even better practice is to try to explain the reasoning behind the test you are proposing. A *reasoned hypothesis* includes an explanation of *why* the researcher thinks changing the independent variable will affect the dependent variable in a particular way. A reasoned hypothesis is often stated in the form of an *if* . . . , *then* . . . , *because* . . . statement. For example, *if* I increase water temperature, *then* the goldfish will open and close its gill covers at a faster rate, *because* as water gets warmer, it contains less dissolved oxygen and the goldfish has to pump more water over its gills to get the oxygen it needs to stay alive.

When writing a reasoned hypothesis, the following format is useful:

1. If (change in independent variable)
2. Then (predicted change in dependent variable)
3. Because (your explanation for the predicted change)

Here are some examples of reasoned hypotheses.

If salt crystals are broken into smaller pieces, *then* they will dissolve faster in water, *because* more surface area will be exposed to the water.

If the vegetation on a slope is removed, *then* the slope will erode more rapidly, *because* plant roots help hold soil in place.

TESTING A HYPOTHESIS

Since a hypothesis states how a researcher thinks changing an independent variable will affect a dependent variable, it can apply only to natural, observable phenomena. If you cannot physically observe a factor described in a hypothesis, either with senses or instruments, then the hypothesis is of little use to a scientist. Wherever possible, the variables in the hypothesis should also be quantifiable, that is, the data should be numerical. Numerical data is less subject to misinterpretation and opinion.

A hypothesis must also be testable. That is, there should be a way either to prove or to disprove the hypothesis. Anyone can make up dozens of possible hypotheses that explain natural phenomena. However, if you cannot test a hypothesis, no way exists of determining whether or not it is correct.

A test of a hypothesis must be designed to prove something that actually happens and can be observed. If you cannot observe the results of a test, nothing has been proven. The absence of an observation is *not* an observation and cannot prove or disprove a hypothesis. This means that a test designed to prove that something *will not* happen is not a valid test. No matter how many times you repeat such a test, you have not proven logically that something will not happen, because it could just happen on the *next* test. For example, suppose you formulate the hypothesis "*If* I flip a coin, *then* it will **NOT** come to rest standing on its edge, *because* a coin on edge is unstable." Then you flip the coin ten times and it always ends up lying flat. Have you proven the hypothesis? Absolutely not! Although *unlikely*, it is entirely *possible* that in some future trial, the coin will come to rest standing on its edge.

Finally, a hypothesis about morals, values, religious beliefs, and similar issues cannot be tested scientifically. A hypothesis about beauty or goodness, for example, cannot be tested, because these are not *observable* qualities—they are *inferred* qualities. For example, the hypothesis "*If* a person is good, *then* that person will not steal, *because* good people do not steal" is not scientifically testable. "Good" is not an observable quality; it is an inference based upon a wide range of observations, and there are varying opinions about the criteria for inferring what is good. Would a person still be good if he or she did not steal but did lie, cheat, or harm other people? Science cannot make moral or value judgments, but science can provide information that people can use when making such judgments.

CONTROLLING TESTS

A test of a hypothesis, planned so that only one variable at a time is changed or tested, is called a ***controlled experiment***. By testing in this way, you know how each variable affects the results. If too many factors are changing at the same time, you will get a result, but you will not know the reason for the result. For example, suppose you design a test of the hypothesis "*If* salt crystals are broken into smaller pieces, *then* they will dissolve faster in water, *because* more surface area will be exposed to the water." In the test, you put coarse salt into a *small* glass of water and fine salt into a *large* glass of water, and the fine salt dissolves faster. You have a result. However, you do not know if the fine salt dissolved faster because of its grain size or because there was more water in which it could dissolve.

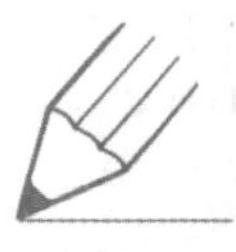

TRY THIS

What factors must be held constant in each of the following experiments to make them controlled experiments?

1. A student wanted to test the idea that dark-colored materials heated up faster than light-colored ones. The student put trays of water into which 10, 20, and 30 drops of food coloring had been added into sunlight and then measured the temperature of the water in each tray.
2. A student heard that dropping a magnet made it weaker. The student compared the number of paper clips the magnet could pick up after being dropped 10 times and 20 times.

REPORTING RESULTS

Once a hypothesis has been tested and the data have been collected and analyzed, the time has come to figure out what they mean. An interpretation of what the test showed must be provided. Was the hypothesis accepted or rejected? Did the data support your explanation of the phenomenon? You can make guesses about the meaning of data beyond the scope of the test, but they must not be claimed as results.

Scientists carefully report the experiments they perform and their results for other scientists. Sharing ideas and experimental results is an important part of science. Experimental results are reported in such a way that other scientists can repeat the experiment and compare their observations with those published in the report. Experimental results are usually published in ***journals***, periodicals containing articles on a particular subject. Literally thousands of journals are dedicated to all areas of scientific research.

PEER REVIEW

Before experimental results are published, they go through a rigorous ***peer review***. This is a process by which scientists work together to review each other's work. Peer review committees evaluate the hypothesis, experimental design, and results of a researcher's work to see if they are valid. They also check to see if the research duplicates work that has already been published. Peer review continues after experimental results are published as other scientists repeat the experiment and compare their results with those published in the report. In this way, errors in one scientist's work can be identified and corrected by the work of another. Review by others helps improve the work that is done.

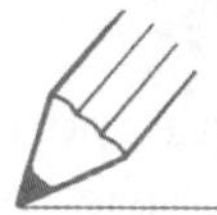

TRY THIS

Now it is your turn to be a peer reviewer! A report of an experiment done by a group of students, and the results they obtained, were submitted for publication to a school science magazine. You are part of a peer review committee that reviews the work of other students. Evaluate the hypothesis, experimental design, and reported results. What suggestions could you make to improve the research project?

Title:
Does Music Help Plants Grow?

Hypothesis:
Classical music will help the plants to grow, and country music will stunt their growth.

Materials:

- 9 small plastic yogurt containers
- 27 bush bean seeds
- Potting soil
- Measuring cup
- Water
- Country music CD or cassette
- Classical music CD or cassette

Procedure:

1. Plant 3 seeds into each yogurt container after filling them almost up to the top with potting soil. Place the seeds just a little bit below the surface.
2. Water the plants with $^1/_4$ cup of water.
3. Put all 9 containers into a spot by a window.
4. Take the plants away from the window at 4:00 P.M. every day. Place the controls into a room where the music that the other 6 plants are listening to cannot be heard.
5. The 3 plants that listened to country music listened to Garth Brooks or Faith Hill, and the classical plants listened to Lorie Line and Mendelssohn, for an hour each day.
6. Record the growth of each plant each night around 9:00 P.M. Since there were 3 of each kind of plant, this fulfills the minimum number of trials: 3.
7. Let each plant grow for 2 whole weeks. Then find out the total growth of the plants.

FORMING LAWS AND THEORIES

Scientists have been reporting the results of their experiments for centuries. Each experiment adds to the storehouse of observed data. Over time, we may observe regularities in data, something that always seems to happen the same way. We may also see relationships between different sets of data.

In science, a ***law*** is merely something that is always observed to happen the same way. A law is a regular behavior, or a relationship between phenomena, that, *as far as we know*, does not change *under the given conditions*. A good example is the law "Like magnetic poles repel; unlike magnetic poles attract." It is an accepted scientific law because magnets are always observed to act this way.

When the results of many experiments show that data are related, there may be an explanation. Scientists may then form a tentative ***theory***, a set of ideas that explains related observations. A theory is not a discovery; only facts can be discovered. Rather, a theory is a model, constructed in the mind of the scientist, that explains discoveries.

Before a tentative theory is accepted in science, it is subjected to intense testing. When many predictions based on a theory are supported by experimental results, the theory may be widely accepted by scientists as the most likely explanation for a set of observations. An accepted theory includes the assumptions underlying the explanation, their consequences, and predictions of observations that follow logically from the theory.

No matter how well a theory explains a set of observations, a new theory might explain them just as well or better, or it may be able to explain a wider range of observations. In science, testing, revising, and discarding old theories for new ones go on continuously. The result of this process is not absolute truth but an increasingly better understanding of how the world works, which is, after all, the goal of science. The evidence that this approach works is the increasing ability of scientists to explain the natural world and to make accurate predictions.

Practice Review Questions

1. Science is best defined as

1. a body of knowledge about the natural world
2. a process by which knowledge of the natural world is obtained
3. both a body of knowledge and the process by which it is obtained
4. the study of matter and energy

2.* Four children can feel and smell an object inside a bag, but they cannot see it. Which of the following is NOT an observation about the object?

1. "It is flat at one end and round at the other."
2. "It smells like peppermint."
3. "It has a bump on it."
4. "I hope it is candy."

Base your answers to questions 3–5 on the diagram and information below.

A student drew these diagrams of a cloud he saw at a distance of several kilometers at 3-minute intervals.

3. Which statement is best described as an observation?

1. The air in the cloud is rising.
2. The air in the cloud is very humid.
3. The cloud is getting darker near the bottom.
4. Rain will soon form in the cloud.

4. Which statement is best described as an inference?

1. The cloud has a flatter bottom than top.
2. The cloud appears to be growing larger.
3. The cloud base height appears to be constant.
4. The air in the cloud is rising.

5. Which statement is best described as a prediction?

1. Rain usually falls from this type of cloud.
2. This type of cloud is common in this area.
3. A bolt of lightning will descend from this cloud.
4. Clouds of this type do not last very long.

* Reproduced from TIMSS Population 2 Item Pool. Copyright © 1994 by IEA, The Hague.

GO ON

6.* Maria collected the gas given off by a glowing piece of charcoal. The gas was then bubbled through a small amount of colorless limewater. Part of Maria's report stated, "After the gas was put into the jar, the limewater gradually changed to a milky white color." This statement is

1. an observation
2. a conclusion
3. a generalization
4. an assumption of the investigation
5. a hypothesis

7. What is the relationship between observations and inferences?

1. They are the same thing.
2. Inferences are conclusions based on observations.
3. Inferences can be made without observations.
4. Observations are the natural result of inferences.

8. A prediction about an event that is changing Earth's surface can best be made if

1. an instrument is used in making one observation
2. the event is noncyclic
3. observations have been made over a long period of time
4. the event has not occurred in the past

9.* Whenever scientists carefully measure any quantity many times, they expect that

1. all the measurements will be exactly the same
2. only two of the measurements will be exactly the same
3. all but one of the measurements will be exactly the same
4. most of the measurements will be close but not exactly the same

10.* Some children were trying to find out which of three lightbulbs was brightest. Which one of these gives the best START toward finding the answer?

1. "One bulb looks brightest to me, so I already know the answer."
2. "All the bulbs look bright to me, so there cannot be an answer."
3. "It would help if we had a way to measure the brightness of a lightbulb."
4. "We can take a vote, and each person will vote for the bulb he or she thinks is the brightest."

11. A student has the idea that studying longer will increase test scores. Which is the best statement of this hypothesis?

1. Test scores affect studying.
2. If test scores increase, then the student studied longer.
3. Studying longer will increase test scores.
4. If a student studies longer, then his or her test scores will increase.

GO ON ➡

12.* The diagrams show different trials Abdul carried out with carts having different-sized wheels. He started them from different heights, and the blocks he put into them were of equal mass.

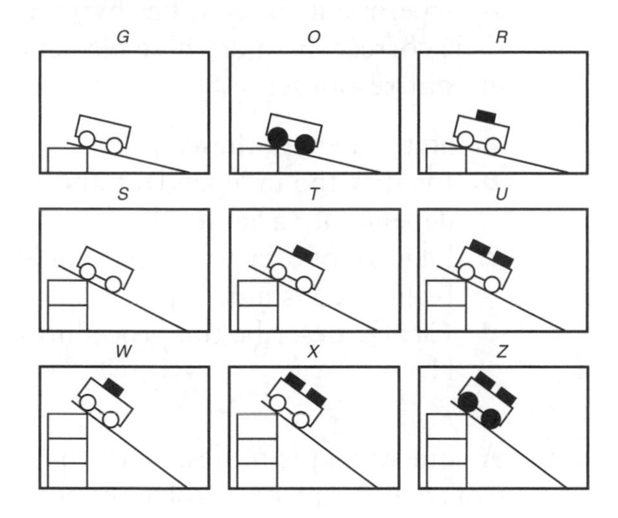

He wants to test the following idea. The heavier a cart is, the greater its speed at the bottom of a ramp. Which three trials should he compare?

1. *G*, *T*, and *X*
2. *O*, *T*, and *Z*
3. *R*, *U*, and *Z*
4. *S*, *T*, and *U*
5. *S*, *W*, and *X*

13.* To find out whether seeds grow better in the light or the dark, you could put some seeds onto pieces of damp paper and

1. keep them in a warm, dark place
2. keep one group in a light place and another in a dark place
3. keep them in a warm, light place
4. put them into a light or dark place that is cool

14.* A cupful of water and a similar cupful of gasoline were placed onto a table near a window on a hot, sunny day. A few hours later, it was observed that both the cups had less liquid in them but that there was less gasoline left than water. What does this experiment show?

1. All liquids evaporate.
2. Gasoline gets hotter than water.
3. Some liquids evaporate faster than others.
4. Liquids evaporate only in sunshine.
5. Water gets hotter than gasoline.

15.* A girl had an idea that plants needed minerals from the soil for healthy growth. She placed a plant in the Sun, as shown in the diagram below.

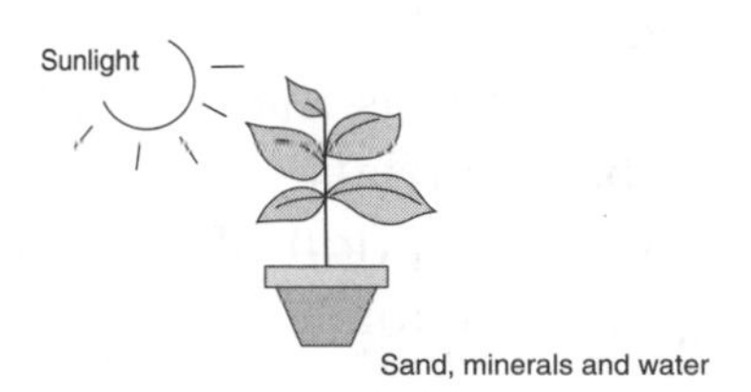

In order to check her idea she also needed to use another plant. Which of the following should she use?

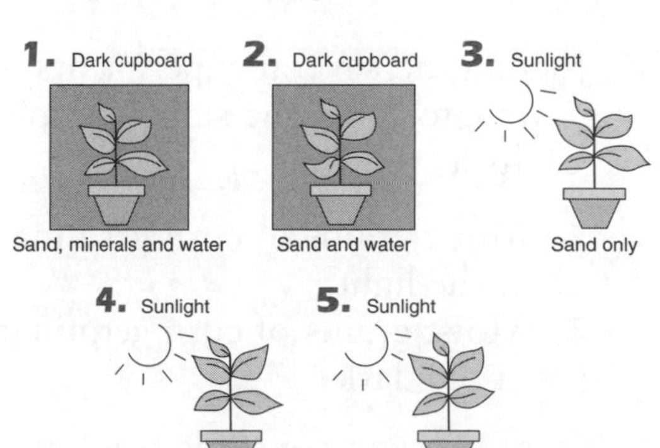

GO ON →

16. Which of the following reasons best explains why a scientific theory may not be accepted?

1. The theory is not able to explain all observations.
2. The theory has not been presented or published.
3. Another theory is able to explain a wider range of observations.
4. All of the above are reasons why a theory may not be accepted.

17. Which of the following best describes a scientific law?

1. a statement voted on for acceptance by 51% of scientists
2. something that is always observed to happen in the same way
3. a hypothesis that has undergone repeated trials
4. an explanation of a natural phenomenon

CONSTRUCTED-RESPONSE QUESTIONS

18.* Juanita did several experiments to germinate corn. She summed up her results as follows:

1. Moist grains of corn germinate in the light.
2. Moist grains of corn germinate in the dark.

What can you conclude from her results?

19. A student plays tennis several times a week. She notices that the tennis ball seems to bounce higher on some courts than on other courts. She wonders if this has something to do with the surface of the court. Design an experiment to see if her hypothesis is correct. Include these elements in your response:

a. State the hypothesis. [1]
b. Identify the independent and dependent variables. [2]
c. Identify two factors that should be held constant. [2]
d. Clearly describe the procedures. [1]

20. A student wants to design a controlled experiment to solve the following problem: When heated on a stove, will tap water or water containing a teaspoon of salt boil faster?

The student plans to place two pots with equal masses of water (one with tap water and one with saltwater) on identical burners in the classroom.

a. Identify the independent (manipulated) variable. [1]
b. Identify the dependent (responding) variable. [1]
c. State two factors that should be kept constant when planning this experiment. [2]

Chapter 3

Engineering Design

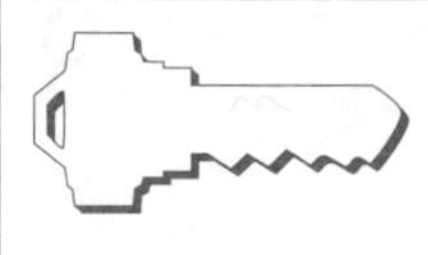

Key Idea: Engineering design is an iterative process involving modeling and optimization (finding the best solution within given constraints); this process is used to develop technological solutions to problems within given constraints.

Engineering design means using scientific knowledge to solve practical problems. The goal of engineering design is to invent, build, and operate efficient and cost-effective structures, machines, processes, and systems. This process involves applying science with a good deal of creativity, ingenuity, and perseverance. For example, the Swiss scientist Daniel Bernoulli discovered that fast-moving fluids exerted less pressure than slower-moving fluids. Aircraft engineers apply this knowledge to design wings. Wings have shapes that force air moving over them to move faster (and exert less pressure) than air moving beneath them. The result is a net pressure exerted upward on the bottom of the wing that provides lift. See Figure 3.1.

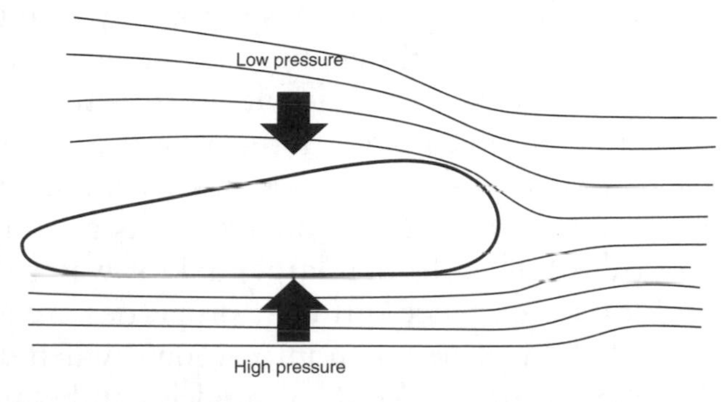

Figure 3.1
Bernoulli's principle.

The problem is that many factors affect how efficient and cost-effective a wing will be. These include how fast the airplane flies, how smoothly air flows over the shape, the amount of lift it produces, the weight and cost of the wing material, and so on. Aircraft engineers may have to design, build, and test hundreds of wings before the best one for a particular aircraft is found. That is why we say engineering design is an ***iterative***, or repetitive, process; one may have to go through the ***design-build-test*** process many times before the best solution to a problem is found. When solving practical problems through engineering design, the use of *models* is often an efficient and cost-effective way to find a solution.

MODELS

A ***model*** is anything that represents the properties of an object or system. A model is both alike and different from a real thing, but it can be used to learn something about the real thing. Observing how a model responds after it is changed may suggest how the real thing would respond if the same thing were done to it. Models are often used to think about things that are too big or too small, or that happen too quickly or too slowly, to be observed or changed directly. For example, models are often used when thinking about atoms. See Figure 3.2. Models may also be used to study things that could be dangerous. That is why carmakers use *models* of people when crash testing vehicles. The usefulness of a model can be tested by comparing predictions made using the model with actual observations in the real world.

Most models are made to ***scale***. That is, the parts of the model are made in the same proportions as the parts are in the original. A doll would look quite odd if its head was one-tenth the size of a real head and its arms were the size of a real arm. A model's scale is the ratio of the size of the model part to the original part. A map drawn to a scale of 1:62,500 means that 1 unit of distance on the map equals 62,500 units of distance on the ground. For example, 1 inch on the map equals 62,500 inches on the ground, or 1 centimeter on the map equals 62,500 cetimeters on the ground.

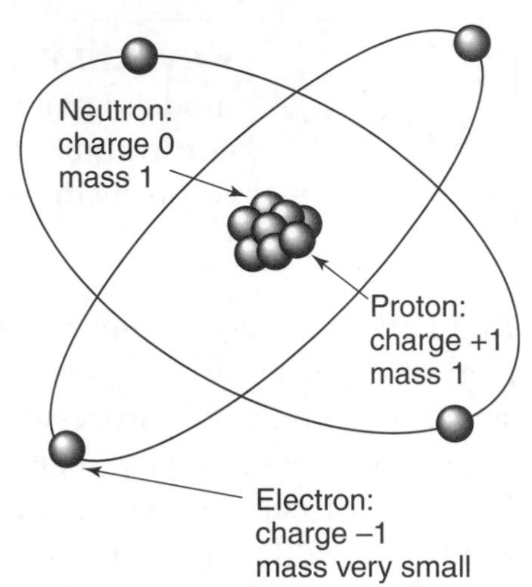

Figure 3.2
A 3-D atomic model.

However, the way things work may change with scale. Some variables change at different rates with changes in scale. For example, as something changes size, its volume changes out of proportion to its surface area. A cube 10 cm on a side has a volume of 1,000 cm^3; a cube 20 cm on a side has a volume of 8,000 cm^3! Therefore, a large tank of water will cool more slowly than a small model tank of water. Since different properties are affected to different degrees by changes in scale, this must be taken into account when designing, building, and testing models.

Many different kinds of models can be made. Some common examples are physical, mental, and mathematical models.

PHYSICAL MODELS

A ***physical model*** is an object, built to scale, that represents in detail another object, event, or process. There are many types of physical models. Some are merely shaped like the real thing, such as a doll or a statue. Some are *mechanical models* that have working parts that permit them to function or move like the real object. Examples of these include an electric train, a radio-controlled airplane, or the solar system model shown in Figure 3.3.

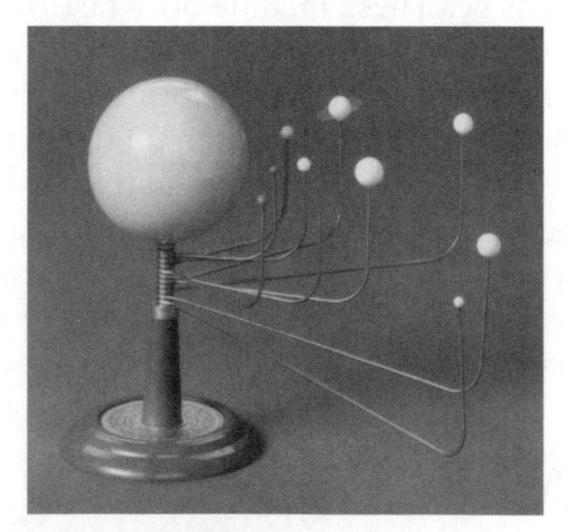

Figure 3.3
Mechanical planetarium.

Sometimes the object may be a *visual model*, something that looks like the real object, such as a globe, a map, a photograph, a drawing, or a set of blueprints. Visual models are often used to represent events or processes. The drawing in Figure 3.4 is a visual model of a natural process—the water cycle. Visual models often help us think about complex natural processes and how changing one part of the process might affect another part of the process.

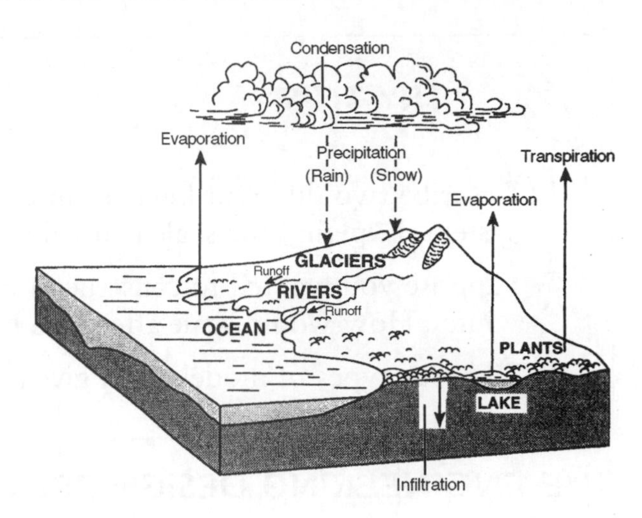

Figure 3.4
Diagram of the water cycle.

MENTAL MODELS

Mental models exist only in your mind. They represent things that cannot be seen. Many scientific concepts cannot be represented by a physical model. You cannot construct a physical model that shows what happens when light enters a black hole. However, when you study physics and earth science, you will develop a mental model of this event.

MATHEMATICAL MODELS

Mathematical models use symbols, such as words or numbers, to represent objects or systems and the ways in which they are related to one another. For example, the scientific concept of density is the relationship between the amount of matter, or mass, that exists in a given volume. The concept of density can be expressed mathematically in words: "Density equals mass divided by volume," or in symbols: $D = M/V$. Mathematical models include equations, graphs, number sequences, geometric figures, and number lines. Although they can be used to represent real objects, events, or processes, they can never be exact in every detail.

Computers have greatly enhanced our ability to use mathematical models to represent complex systems. The meteorologist shown in Figure 3.5 is using a computer to display a mathematical model of a tornado. Changing the model to see what happens would take months to recalculate by hand. With a computer, though, the meteorologist can see what happens in just seconds.

Many different models can be used to represent the same thing. For example, a dollhouse, a picture, and a set of architectural plans can all represent a real house. The kind of model used depends on its purpose. The usefulness of a model may be limited if it is too simple or if it is needlessly complicated.

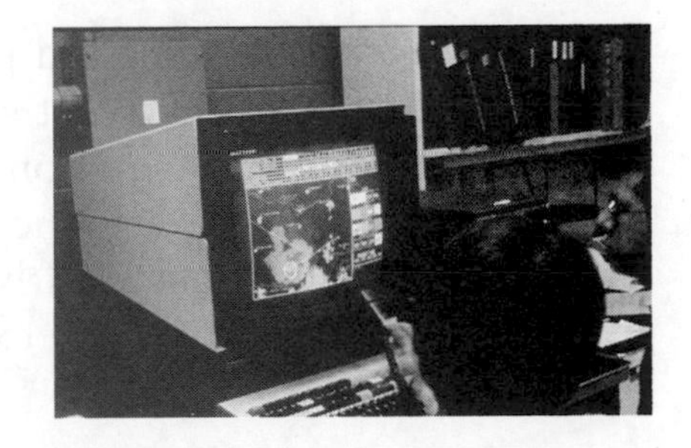

Figure 3.5
A mathematical model on the computer.

TRY THIS

1. Describe two different kinds of models you could use to represent the solar system. Explain how scale is involved in each of these models.
2. Suppose you wanted to construct a model of a lake to investigate cooling in winter. How would scale affect the behavior of your model?
3. List four types of models, and give an example of each.

THE ENGINEERING DESIGN PROCESS

Technology is the application of scientific knowledge to industrial, agricultural, or business-related problems. The engineering design process is used to develop technological solutions to these real-world problems within given constraints.

IDENTIFYING NEEDS

The first step in the process is to identify the need or opportunity for a technological solution. Not every problem can be solved by technology, but you often find parts of a problem that might be solved by technology. For example, by 1970, people recognized that most air pollution came from cars and factories. Investigation of the problem revealed many possible technological solutions. These included developing cleaner burning fuels, finding a way to remove harmful gases from exhaust, and designing engines that would burn fuel more completely. In 1970, Congress passed the Clean Air Act, which set strict limits on emissions from factory smokestacks and automobile tailpipes. Companies now had a reason to clean up car exhaust, but how? They needed ideas for how to solve the problem.

OBTAINING IDEAS

Ideas come from people, and people communicate their ideas mainly by writing and speaking. Therefore, the first step in obtaining ideas is to search printed, electronic, and human sources of information for ideas that people have already had. One such search revealed that in 1962, a French mechanical engineer, Eugene Houdry, had patented a "catalytic muffler" that greatly reduced the amount of carbon monoxide and unburned hydrocarbons in automobile exhaust. Today, the device is standard on all American cars.

PROPOSING TECHNOLOGICAL SOLUTIONS

Most problems have more than one solution, and most solutions have constraints. For example, automobile emissions can also be reduced if cars burn less fuel. There are many ways to make cars more fuel-efficient. For example, their shape can be changed to reduce drag. However, if the car is given an aerodynamic shape, it often has less passenger space. The less a car weighs, the less energy (fuel) is needed to move it. However, less weight often means less strength, and who wants a car made of parts that break easily?

OPTIMIZATION AND TRADE-OFFS

No perfect engineering design solution exists. Solutions that are best in one respect, such as strength or safety, may be inferior in other ways, such as cost or appearance. Finding the most functional and efficient solution to a problem is called ***optimization***. Which solution is considered optimum often depends on the trade-offs one is willing to accept. ***Trade-offs*** are features that must be sacrificed to get others. For example, adding lead to gas prevents engine knock. However, leaded gas ruins a car's catalytic converter. Drivers have had to accept more frequent engine knock as a trade-off for cleaner exhaust.

Technologies often have drawbacks as well as benefits. A technological solution to one problem may create other problems. For example, nuclear power plants produce electricity inexpensively. However, they also produce hazardous waste products that are difficult to dispose of safely.

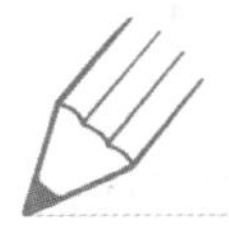

TRY THIS

Getting people from one place to another is a common problem for which there are many technological solutions. List some alternatives for getting a student from home to school. For each alternative, list some trade-offs associated with that solution.

DEVELOPING PLANS

Once a technological solution to a problem has been identified, it must be designed, built, and tested. Plans must be developed, including drawings with measurements and details of construction. It is important to realize that you may not actually be able to make or do everything you design. That is why a model of the solution is usually built first. Building a model is helpful for spotting and ironing out glitches.

TESTING A DESIGN

Even the best designs can fail. Designs fail for many different reasons. Models are useful for testing a design because models are usually easier to modify and less expensive to build than the real thing. If you test a model ahead of time, you may be able to reduce the likelihood of failure of the real system later on.

Testing a design is not usually a one-step process. When a design fails, the reason for the failure is identified, and the design is changed. Then the modified design is built and tested again. A design may have to go through several test-evaluate-modify-retest cycles before it is optimized.

Practice Review Questions

1. Which of the following would not be considered a model?

   1. a globe
   2. a radio-controlled airplane
   3. $D = M/V$
   4. a microscope

2.* 1 centimeter on the map represents 8 kilometers on the land.

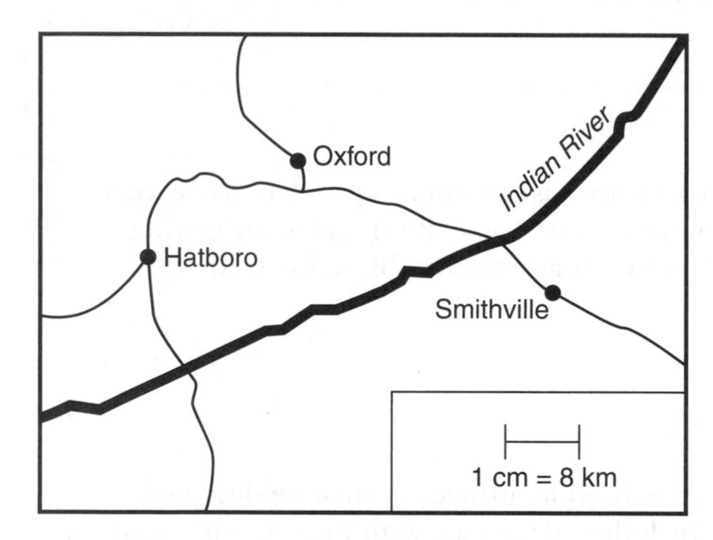

About how far apart are Oxford and Smithville on the land?

   1. 4 km
   2. 16 km
   3. 32 km
   4. 50 km

3. Which of the following statements about technology is most accurate?

   1. Technology can always provide successful solutions for problems.
   2. Technology often has drawbacks as well as benefits.
   3. Technology has had little effect on the course of history.
   4. Technology is a relatively recent development.

4. All of the following are examples of a trade-off *except*

   1. giving a car an aerodynamic shape makes it less roomy
   2. paving over land to create roads decreases the amount of water that seeps back into the ground to supply wells
   3. composting leaves decreases a town's solid wastes and supplies homeowners with fertile soil for their gardens
   4. burning garbage generates energy and reduces solid wastes, but the process releases harmful gases into the atmosphere

GO ON

5. A car company has built a wind tunnel to test the amount of wind resistance on different car body shapes. Decreasing wind resistance enables a car to get better gas mileage. The company is trying to design a shape that minimizes wind resistance while maximizing interior space. This is an example of

1. peer review
2. symbolic representation
3. optimization
4. a mental model

6. In designing engineering solutions to problems, which of the following should be taken into account?

1. scientific laws and engineering principles
2. properties of materials and construction techniques
3. cost, safety, appearance, and environmental impact
4. all of the above

7. Models are useful when testing an engineering design because

1. models always behave exactly like the original
2. models are easier to modify and less expensive than the original
3. models never fail and always provide successful solutions to problems
4. models are always technologically superior to the original

* Reproduced from TIMSS Population 2 Item Pool. Copyright © 1994 by IEA, The Hague.

CONSTRUCTED-RESPONSE QUESTIONS

8.* It takes 10 painters 2 years to paint a steel bridge from one end to the other. The paint that is used lasts about 2 years. So when the painters have finished painting at one end of the bridge, they go back to the other end and start painting again.

a. Why MUST steel bridges be painted?

b. A new paint that lasts 4 years has been developed and costs the same as the old paint. Describe two consequences of using the new paint.

9.** Machine *A* and Machine *B* are used to pump water from a river. The table shows the volume of water each machine removed in one hour and how much gasoline each of them used.

	Volume of Water Removed in 1 Hour (liters)	Gasoline Used in 1 Hour (liters)
Machine *A*	1,000	1.25
Machine *B*	500	0.5

a. Which machine is more efficient in converting the energy in gasoline to work?

Answer: ______________________

b. Explain your answer.

____

____

** Reproduced from IEA TIMSS 1999 Science Released Items for Grade 8, TIMSS International Study Center, Boston College, MA

Unit 2

The Living Environment

Standard 4: Students will understand and apply scientific concepts, principles, and theories pertaining to the physical setting and living environment and recognize the historical development of ideas in science.

How many different kinds of animals are there? How do plants get food? Where did the dinosaurs go? How does my body work? What do all of these questions have in common? They have to do with the living environment—the living things that surround and interact with us. The search for answers to questions like these is the goal of biology, the branch of science that deals with the living environment.

Biology tries to figure out what it means to be alive and how living things do some of the things they do. Biologists study the insides of plants and animals to figure out how they are put together and how they work. They study the relationships between living things and the environments in which they live. They also study how living things change during their lifetime and over long periods of time.

The chapters in this unit will focus on the characteristics of life, the structure of living things, how living things maintain and continue life, how they change over time, and how living things live in their environments and depend on both the living and nonliving environment for their survival.

Chapter 4

The Unity and Diversity of Life

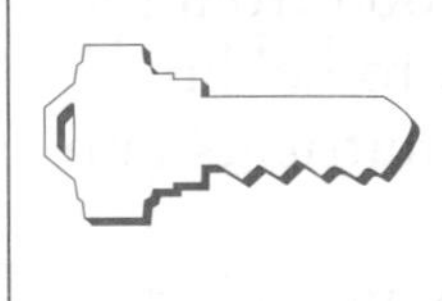

Key Idea: Living things are both similar to and different from each other and from nonliving things. Living things are similar to each other yet different from nonliving things. The cell is a basic unit of structure and function of living things (cell theory). For all living things, life activities are accomplished at the cellular level.

THE CONCEPT OF LIFE

Most people have a pretty good sense of what the word *living* means. However, it is hard to come up with an exact definition of life. One dictionary defines *life* as "the quality that distinguishes a vital and functioning being from a dead body" and *dead* as "deprived of life." In other words, life is what something has when it is not dead, and dead is what that same thing is when it lacks life. This kind of circular reasoning gets us nowhere.

If you walked through a forest in springtime and saw trees sprouting new leaves, you would consider them living. Suppose you came across a tree that had no leaves and whose bark had fallen off. Would you consider it to be living? You would probably decide that the tree was dead. The growth of new leaves is a sign of life. ***Biologists***, or scientists who study life, have not been able to agree on a simple definition of life. However, they have been able to agree on the signs of life, that is, the characteristics that all living things have in common that set them apart from nonliving things. One way in which all living things are alike is that they all perform certain activities in order to stay alive, called ***life functions***. Another characteristic living things share is that they are all made of tiny individual units called ***cells***. When taken together, being made of cells and performing life functions characterize life, and this defines our concept of life. Life functions are what we use to decide whether a given object is living, dead, or nonliving.

THE UNITY OF LIFE

Living things are both similar to and different from one another and from nonliving things. The ways in which all living things are similar to one another is called the ***unity of life***. Performing life functions and being made of cells are the common threads that tie all living things together and make all life similar.

LIFE FUNCTIONS

Every living thing performs *all* of the life functions. No matter how different from one another living things may be, they must all carry on these activities to stay alive. The life functions include the following eight processes.

Nutrition is the intake and use of food by living things. Food contains nutrients. *Nutrients* are substances that an organism can use to obtain energy or to supply the materials needed for the growth and repair of the body. The taking in of food from the environment is called *ingestion*. Most nutrients, however, are too complex to be directly used by the organism. Therefore, the organism must first break down the nutrients into simpler substances usable by cells, a process called *digestion*.

There are two basic types of nutrition. Green plants and some bacteria and other one-celled organisms are able to make their own food. They take in simple substances from their environment and use the Sun's energy to change them into food. Organisms that cannot make their own food use plants and other animals for food.

Transport is the process by which the simple substances obtained from nutrients are absorbed by the organism and circulated throughout the organism. Transport also includes the movement of wastes and other products of life processes from one place to another within the organism.

Respiration includes the processes by which the energy stored in food is changed into a form that can be directly used by the living cells in an organism. Respiration involves a complex series of chemical reactions. When food is oxidized (chemically burned), water and carbon dioxide are formed. The chemical bonds in food contain more energy than the chemical bonds in water and carbon dioxide. When food is oxidized, the "excess" energy is released and used by the living cells in the organism. The water and carbon dioxide are wastes.

Excretion is the process by which the waste products of life functions are removed from an organism. A buildup of wastes from respiration and other life functions could be harmful to an organism. Some wastes are poisons. Therefore, excretion is essential if an organism is to remain alive.

Synthesis involves the processes by which large molecules are built from small molecules. During synthesis, the small molecules produced by the digestion of food are built up into larger molecules that can be used to build cells, to repair or replace worn-out parts, or to grow. The process of making materials part of the body of an organism is called *assimilation*.

Regulation includes all of the processes by which an organism responds to changes within itself and in its surroundings. These responses control and coordinate the life functions. Most organisms respond in ways that protect them. For example, your body is sensitive to changes in temperature. If it gets too cold, your body may respond by shivering. Shivering muscles produce heat.

An organism may also respond by making itself move, or *locomotion*. When you accidentally touch a hot stove, your hand almost immediately jerks away. If someone throws a snowball at you, you respond by ducking out of the way.

Growth is the process by which living things increase in size and in the amount of material they contain. It is one result of the assimilation of nutrients. In one-celled organisms, growth is simply an increase in the size of the cell. In organisms made of many cells, growth involves increasing both the size and the number of cells in an organism.

Reproduction is the process by which living things produce offspring, or new organisms of their own kind. In reproduction, new cells arise from preexisting cells. Living things reproduce to perpetuate their own kind and pass their characteristics on to their offspring.

These life functions are all accomplished through a huge number of chemical reactions, which together are called ***metabolism***. All of the life functions require energy. Living things use energy to do work, grow, and maintain themselves. Energy is central to their ability to sense and react to the external environment. Living things also use energy to maintain livable conditions inside themselves no matter what their surroundings are like, a condition known as ***homeostasis***.

Nonliving objects may perform one or a few of these functions. For example, an icicle can grow as more and more water freezes on its surface. A thermostat regulates the temperature in your house. A bus transports passengers and "excretes" gases from its tailpipe. Yet nonliving things never perform *all* of the life functions.

In some cases, it may not be clear whether an object is living or nonliving. One example is a virus. Inside a living thing, viruses can carry out one life process: they can reproduce. However, does the ability to reproduce mean that something is alive? Outside of a living thing, viruses do not carry out all of the life processes. For example, they do not grow or respond to changes, and they can be stored in a jar like chemicals. For this reason, many biologists do not consider viruses living things.

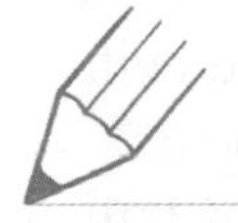

TRY THIS

Explain why an automobile is not considered a living thing, even though it moves; "ingests" fuel; oxidizes fuel to release energy; "excretes" exhaust gases; has systems that transport fuel, air, and coolant; and has devices that regulate its operation, such as thermostats to regulate temperature and fuel injection systems to regulate the flow of fuel and air to the engine.

CELLS

The other characteristic that all living things share is that they are all made up of small structures called cells. Some living things are made up of just one cell. Others are made up of billions of cells. Whether one celled or many celled, a whole living thing is called an ***organism***. As biologists have studied living things, it has become clear that in all living things, the life functions are actually carried out by the cells. For this reason, ***cells*** are said to be the smallest unit of structure and function in a living thing.

THE DEVELOPMENT OF THE CELL THEORY

Our current ideas about cells can be traced back to the mid-1600s. In 1665, a scientist named Robert Hooke used a microscope to look at thin slices of cork and other plant materials. Through the microscope, he saw boxlike structures, which he called *cells*. What Hooke actually saw were the cell walls of dead plant cells.

At about the same time, a Dutch fabric merchant named Anton van Leeuwenhoek made a hobby of building simple, but very powerful, microscopes. Leeuwenhoek's curiosity led him to observe almost anything that could be placed under his lenses. In the process, he discovered living things that no one else had ever seen.

As the years passed, scientists used microscopes to look at material from all kinds of living things. They found cells in the roots, stems, and leaves of plants. They found cells in skin, bones, muscle, blood, and all kinds of other materials from animals. In 1831, Robert Brown noted that all plant cells contained a small, round mass that he called the *nucleus*. This was the first of many structures that would be found within cells. By 1838, Matthias Schleiden put forward the idea that all plants were made of cells. The following year, Theodor Schwann proposed that all animals were also made of cells. In 1855, Rudolph Virchow added the final part of the theory by stating that all cells form from already existing cells.

By the end of the nineteenth century, biologists had discovered many of the structures found within cells. They had also observed and described the sequence of events in ***cell division***, the process by which one cell divides to form two or more cells.

A ***theory*** offers an explanation for a large number of observations. By 1900, biologists had put together their findings and ideas in the form of the ***cell theory***. The cell theory has three main parts.

1. **All living things are made up of cells.** Some organisms consist of a single cell, such as an amoeba or bacteria. Others consist of billions of cells, like you.
2. **All cells perform the life functions.** The life functions of a many-celled organism are the combined effects of the life functions of its individual cells. For example, the liquid wastes you excrete are the liquid wastes collected from all of the cells in your body.
3. **All cells arise from other living cells.** Cells arise from other living cells by the process of cell division or by reproduction. Many-celled organisms reproduce by the reproduction of certain cells. This idea had an important part in the theory of evolution (which will be discussed later).

These simple statements emphasize the unity of life, the basic sameness of all living things. Though living things may differ in many ways, they all share a common unit. They are all made of cells.

CELL STRUCTURES

Today's microscopes show that cells contain many small structures. Each structure helps the cell to perform a life function. Although there are many different kinds of cells, many share certain basic structures. Figure 4.1 shows a typical plant cell and a typical animal cell. Notice that both have a nucleus, cytoplasm, and a cell membrane. The ***nucleus*** is the control center for the cell's activities. It directs everything the cell does. Structures inside the nucleus called ***chromosomes*** store genetic material containing the directions for all cell activities. The nucleus is surrounded by a thin membrane that seems to control the flow of substances into and out of the nucleus. The entire nucleus is embedded in a jellylike fluid called ***cytoplasm***. Many other cell parts are scattered throughout the cytoplasm, and it is here that most of the life functions occur. A saclike lining called the ***cell membrane*** surrounds the cytoplasm. The cell membrane holds the contents of the cell together and controls the movement of substances into and out of the cell.

Notice that the animal cell also has a part you will not find in plant cells—the *centrioles*. ***Centrioles*** are located near the nucleus and are involved in cell division in animal cells.

Plant cells also have structures you will not find in animal cells. Plant cells are surrounded by a rigid, nonliving ***cell wall*** that protects the cell and gives the cell its shape. The cell wall is made mostly of cellulose and has many small openings through which substances can pass freely to and from the cell membrane. Many plant stem and leaf cells contain small, green bodies called ***chloroplasts***, cell structures that trap the Sun's energy and use it in a process that makes food. Chloroplasts are green because they contain a special green substance called *chlorophyll*.

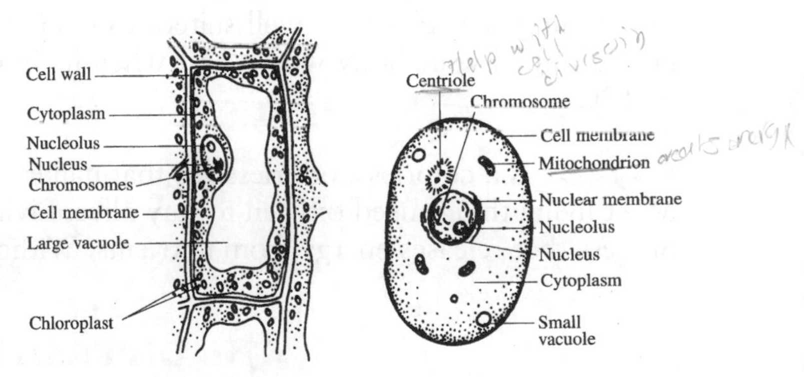

Figure 4.1
Typical plant (*left*) and animal (*right*) cells.

TRY THIS

1. What are some cell structures found in both plant and animal cells?
2. What cell structures are found only in plant cells? Only in animal cells?
3. Why are plant cells usually more rigid than animal cells?

THE NEEDS OF LIVING THINGS

All living things are composed of matter and need energy in order to stay alive. Everything an organism needs to stay alive is found in its environment. Some of the most basic needs of all living things include the following four items.

Water Life functions involve chemical reactions among a wide variety of substances. The most important is water. Water provides the medium in which all of these chemicals can dissolve and interact. It is the chemical soup in which the chemical reactions of metabolism take place. Most living things are 60–90% water, though jellyfish can be up to 95% water.

Light Sunlight is the main source of energy for all living things. Green plants use sunlight as an energy source to convert carbon dioxide and water into more complex chemical compounds (nutrients). In the process, the light energy is stored in the chemical bonds of these complex molecules. This energy is later released by respiration in other organisms that take the plants in as food.

Temperature The Sun also supplies the energy that warms Earth. All organisms live in a thermal environment. The temperature of the environment influences life in many ways. It affects the rate at which all of the chemical reactions of metabolism take place. It affects the rate of evaporation and thus the need of all living things for water. Temperature also governs climate and thereby the places where organisms can live. Every living thing is adapted, or well suited, to living within a certain range of temperatures. For example, your body works best when its internal temperature is between 36.1°C and 37.8°C.

Oxygen is a colorless, odorless gas that makes up about 21% of Earth's atmosphere. Most living things need oxygen to stay alive. Living things need oxygen for respiration, the process that releases energy from nutrients. Without energy, living things cannot stay alive.

THE DIVERSITY OF LIFE

About 1.5 million different kinds of things live on Earth, and more are being identified each year. The unity of life is that all of these living things are made of cells and perform the same life functions. The ***diversity*** of life is the great variety of ways in which different organisms perform those life functions.

CLASSIFICATION

An important part of scientific thinking is organizing observations. One way that scientists organize their observations of living things is by classification. ***Classification*** is the sorting of living things into groups based on their similarities and differences. Classifying living things makes studying them easier and understanding the relationships among them easier. The branch of science that deals with classifying living things into groups is known as ***taxonomy***.

Although people have always classified living things, for centuries there was no one system that everyone agreed on. Different people had different opinions about how organisms should be grouped. Some people used vague reasons for putting living things into groups, such as *domestic animals* or *water animals*.

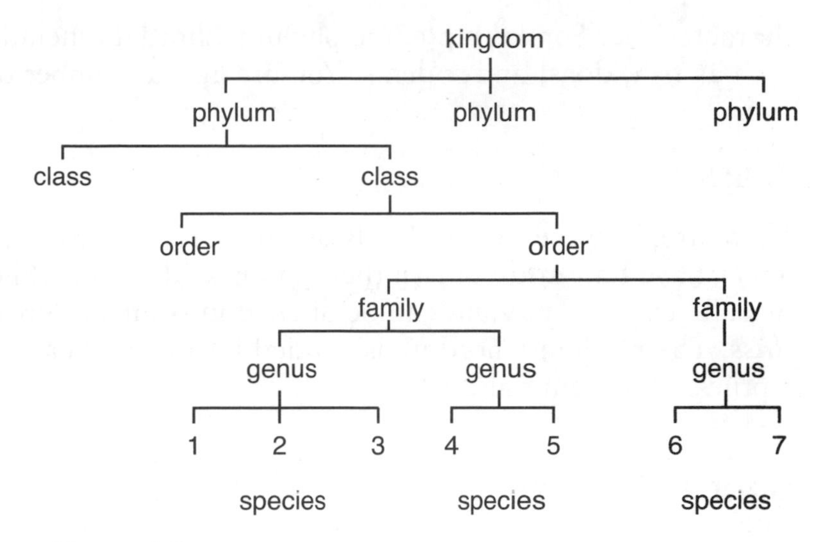

Figure 4.2
Categories for classifying living things.

Our modern classification system can be traced back to the work of the Swedish doctor and naturalist Carolus Linnaeus (1707–1778). Linnaeus's innovation was to group organisms according to the characteristics they have in common and to identify each organism using a two-name, or binomial, system. He also developed a system for grouping organisms into increasingly general categories: kingdom, class, order, genus, and species—phylum and family were added later. See Figure 4.2.

KINGDOM

Linnaeus divided all living things into two great groups, or ***kingdoms***: plants and animals. As time went on, the number of known living things swelled. Microscopes revealed tiny organisms that did not fit clearly into either kingdom. As a result, many scientists today recognize five kingdoms. See Figure 4.3. Members of each kingdom have characteristics in common. For example, suppose you found a living thing with green leaves, a stem, and roots. Even if you did not know its name, you would know enough to classify it as a member of the plant kingdom.

PHYLUM

All animals share certain characteristics. They take in food and move about by walking, crawling, flying, or swimming. However, some animals have backbones, some have shells, and others have no hard parts at all. These characteristics can be used to divide the animal kingdom into smaller groups called *phyla* (singular, phylum.) A ***phylum*** is a major group within a kingdom whose members share at least one special

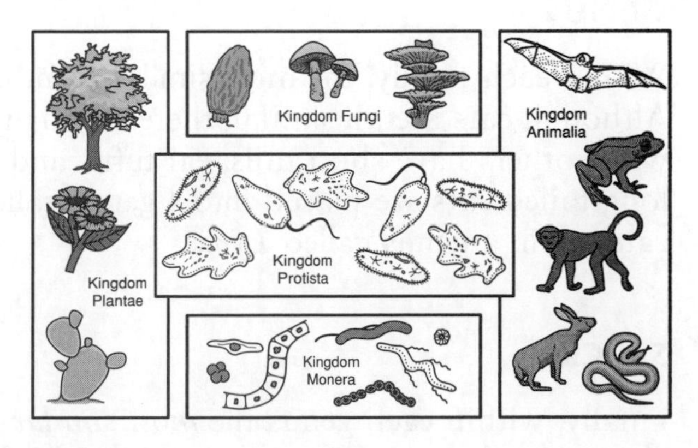

Figure 4.3
The five kingdoms.

characteristic. For example, the phylum Chordata includes all animals with a backbone, such as fish, dogs, and rodents. You are also a member of the phylum Chordata.

CLASS

There are, however, many kinds of chordates. For example, some animals with backbones have hair and nourish their young with milk, while others have feathers and lay eggs. Thus, each phylum can be divided into smaller groups, each of which is called a *class*. The phylum Chordata is divided into a number of classes, including birds, fish, reptiles, and mammals.

ORDER

Members of a class that have a lot in common are put into even smaller groups, each of which is called an ***order***. For example, some mammals, such as cats, dogs, bears, and skunks, have a well-developed sense of smell, sharp claws, and powerful jaws with large teeth specialized for seizing, cutting, and tearing meat. These members of the class are grouped into the order Carnivora. Other mammals, such as African and Asian elephants, have a muscular trunk and greatly enlarged upper incisor teeth, or tusks. These members of the class are grouped into the order Proboscidea.

FAMILY

Within an order, the most similar members are grouped into a ***family***. For example, cats, dogs, and bears share certain characteristics and are therefore all members of the order Carnivora. However, a lion is clearly more similar to a leopard than it is to a dog. So all of the different catlike Carnivora are grouped together in the family Felidae. A fox is clearly more similar to a wolf than it is to a bear, so all of the different doglike Carnivora are grouped together in the family Canidae.

GENUS

Within each family, the most similar members are grouped into a ***genus*** (plural, genera). Although cats are all similar, there are many kinds of cat. Some cats have long tails, while others have short tails, ear tufts, and a ruff of fur on their cheeks and throat. So, long-tailed cats are placed into a genus called *Felis* (Latin for cat), and the short-tailed cats are in a genus called *Lynx*.

SPECIES

Finally, within each genus the *most similar* organisms are grouped into ***species.*** For example, all of the long-tailed cats can be further subdivided based on differences in color, size, thickness of hair, and the like. The mountain lion is *Felis concolor*, the

African lion is *Felis leo*, and the house cat is *Felis catus*. Organisms within a species are so similar that they can mate and reproduce more of their own kind.

By using this system, every organism can be identified by the kingdom, phylum, class, order, family, genus, and species to which it belongs. The scientific name of an organism is usually a combination of its genus name and its species name. The first letter of the genus is capitalized, the first letter of the species is not. Both names are written in italics or are underlined. Look at Table 4.1 below. If you put together the genus name *Felis* and the species name *catus*, you have the scientific name *Felis catus*. To scientists around the world, *Felis catus* means the common house cat. What would be the scientific name for a tiger?

You can think of the scientific name of a living thing as its biological address. Your address identifies you out of all the people on Earth. A typical address includes your country, state, city, zip code, street, house or apartment number, and name. The country contains the largest number of people. As you move down the categories from state, to city, to zip code, to street, and finally to house or apartment number, the number of people with that same description gets smaller and smaller. Biological classification works in much the same way. A kingdom contains a huge number of different living things. As you move down the groups to species, the number of different kinds of living things in each group gets smaller and smaller.

Linnaeus classified living things based on their outward appearance. Today, scientists group living things according to how closely they are related. What do we mean by closely related? Most modern systems of classification are based on the ***theory of evolution***. The theory of evolution proposes that long ago, very simple forms of life

Table 4.1 **The Classification of Some Common Living Things**

Category	Human	House Cat	Tiger	Bread Mold	Pond Algae	Red Oak
Kingdom	Animalia	Animalia	Animalia	Fungi	Protista	Plantae
Phylum	Chordata	Chordata	Chordata	Zygomycota	Chlorophyta	Tracheophyta
Class	Mammalia	Mammalia	Mammalia	Zygomycetes	Euconjugatae	Angiospermae
Order	Primates	Carnivora	Carnivora	Mucorales	Zygnematales	Fagales
Family	Hominidae	Felidae	Felidae	Mucoraceae	Zygnemataceae	Fagaceae
Genus	*Homo*	*Felis*	*Panthera*	*Rhizopus*	*Spirogyra*	*Quercus*
Species	*sapiens*	*catus*	*tigris*	*stolonifer*	*crassa*	*rubra*
Scientific name	*Homo sapiens*	*Felis catus*	*Panthera tigris*	*Rhizopus stolonifer*	*Spirogyra crassa*	*Quercus rubra*

arose by natural processes. Over millions of years, these primitive life-forms gradually changed, or *evolved*, into all of the different types of organisms on Earth today. It follows from this theory that all present forms of life evolved from earlier forms of life. Thus, any two organisms could be traced back to a common ancestor. The more recently the common ancestor of the two organisms existed, the more closely related the modern organisms are. See Figure 4.4.

Closely related organisms are likely to have similar forms and structures. That is why biologists classify living things into groups based on their physical structure. Other factors, such as genetic, embryonic, and chemical similarities and differences, are also used.

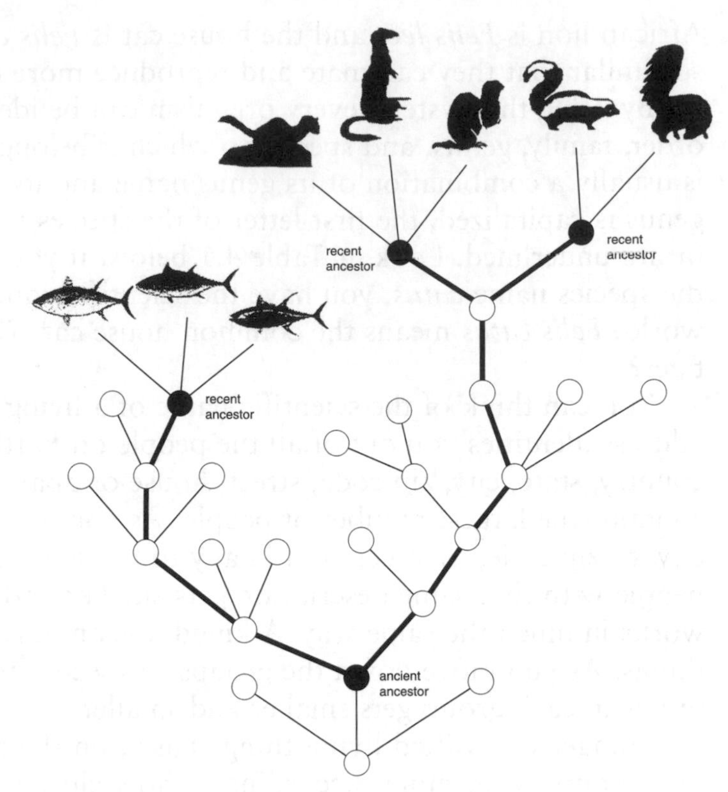

Figure 4.4
Closely related and distantly related common ancestors.

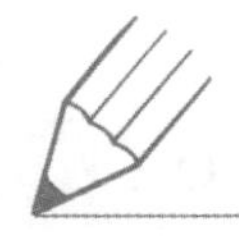

TRY THIS

These three organisms share a common ancestor. Which two organisms most likely share the most *recent* common ancestor?

Category	Mink	Fur Seal	Sea Otter
Kingdom	Animalia	Animalia	Animalia
Phylum	Chordata	Chordata	Chordata
Class	Mammalia	Mammalia	Mammalia
Order	Carnivora	Carnivora	Carnivora
Family	Mustelidae	Otariidae	Mustelidae
Genus	*Mustela*	*Callorhinus*	*Enhydra*
Species	*vison*	*ursinus*	*lutris*
Scientific name	*Mustela vison*	*Callorhinus ursinus*	*Enhydra lutris*

Practice Review Questions

1. One characteristic of *all* living things is that they

1. ingest food
2. respond to changes by moving
3. are made of cells
4. are made of many cells

2. In multicellular living things, cell division is required for tissue repair and

1. excretion
2. respiration
3. growth
4. circulation

3. Which life function involves changing the energy stored in food into a form that can be directly used by cells?

1. excretion
2. respiration
3. synthesis
4. transport

4. Which life function is shown by the diagrams below of the same cell?

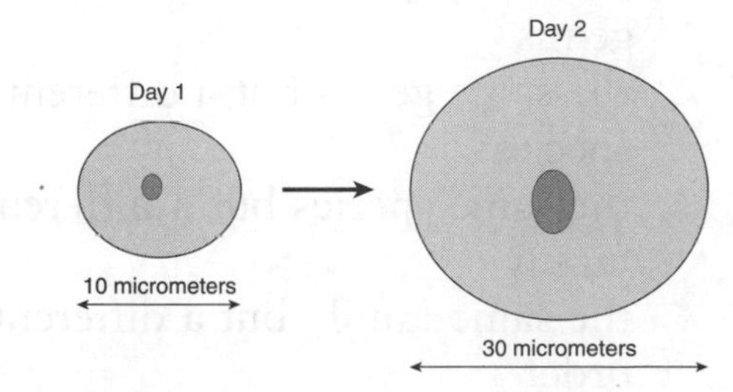

1. excretion
2. reproduction
3. growth
4. transport

5. A one-celled living thing absorbs substances from its environment and circulates them throughout its cytoplasm. Of which life function is this an example?

1. transport
2. synthesis
3. reproduction
4. respiration

6. Living things combine simple molecules to form complex molecules by the process of

1. ingestion
2. synthesis
3. excretion
4. respiration

7. Which term includes all the activities that sustain life?

1. homeostasis
2. synthesis
3. metabolism
4. excretion

8. While waiting for the school bus one winter morning, a child's body temperature decreases. The child responds by shivering, which raises body temperature. This process is an example of

1. regulation
2. classification
3. nutrition
4. growth

GO ON

9. The idea that all cells come from pre-existing cells was proposed by

   1. Schleiden and Schwann
   2. Hooke
   3. Virchow
   4. Van Leewenhoek

10. According to the cell theory, which statement is correct?

    1. Cells can arise from nonliving material.
    2. Cells are parts of living things but are not themselves alive.
    3. Some living things are not composed of cells.
    4. Cells perform all the life functions.

11. The diagram below shows a cell.

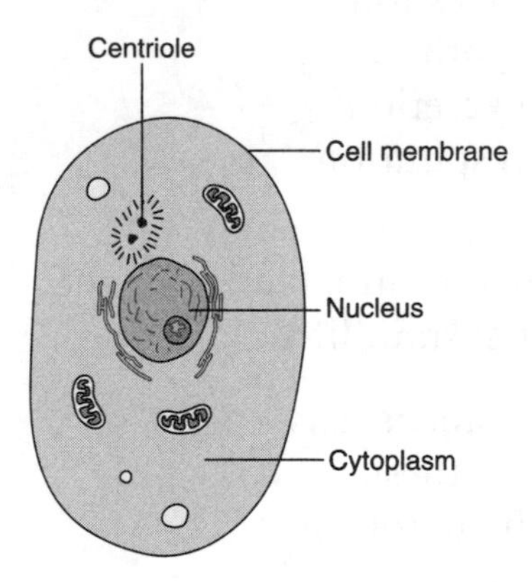

In which type of organism would this cell be found?

1. plant
2. animal
3. virus
4. fungus

12. Which structure appears in the cells of a carrot but not in the cells of a rabbit eating the carrot?

    1. nucleus
    2. cytoplasm
    3. cell membrane
    4. cell wall

13. Biological classification is based primarily on

    1. size of organisms
    2. habitat in which organisms are found
    3. evolutionary relationships between organisms
    4. an organism's nutrition

14. The correct order for classifying an organism from the broadest to the most specific category is

    1. kingdom, genus, order, family
    2. class, genus, kingdom, species
    3. order, kingdom, class, family
    4. kingdom, phylum, genus, species

15. A mountain lion is classified as *Felis concolor*, and a house cat as *Felis catus*. These two animals belong to

    1. the same species but a different genus
    2. the same genus but a different species
    3. the same species but a different family
    4. the same family but a different order

Chapter 5

Organization in Living Things

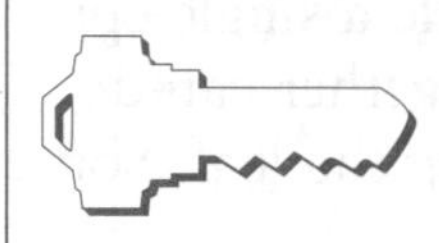

Key Idea: Living things are both similar to and different from each other and from nonliving things. Human beings are an interactive organization of cells, tissues, organs, and systems.

Some living things are ***unicellular***; their bodies are made of just one cell. Bacteria and algae are unicellular. In unicellular organisms, specialized cell structures carry out life functions. Other living things are ***multicellular***; their bodies are made of many cells. Most plants and animals, such as trees, tulips, cats, dogs, and people, are multicellular organisms. In multicellular living things, cells work together in groups that form body parts that are specialized to carry on life functions.

LEVELS OF ORGANIZATION IN LIVING THINGS

CELLS

All living things are made of basic units called ***cells***. Cells are called basic units because they are used for building body parts and for carrying out life functions. Just as a brick is the basic unit of a brick wall, cells are the basic units of the body.

Cells come in different sizes. Most plant and animal cells are microscopic, measuring between 5 and 50 micrometers in diameter. (A micrometer is 1/1000 of a millimeter!) The average human egg cell is 100 micrometers in diameter, but some egg cells are very large. Think of how large an ostrich egg can be!

Cells also come in different shapes, such as spheres, cubes, and rectangles. Cells may be long or short, thick or thin, wide or narrow. The shape of a cell is related to the job or function it performs. See Figure 5.1.

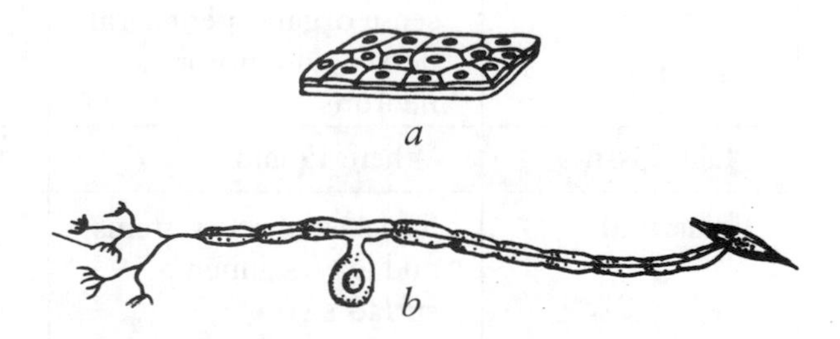

Figure 5.1
Cells have shapes suited for their functions. (*a*) Epithelial cells are flat and thin like a floor tile, a good shape for covering a lot of surface area with the least amount of material. (*b*) Nerve cells are long and thin like a wire, a good shape for covering large distances with the least amount of material.

TISSUES

In multicellular organisms, cells form groups and have become specialized, or adapted, to perform certain tasks. A group of cells that are similar and act together to perform a life function is called a ***tissue***. For example, muscle cells work together to cause movement. A single microscopic cell cannot move an entire human arm. However, many muscle cells working together as muscle tissue can do the work. Some of the different types of animal tissues are shown in Table 5.1 below.

Plants are also made of many different kinds of cells. For example, inside a plant leaf, each kind of cell is specialized to do a specific job. Epidermis cells cover and protect the leaf. Xylem cells and phloem cells transport materials throughout the leaf. Some fundamental cells contain chloroplasts and are where photosynthesis takes place. While a single epidermis cell is too tiny to cover an entire leaf, many epidermis cells joined together can do the job. It is the epidermis *tissue* that does the job of covering and protecting the leaf. Some of the different types of plant tissues are shown in Table 5.1.

Table 5.1 **Types of Tissues**

Animal Tissues	Where Found	Function of Tissue
Epithelial	Skin, linings of airways and digestive tract, linings and coverings of organs and glands	Provide a barrier between organs and the external environmental to protect organism from microorganisms, injury, and fluid loss. Secretion and absorption.
Connective	Tendons, ligaments, bones, cartilage, blood	Give shape to organs and hold them in place. Often made of cells separated by a nonliving substance, e.g., bone cells embedded in minerals, cartilage cells embedded in flexible proteins, blood cells suspended in liquid.
Muscle	Skeletal muscles, organs, heart	Contract to produce force and cause motion such as locomotion or movement in internal organs.
Nerve	Brain, spinal cord, sense organs, peripheral nerves and motor neurons	Transmit communications throughout the body.
Plant Tissues	**Where Found**	**Function of Tissue**
Epidermal	Surfaces of roots, stems and leaves; inner surfaces	Provide barrier between organs and the external environment to protect organism from microorganisms, injury, and fluid loss. Secretion and absorption.
Vascular	Tubes running inside roots, stems, and leaves	Transport materials throughout the plant. Xylem conducts water and dissolved minerals upward from plant roots; phloem transports food to all plant parts.
Fundamental	Basic tissue that makes up most of the plant body	Provide support for stems and leaves, tissue where photosynthesis takes place.

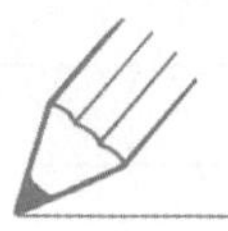

TRY THIS

1. What is the difference between a cell and a tissue?
2. List the types of tissues that compose your hand.

ORGANS

Tissues join together to form organs. An ***organ*** is a group of tissues working together to perform a life function. Organs are the main working parts of plants and animals. The heart is an organ that pumps blood throughout the body. The heart is mostly muscle tissue, but it also contains epithelial tissue, nerve tissue, and blood tissue. The lungs are also organs. So are the stomach, liver, eyes, ears, brain, and bladder. Many plants have roots, stems, leaves, and reproductive structures. These organized groups of tissues are responsible for a plant's life functions.

ORGAN SYSTEMS

A group of organs that work together to carry out a specific set of life functions is called an ***organ system***. For example, the mouth and stomach are parts of the digestive system. They work together to carry out the process of nutrition, taking in food and breaking it down into a usable form. Many organisms have similar organs and specialized organ systems. Think of how many different living things have a mouth and a stomach as part of their digestive system. An ***organism*** is nothing more than a group of organ systems that work together to perform all of the life functions. As Figure 5.2 shows, groups of cells form tissues. Groups of tissues that work together form organs. Different organs that work together form organ systems. All of the organ systems work together to make the organism. Table 5.2 lists the major human organ systems.

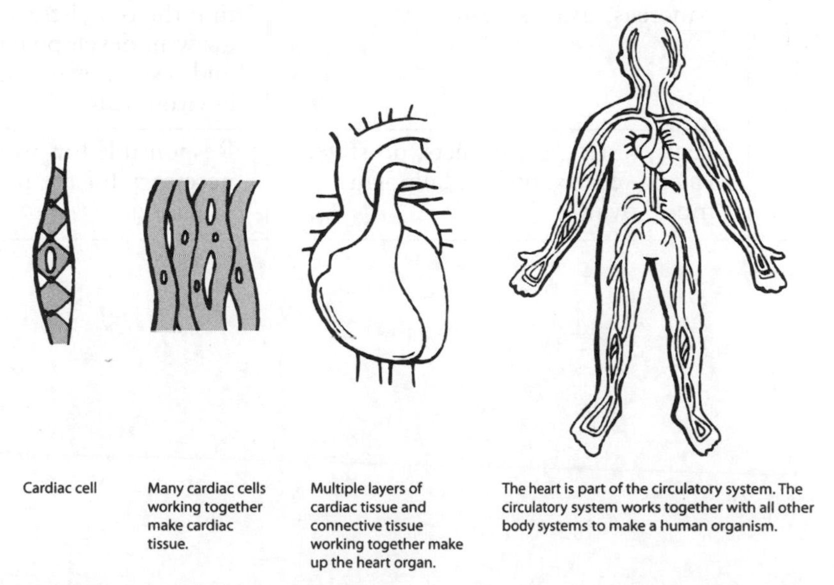

Figure 5.2
Cells, tissues, organs, and organ systems.

Table 5.2 **Major Human Organ Systems**

Organ System	Organs	Function
Digestive	Mouth, esophagus, stomach, liver, pancreas, gallbladder, small intestine, large intestine, rectum, anus	Mechanical and chemical breakdown of food
Respiratory	Lungs, air sacs, windpipe, bronchial tubes, epiglottis, diaphragm	Gas exchange—supplies oxygen and removes carbon dioxide
Circulatory	Heart, veins, arteries, capillaries, lymph nodes, red blood cells, white blood cells	Moves substances to and from cells where they are needed or produced, responding to changing demands
Excretory	Kidneys, urinary bladder, urethra, ureters, skin	Disposal of dissolved waste molecules, elimination of liquid and gaseous wastes, and removal of excess heat energy
Skeletal	Bones, bone marrow, cartilage	Provides support, protects organs, produces blood cells, works together with muscles to accomplish locomotion, stores minerals and fats
Muscular	Muscles, tendons, ligaments	Produces motion; maintains stability and posture; produces body heat; and helps move materials within the body by contraction of muscles in organs, blood vessels, tubes, and sacs
Nervous	Brain, spinal cord, nerves, sensory organs	Obtains information about surroundings, processes and transmits information to control and coordinate body functions and responses to changes in the environment
Endocrine	Adrenal, pituitary, thyroid, pancreas, ovaries, testes	Produces and releases chemicals into the bloodstream that regulate growth, development, reproduction and responses to changes in the environment
Reproductive	Penis, testes, sperm ducts, prostate, vagina, cervix, uterus, fallopian tubes, ovaries	Responsible for producing sex cells necessary for the production of offspring

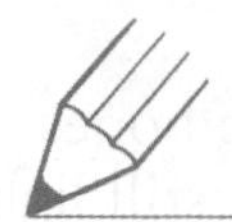

TRY THIS

1. Two organs are considered part of the same body system if the organs
   1. are located next to each other
   2. work together to carry out a life function
   3. work independently of each other
   4. are made up of cells
2. Which list of words correctly matches the diagram?

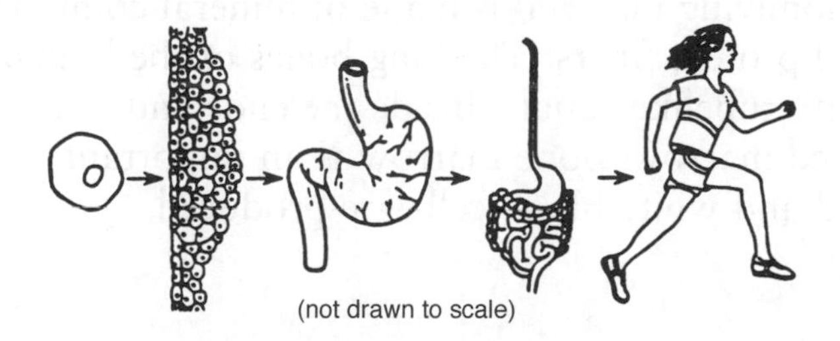

(not drawn to scale)

   1. cell → tissue → organ → system → organism
   2. cell → system → tissue → organism → organ
   3. cell → organism → tissue → system → organ
   4. cell → organism → organ → system → tissue
3. What is the difference between a tissue and an organ?

Performance Indicator

Explain the functioning of the major human organ systems and their interactions.

THE MAJOR HUMAN ORGAN SYSTEMS

Each human organ system is composed of organs and tissues that perform specific life functions and interact with each other. Tissues, organs, and organ systems help to provide all cells with nutrients and oxygen and to remove wastes.

Organ systems often work together. For example, the skeletal and muscular systems provide support, shape, and protection for body organs. They also work together to move body parts and produce locomotion. ***Locomotion*** is the movement of the body from place to place. Locomotion is necessary to escape danger, obtain food and shelter, and reproduce.

THE SKELETAL SYSTEM

The 206 bones of the human body form the ***skeletal system***. See Figure 5.3. The bones support the body, give it shape, and protect internal organs. They are strong yet still light enough to allow for ease of movement. The different sizes and shapes of bones allows the body to do many different things in many different ways.

BONES

The ***bones*** of the skeletal system are living and grow with the body. Bones are made of several kinds of tissue. Beneath a tough outer tissue, living bone cells nestle in spaces between hard, nonliving material. The nonliving material is made of mineral compounds (usually of calcium and phosphorus) and protein fibers. The long bones of the human body have spongy ends and are filled with tubelike canals. Inside the ends and inner spaces of these bones is a soft tissue called marrow. Bone marrow is an important substance. In some bones, it is where red and white blood cells are produced.

CARTILAGE

The ends of many bones also have layers of cartilage. ***Cartilage*** is a tough tissue made of living cells and nonliving material that is not as hard and rigid as bone. Cartilage is slick and flexible because its nonliving material is mostly protein fibers with little or no minerals deposited between the living cells. In adults, cartilage is found at the ends of ribs, at joints, and in the nose and outer ear. Feel the outer edge of your ear. It is contains flexible cartilage.

JOINTS

Joints are places where bones are connected to one another. Some joints, like those in the skull, are immovable. However, most joints can move. The elbows, knees, shoulders, and hips are joints. Without joints, you would be unable to move. At joints, cartilage covers the bones where they meet. The cartilage acts like a cushion and protects bones from rubbing against one another.

Figure 5.3 shows some of the different kinds of joints in your body. The joint in your elbow is a hinge joint. *Hinge joints*, like door hinges, allow back-and-forth motion. *Ball and socket joints*, like those at the shoulder and hip, can move in all directions and allow the widest range of motion. In this kind of joint, the end of one bone is ball shaped and fits into a cuplike hollow, or socket, of another bone. Your neck has a *pivot joint* that allows you to turn your head from side to side as well as up and down. Ankles and wrists have *gliding joints*; they allow limited movement in all directions.

LIGAMENTS AND TENDONS

At movable joints, bones are connected to one another by special tissues called ligaments. ***Ligaments*** are strips of tough tissue that hold the bones together but that also stretch in order to allow the bones to move. Muscles attach to bones with strong, cordlike fibers called ***tendons***.

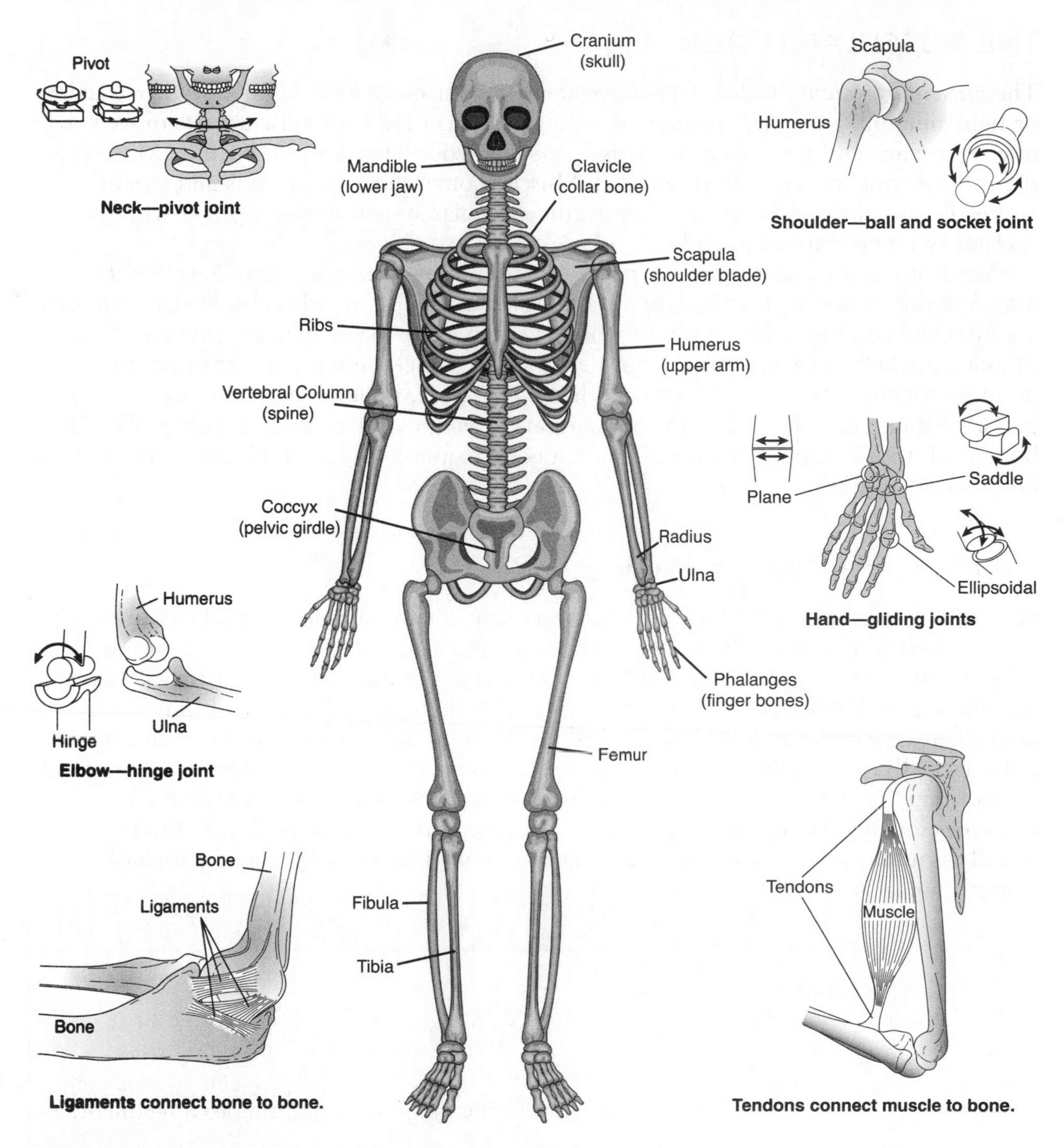

Figure 5.3
The Skeletal System.

THE MUSCULAR SYSTEM

The ***muscular system*** produces the movements of your body. Muscles work by contracting, or tightening up, and then relaxing, or resting. Different kinds of muscle tissue move your internal organs and your skeleton. Muscle tissue in your heart keeps it beating. Muscles in the wall of your stomach churn up food. Muscles connected to your arms and legs help you walk and pick up things. In this way, the muscular system works together with the skeletal system to move the body.

Most of the muscular system is made of skeletal muscles. See Figure 5.4. *Skeletal muscle* tissue is made of striped, or striated, fibers made of muscle cells. Skeletal muscles are attached to bones. The bones serve as levers to exert forces that are created when a muscle attached to them contracts, or gets shorter. Skeletal muscles exert forces by pulling on bones, never by pushing on them. Therefore, skeletal muscles often work in pairs, with one muscle pulling the bone in one direction and another muscle pulling the bone back to its original position. This is a good example of how different systems of the body work together.

Voluntary and Involuntary Muscle

Skeletal muscles are ***voluntary muscles*** because you can consciously make them work or rest. You can decide whether or not to stick out your tongue, wave your hand, or wiggle your toes. ***Involuntary muscles*** are not under conscious control; they work automatically. *Cardiac muscle* is involuntary muscle tissue that is found only in the heart. Cardiac muscle tissue is striped like skeletal muscle, but it contracts and relaxes automatically. The walls of blood vessels, the stomach, and many other organs contain smooth muscle tissue. *Smooth muscle* tissue is not striped and is involuntary. You do not consciously control smooth muscles. The sound of your stomach grumbling is usually liquids and gases being forced through your intestines by contractions of smooth muscles.

TRY THIS

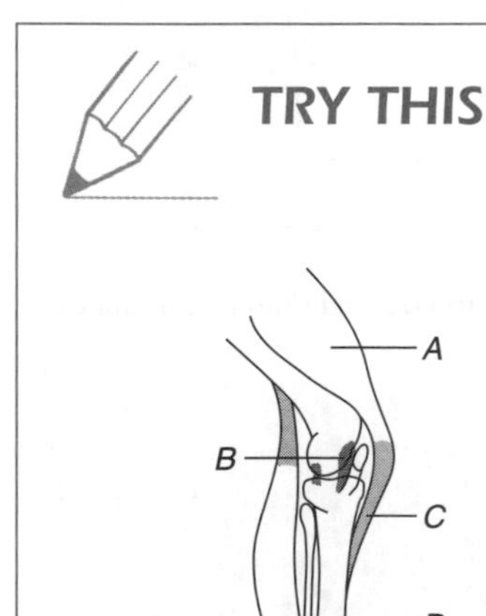

1. Which letter in the diagram is pointing to a ligament?

   1. *A* 2. *B* 3. *C* 4. *D*

2. Using one or more complete sentences, state the name and function of part C.

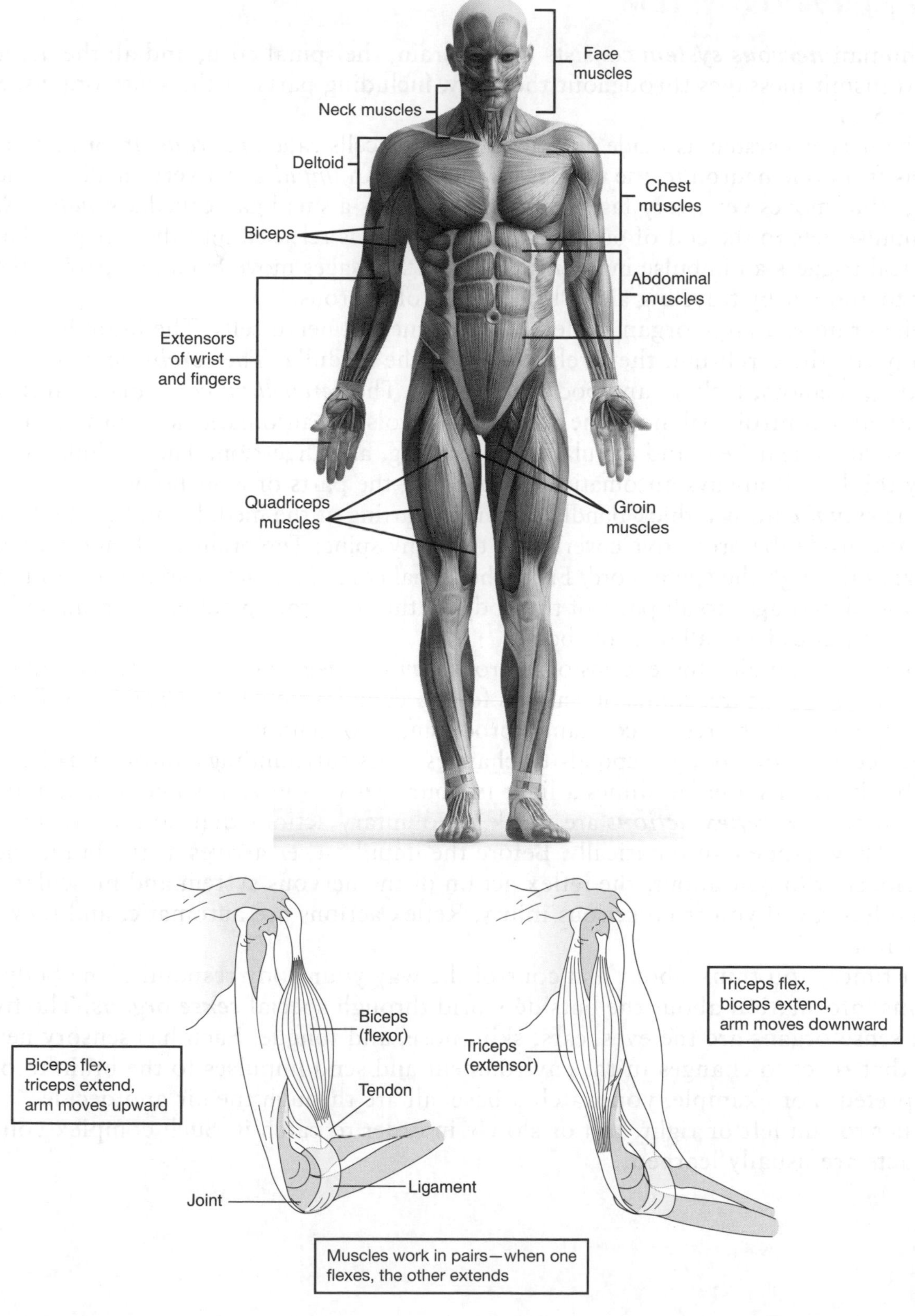

Figure 5.4
The Muscular System.

THE NERVOUS SYSTEM

The human ***nervous system*** consists of the brain, the spinal cord, and all the nerves that transmit messages throughout the body, including parts of the sense organs. See Figure 5.5.

The nervous system is made of individual nerve cells called *neurons*. Information travels from one neuron to the next as an impulse. An *impulse* is a very small electrical charge that moves very, very fast. Between neurons is a small gap called a *synapse*. When an impulse gets to the end of one neuron, a chemical is released into the synapse. This chemical triggers an impulse in the next neuron. Messages move from one part of the body to another by traveling along lengthy sets of neurons.

The brain is a large organ made of interconnected nerve cells. The brain has three main parts: the cerebrum, the cerebellum, and the medulla. The *cerebrum* controls thinking, memory, feeling, and body movement. The *cerebellum* coordinates muscular activity and controls balance. The *medulla* controls the automatic activities of the body such as heartbeat and circulation, breathing, and digestion. The medulla does not really think; it is always automatically operating the parts of your body.

The *spinal cord* is a thick bundle of nerves starting at the medulla and going down the back inside the protective covering of the bony spine. The brain sends and receives messages through the spinal cord. From the spinal cord, 31 pairs of spinal nerves branch off to send messages to all parts of the body. In this way, the spinal nerves send and receive impulses from all over the body.

The body contains three kinds of neurons: *sensory neurons, motor neurons,* and *interneurons*. The three kinds of neurons form a complex network of pathways along which the nervous system is constantly processing information.

Sometimes your body responds to changes in its surroundings automatically. You blink when someone shines a light in your eyes or sneeze if something irritates your nose. Such ***reflex actions*** are quick, involuntary actions that do not involve the brain. They happen automatically. Before the impulse ever arrives at the brain and is consciously thought about, the reflex action of the nervous system and muscular system has saved you from serious injury. Reflex actions are automatic, and they occur fast.

At times, you think about and control the way your body responds. The body obtains information about the outside world through special ***sense organs***. The five main sense organs are the eyes, ears, skin, nose, and tongue. Each has sensory neurons that react to changes in the environment and send impulses to the brain to be interpreted. For example, you watch a baseball arc through the air and decide whether to run left or right, fast or slowly in order to catch it. Such complex voluntary acts are usually learned.

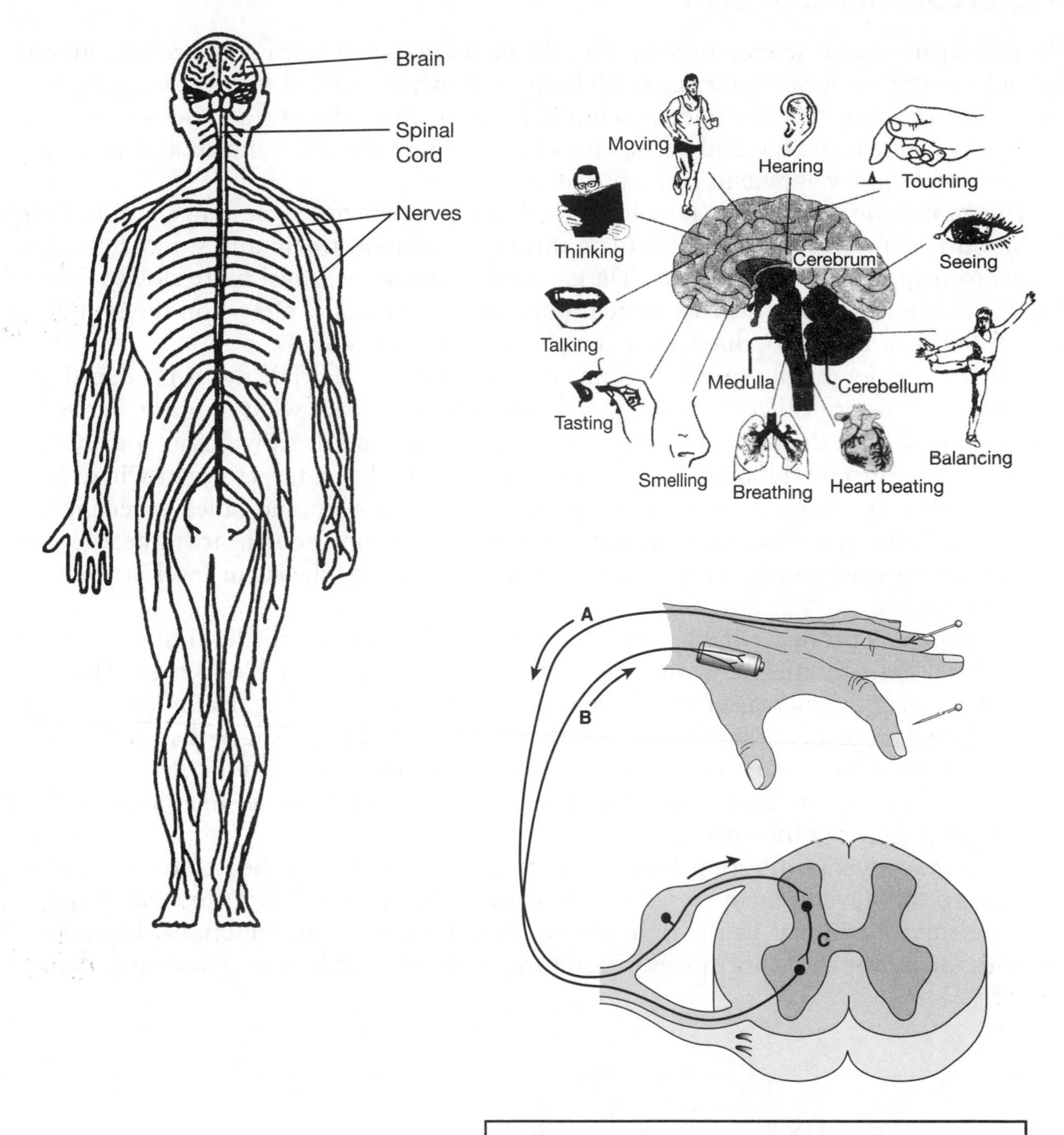

Reflex Arc—finger touches pin and sensory neuron *(A)* sends pain impulse to spinal cord and brain. Inside the spinal cord, interneurons *(C)* send impulses to the motor neurons *(B)*, which trigger muscle to pull finger away.

Figure 5.5
The Nervous System.

THE ENDOCRINE SYSTEM

The endocrine system works together with the nervous system to send and receive messages throughout the body. Some messages tell body parts what to do. Others cause changes inside the body that regulate body functions. Together, the endocrine and nervous systems control and coordinate the body's responses to changes in the environment and also regulate growth, development, and reproduction.

The ***endocrine system*** is made of glands that release chemical messengers directly into the bloodstream. The chemicals circulate through the bloodstream and stimulate target organs to respond in a certain way. These chemical messages travel more slowly than the electrical impulses of the nervous system but can have dramatic effects on the body. The major glands of the endocrine system are shown in Figure 5.6.

Glands are organs that produce fluids that the body needs. The fluids produced by endocrine glands are called hormones. ***Hormones*** help regulate what happens in tissues and organs all over the body. In general, hormones speed up or slow down your body. Once released into the bloodstream, hormones are carried to a target organ where the hormone acts. For example, the adrenal glands produce a hormone called adrenaline. *Adrenaline* helps your body deal with emergencies by causing your heart to beat faster, your blood pressure to rise, your airways to open wider, and the sugar level in your blood to rise.

The pancreas produces the hormone *insulin,* which regulates the level of sugar in the blood. After eating a meal high in sweets, the level of sugar in the blood rises. The pancreas detects the increased blood sugar and releases insulin. The insulin causes sugar in your blood to be taken into body cells that need sugar. Excess sugar is stored in the liver or removed from the blood in urine. Without insulin, sugar cannot get into cells where it is needed. Having too much insulin lowers the sugar level in the blood, and cells cannot get the energy they need to operate.

More than twenty different hormones affect everything from heart rate and blood pressure to the development of sex characteristics. See Figure 5.6. Your body must have the right hormones at the right times to carry on basic life functions. However, the body does not store hormones. Therefore, a person's endocrine glands must work properly.

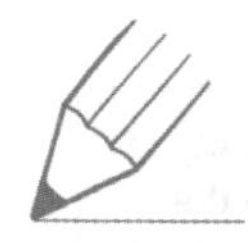

TRY THIS

1. How does the way in which the endocrine system sends messages throughout the body differ from the way in which the nervous system sends messages throughout the body?
2. Name two different hormones. For each hormone, identify the gland that produces it and how the hormone affects the human body.

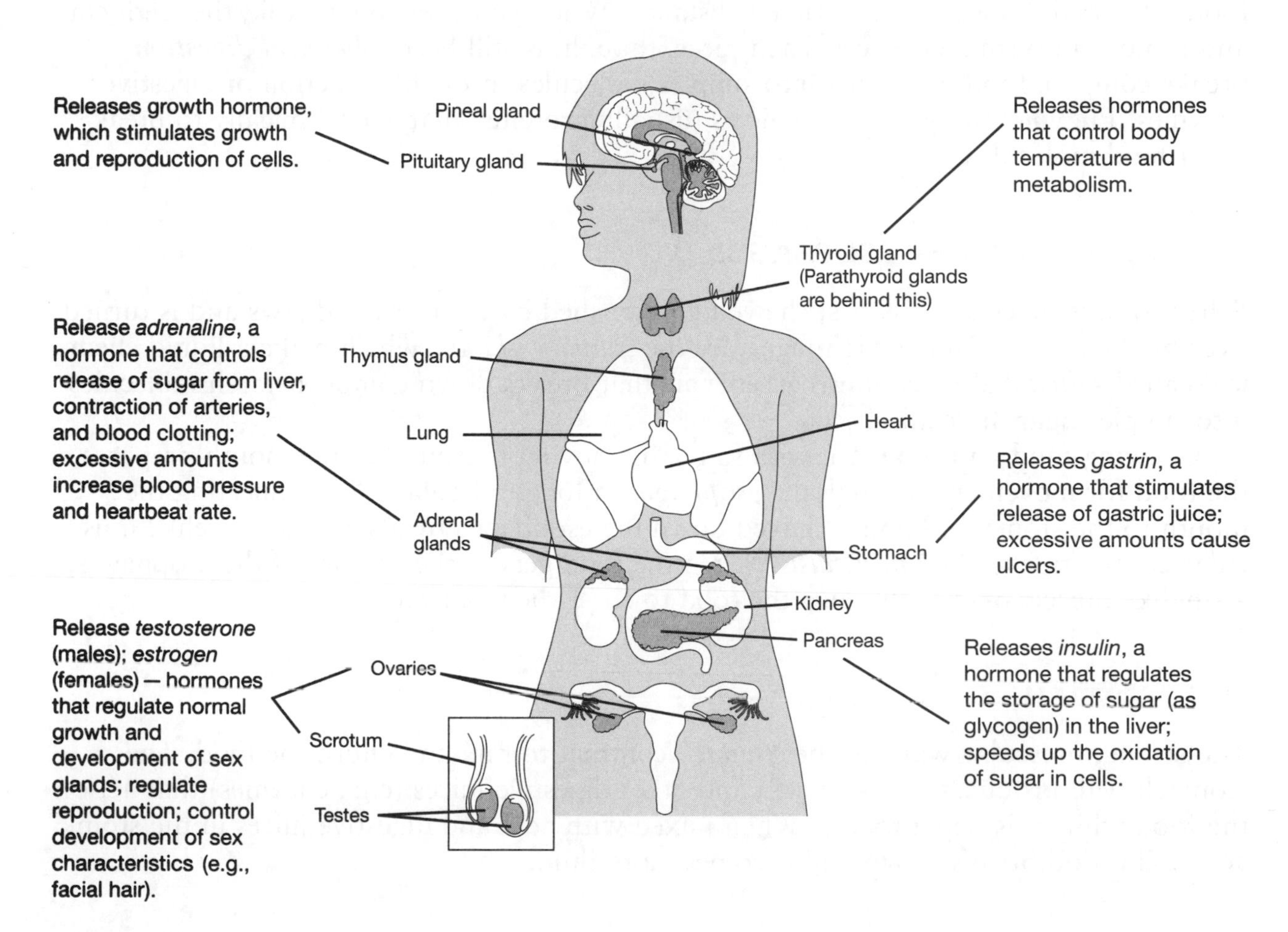

Figure 5.6
The Human Endocrine System.

THE DIGESTIVE SYSTEM

The digestive system consists of organs that make food available and ready for use by body cells. Taking food into your body is called ***ingestion***. Breaking food down into molecules that are small enough to be absorbed and used by the body's cells is called ***digestion***. Figure 5.7 shows the organs of the human digestive system.

The digestive system breaks down food in two ways. *Mechanical digestion* breaks food into smaller pieces of the same substance. When you chew on a steak, the teeth cut and grind it into smaller pieces. Each piece, though, is still beef. *Chemical digestion* breaks complex food molecules into simpler molecules through the action of digestive enzymes. ***Enzymes*** are special chemicals (usually proteins) that act chemically to break apart food molecules.

THE MOUTH AND THE ESOPHAGUS

When food is ingested, it is first chewed and crushed by the teeth and jaws and is turned over by the tongue. Under the tongue are the *salivary glands,* which make a liquid chemical called *saliva.* Saliva contains an enzyme that breaks down complex starch molecules into simple sugar molecules.

Chewing food and mixing it with saliva by moving it around in the mouth prepares the food for movement through the *esophagus,* a long muscular tube connecting the mouth to the stomach. Food is moved down the esophagus by a series of wavelike muscular contractions called *peristalsis.* When the food gets to the bottom of the esophagus, a ringlike muscle opens, allowing the food to enter the stomach.

THE STOMACH

The strong, muscular walls of the *stomach* contract to mix and churn the food. The stomach wall produces a strong acid and other digestive juices (e.g., enzymes) that act on the food while it is being mixed. When mixed with acid and digestive juices in the stomach, solid food breaks down and becomes more fluid.

THE SMALL INTESTINE

After leaving the stomach, food enters the *small intestine.* Peristaltic movements further break down food into smaller pieces and speed up absorption by constantly bringing freshly digested food in contact with the intestinal walls. Most chemical digestion and almost all absorption of food into the bloodstream take place in the small intestine.

THE LARGE INTESTINE

When food materials that have not been either digested or absorbed reach the end of the small intestine, a ring of muscle opens and they enter the *large intestine. No digestion occurs in the large intestine.* One of the main jobs of the large intestine is to reabsorb water from the food mass. This helps the body conserve water. Another important job is to eliminate digestive wastes from the body. As undigested and indigestible food move through the large intestine and as water is absorbed, the food become solid wastes, or *feces*. Feces include cellulose from cell walls of plants eaten as food, bacteria, mucus, bile, and worn-out intestinal lining cells. Feces are stored in the last part of the large intestine, the *rectum*, and leave the body through a muscular ring called the *anus*.

Teeth cut and grind food into smaller pieces. Enzymes from the salivary glands break down starch molecules into simpler sugar molecules.

Mouth

Salivary gland

Esophagus

Food moves down the esophagus by a series of wave-like muscular contractions called *peristalsis*.

Bile breaks up fats and oils into tiny droplets, speeding digestion by increasing surface area acted on by enzymes.

The strong, muscular walls of the stomach contract to mix and churn the food.

Liver

Stomach

Gallbladder

Pancreas

Pancreatic juice neutralizes acidic food and helps digest carbohydrates, proteins, and fats.

The large intestine reabsorbs water and vitamins made by bacteria in intestine, eliminates wastes.

Large intestine

Small intestine

The small intestine is very long, so food has a long time to be absorbed. The lining has many folds that increase the surface area through which food can be absorbed.

Appendix

Rectum

Anus

The folded lining of the small intestine is covered with millions of tiny, fingerlike projections called *villi* (sing. villus). Villi increase surface area and contain a network of tiny blood vessels that absorb food molecules.

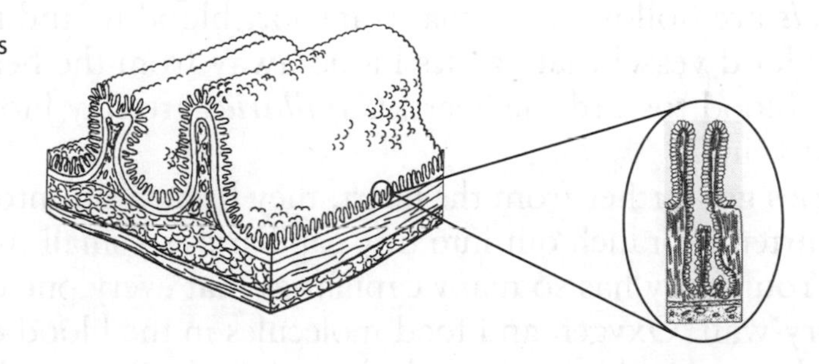

Figure 5.7
The Digestive System.

THE CIRCULATORY SYSTEM

The job of the ***circulatory system*** is to transport substances, such as food, oxygen, and water, to the body cells and to carry away wastes produced by the cells. The human circulatory system consists of the *heart,* a pump that forces *blood* through the body, and the *blood vessels,* a network of tubes through which the blood moves. See Figure 5.8.

THE HEART

The fist-sized human ***heart*** is located between the lungs in the chest cavity. The heart is made mostly of muscle tissue and is the most powerful organ in the body. The heart contracts and relaxes in a regular rhythm known as the *heartbeat.* The beating of the heart is automatic. Every time the heart contracts, a surge of blood is forced into the arteries leading away from the heart, causing them to bulge. This bulge is called a *pulse.* The surge of blood in the arteries can be felt as a pulse at your wrist. If you hold two fingers on the wrist bone below your thumb, you can count the number of times your heart beats in a minute.

THE BLOOD

Blood is a complex fluid mixture that includes cells, dissolved food and gases, cellular wastes, proteins, enzymes, and hormones. About half of blood is *plasma,* a fluid that is mostly water and dissolved proteins. The rest of blood consists of blood cells. ***Red blood cells*** carry oxygen from the lungs to body cells and carry carbon dioxide from body cells to the lungs. Red blood cells get their red color from hemoglobin, a pigment that contains iron. Hemoglobin attracts and holds on to oxygen in the bloodstream but also easily gives up oxygen to body cells.

White blood cells are larger than red blood cells and often have large, irregularly shaped nuclei. White cells protect the body against infection by bacteria and other microorganisms.

BLOOD VESSELS

Blood vessels are hollow tubes that transport blood to and from all parts of the body. An ***artery*** is a blood vessel that carries blood away from the heart. A ***vein*** is a blood vessel that carries blood toward the heart. ***Capillaries*** are tiny blood vessels that interconnect arteries and veins.

As arteries get farther from the heart, they branch off into smaller and smaller arteries. Very small arteries branch out into tiny capillaries so small that their walls are only one cell layer thin. Your body has so many capillaries that every one of its trillions of cells lies next to a capillary wall. Oxygen and food molecules in the blood enter cells and wastes leave the cells and enter the blood through these thin capillary walls. Capillaries then merge back into larger and larger blood vessels called veins that carry the blood back to the heart. Large veins rush the blood back to the heart, where it is pumped into the lungs to pick up more oxygen.

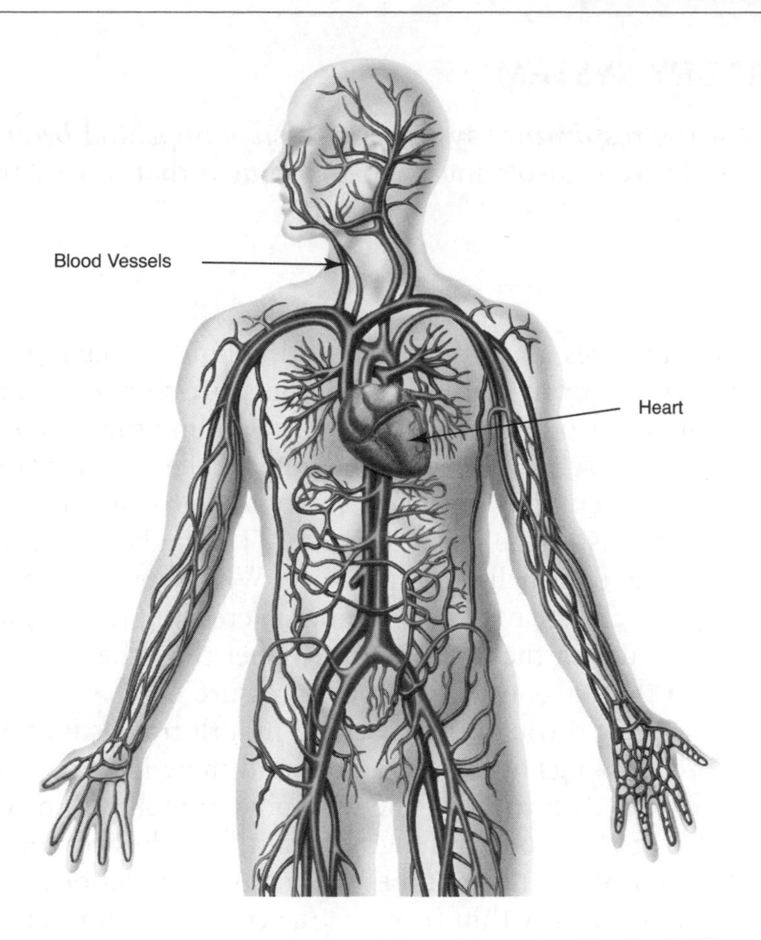

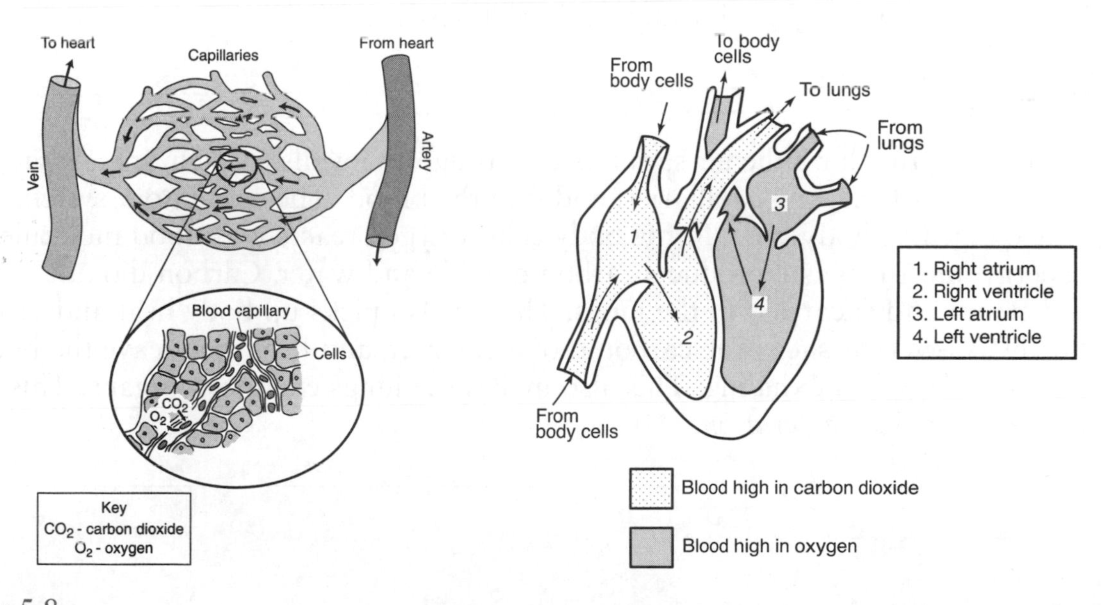

Figure 5.8
The Human Circulatory System. Circulation in the human heart. Oxygen-poor blood from the body flows into the *right atrium (1)* and then into the *right ventricle (2)*. When the muscular right ventricle contracts, blood is forced into the lungs. In the lungs, the blood gets filled with oxygen. From the lungs, oxygen-rich blood flows into the *left atrium (3)* and then into the *left ventricle (4)*. The left ventricle is a thick muscle that is powerful enough to send the blood surging out through the body's largest blood vessel, the aorta, and throughout the body.

THE RESPIRATORY SYSTEM

The main organs of the ***respiratory system*** are the *windpipe* and *bronchi*, through which air enters the *lungs*. There is also a muscular *diaphragm* that moves air into and out of the lungs. See Figure 5.9.

BREATHING

The respiratory system moves air into and out of the lungs by changing air pressure inside the lungs. Air moves from regions of high pressure to regions of low pressure. When you *inhale,* muscles around the ribs contract, pulling the ribs upward and outward. A strong sheet of muscle underneath the lungs, called the *diaphragm,* contracts and pulls downward. This enlarges the chest cavity. As air inside the lungs expands to fill this volume, air pressure inside the lungs decreases. The higher-pressure air from outside then rushes into the lungs to equalize the pressure. When you *exhale,* the rib muscles and diaphragm relax and the chest cavity gets smaller, increasing the pressure inside the lungs. Since the air pressure in the lungs is now higher than the air pressure outside the body, air rushes out of the lungs to equalize the pressure.

Air that enters the body through the nose and mouth travels into a large breathing tube called the *windpipe.* Rings of cartilage surround the windpipe and keep it from collapsing when the body bends or twists. The windpipe branches into two tubes called the *bronchi.* One bronchus goes into each lung, where it branches off into smaller and smaller tubes. At the end of the tiniest tubes are many microscopic, balloon-like air sacs. The walls of the air sacs are very thin (usually one cell layer thin) and are always kept wet. A network of tiny capillaries surrounds each air sac.

GAS EXCHANGE

Oxygen from the air filling the air sacs moves through the walls of each tiny air sac, through the walls of the tiny capillaries, and into the blood. The circulatory system then carries the oxygen to body cells. In the body cells, oxygen reacts with food molecules to release energy and the waste products carbon dioxide and water. Carbon dioxide enters the bloodstream and is carried to the lungs. The air also picks up body heat and water from the always-wet air sacs. The carbon dioxide, water, and heat then leave the body, along with the air, when exhaling. This also makes the lungs excretory organs. This entire process is called *respiration.*

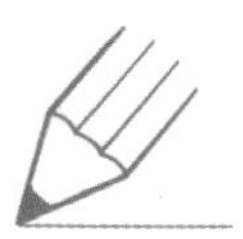

TRY THIS

1. How does the air you exhale differ from the air you inhale?
2. What changes would you expect to occur in your blood if you were to hold your breath for a long time?

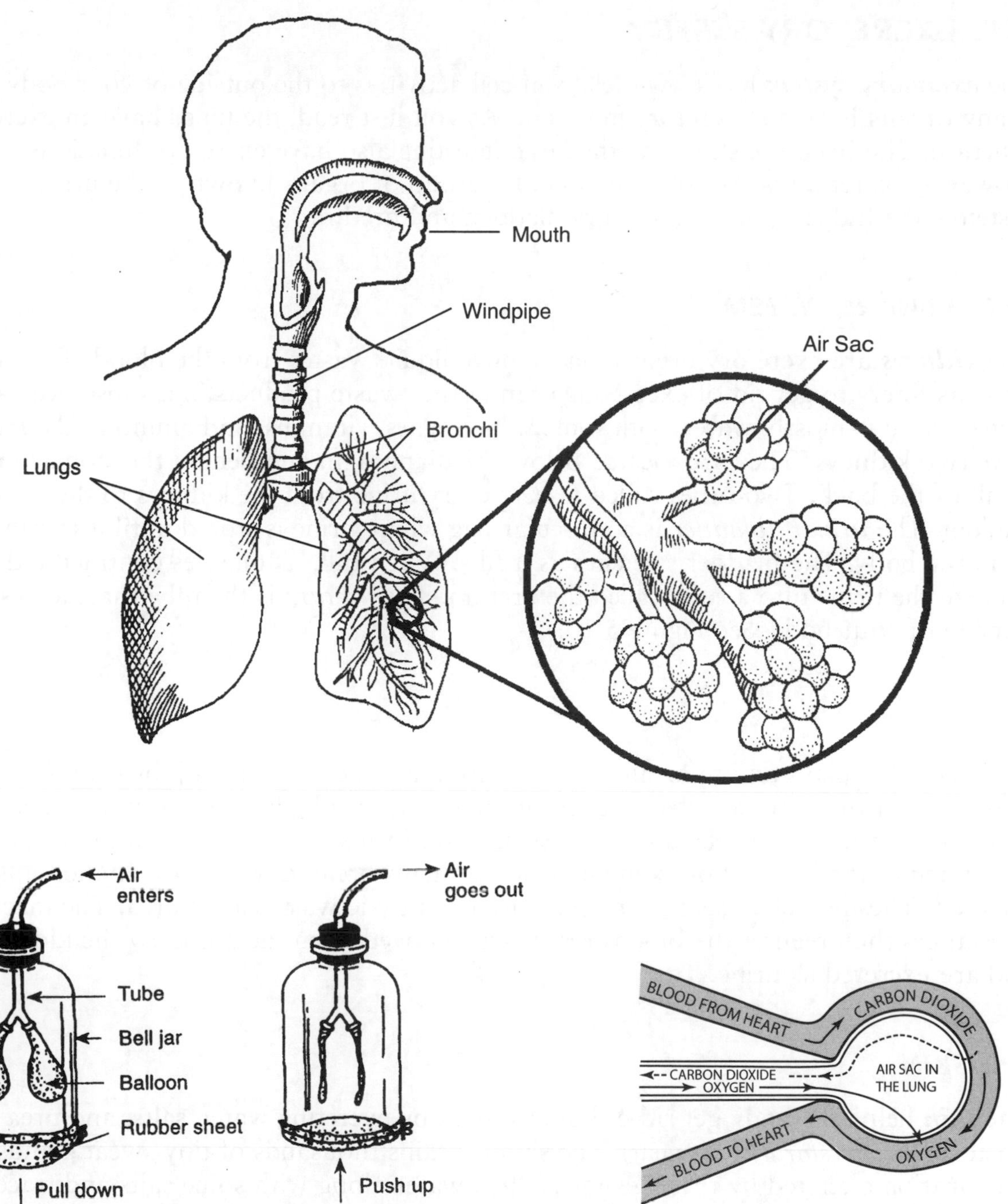

Figure 5.9
The Human Respiratory System.

THE EXCRETORY SYSTEM

The ***excretory system*** moves wastes from cell activities to the outside of your body. Many organs have an excretory function. As you just read, the lungs have an excretory function. The liver, the skin, and the large intestine also have excretory functions. However, excretion is the sole function of a group of organs known as the urinary system—the kidneys, ureters, urinary bladder, and urethra.

THE URINARY SYSTEM

The ***kidneys*** are excretory organs that remove liquid wastes from the blood. The kidneys serve as filters to get rid of excess sugar and other waste products in a substance called *urine*. Urine is mostly water, with some salts, sugars, vitamins, and amino acids. Humans have two kidneys. They are located below the diaphragm and behind the stomach in the small of the back. Two tubes called *ureters* carry urine from the kidneys to the urinary bladder. The *urinary bladder* is a muscular bag where urine is stored until it is removed from the body. The bladder stretches as it fills. When full, its muscles contract and squeeze the urine into a tube called the urethra. The *urethra* is the tube that carries urine outside of your body. See Figure 5.10.

THE LIVER

The ***liver*** also plays an important role in excretion. One of its main jobs is to *detoxify* or remove harmful or toxic substances from the blood. Inside the liver, complex chemical reactions change these substances to less-harmful forms or make them inactive. Thus, the liver purifies the blood. For example, when proteins break down, toxic ammonia is released. The liver changes the ammonia to a less-toxic waste called urea. The inactive substances then reenter the bloodstream, are removed from the blood by the kidneys, and are excreted as urine.

THE SKIN

The ***skin*** helps the body get rid of liquid wastes by excreting water, salts, and urea as a waste called *perspiration* (sweat). The skin contains thousands of tiny sweat glands. The perspiration excreted by these glands is 99% water along with some salts and traces of urea. The skin also helps the body excrete heat to regulate body temperature. Blood vessels in the skin open wider when the body is overheated. This allows more blood to flow near the surface of the skin and radiate heat to the air. Evaporation of the water in perspiration also cools the surface of the skin. When the body is too cold, blood vessels in the skin constrict. Less blood flows through the skin, and less heat is lost. With less blood flowing through sweat glands, perspiration decreases.

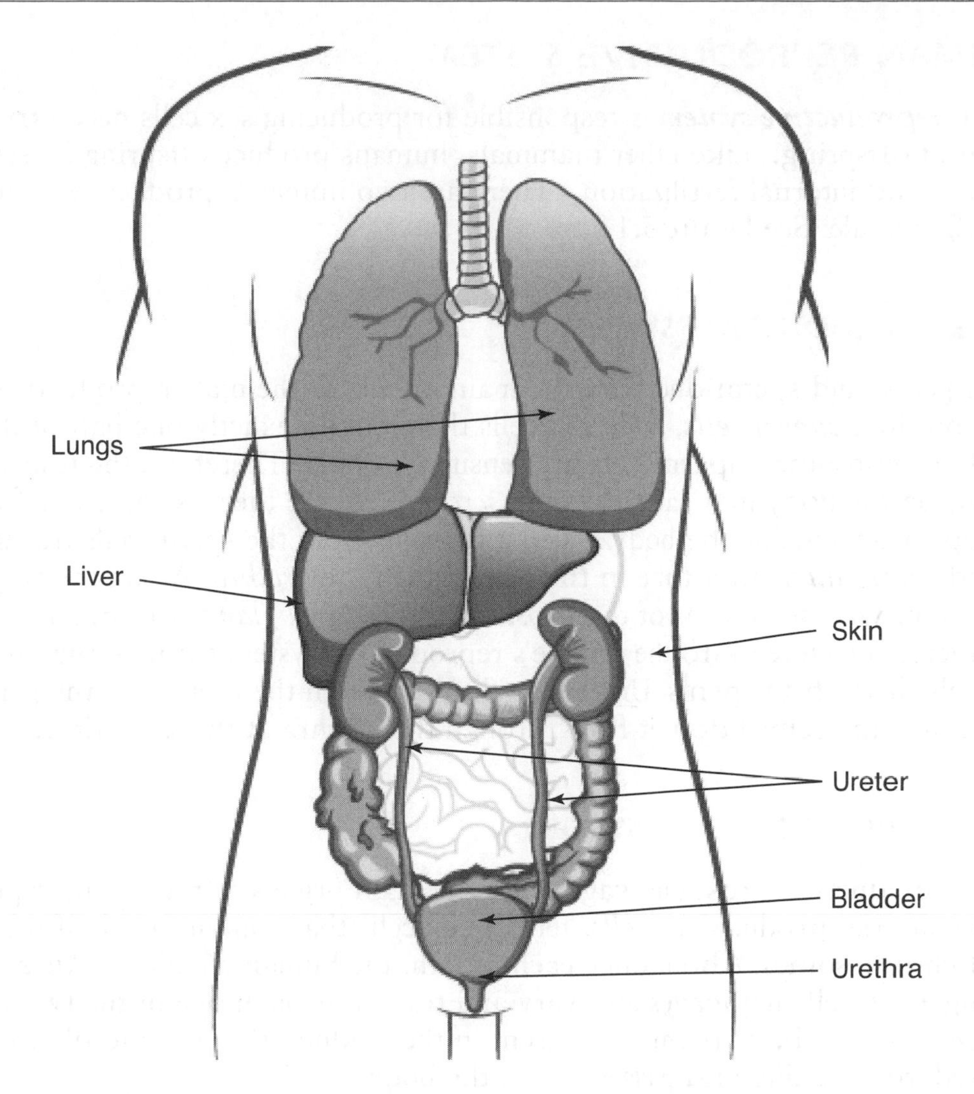

Figure 5.10
The Human Excretory System.

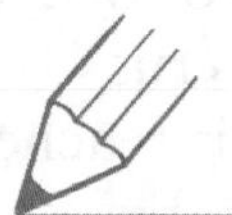

TRY THIS

1. Describe two ways in which perspiring is useful to your body.
2. Explain how the body eliminates: **1.** solid wastes, **2.** liquid wastes, and **3.** gaseous wastes

THE HUMAN REPRODUCTIVE SYSTEM

The human ***reproductive system*** is responsible for producing sex cells necessary for the production of offspring. Like other mammals, humans produce offspring by sexual reproduction and internal fertilization. There are two human reproductive systems—the male and the female. See Figure 5.11.

THE MALE REPRODUCTIVE SYSTEM

The testes, penis, and sperm ducts are the main organs of the male reproductive system. The *testes* produce *sperm cells,* male sex cells that contain exactly one-half of the normal number of chromosomes. Sperm cells are sensitive to heat. Therefore, the testes are located outside the body in a sac called the *scrotum,* where there is some relief from the higher temperatures inside the body. During reproduction, the sperm cells travel through tubes called *sperm ducts* to a tube in the penis called the *urethra.* Along the way, the sperm cells mix with secretions of the *prostate* and *seminal glands* to form a fluid called *semen.* Semen is delivered into the female's reproductive system through the urethra, which travels through the penis. Urine also flows through the urethra of the penis. However, urine and semen do not flow through the urethra at the same time.

THE FEMALE REPRODUCTIVE SYSTEM

The ovaries, oviducts, uterus, and vagina are the main organs of the female reproductive system. The *ovaries* produce *egg cells,* female sex cells that contain one-half the normal number of chromosomes. About once each month, the human female *ovulates,* or produces a single egg cell that leaves an ovary and travels through one of the two oviducts to the *uterus,* or womb. If sperm are present in the oviduct, the egg is fertilized. If it is not fertilized, the egg dies and passes out of the body.

In *fertilization,* a male sperm cell (with one-half the normal number of chromosomes) and female egg cell (with one-half the normal number of chromosomes) unite to form a new cell that has a full set of chromosomes. This new cell, called a *fertilized egg* or *zygote,* has the genetic information necessary to develop into a human being.

Each month, the uterus prepares to receive a zygote. Its inner walls thicken to form a soft, blood-rich surface. If an egg cell is fertilized, it becomes attached to the uterus where, over a period of about nine months, it develops into a new baby. When it reaches full size, the uterus begins to contract and squeezes repeatedly to push out the baby. Slowly, the baby emerges through the *vagina,* or birth canal. After birth, the baby gets nourishment from the female's *mammary glands,* or breasts. Mammals get their name from these mammary glands, which produce milk after the birth of a child.

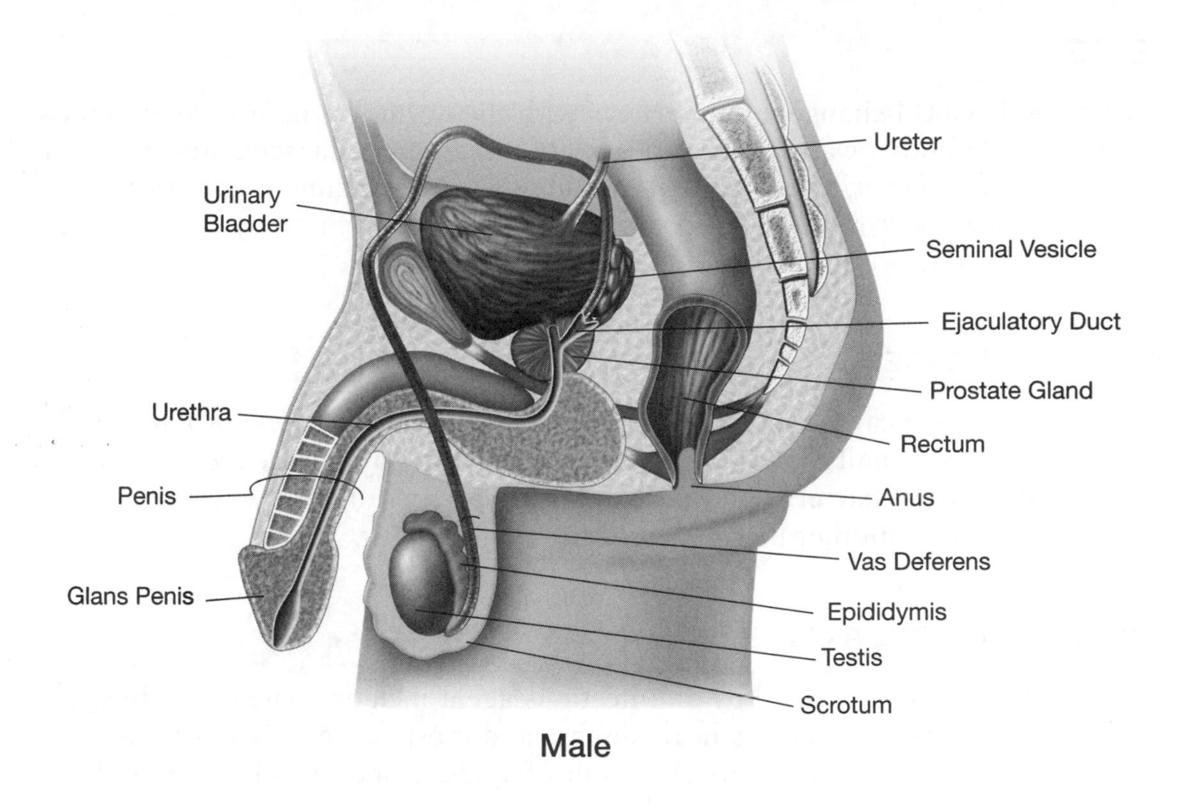

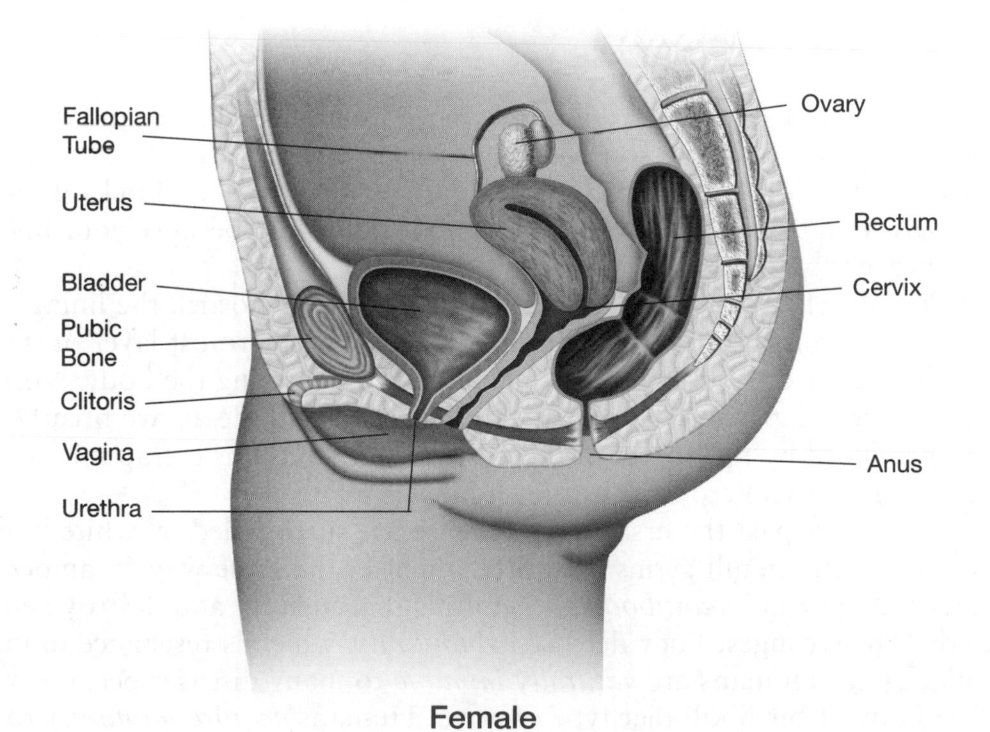

Figure 5.11
The Human Reproductive System.

DISEASE

Disease is any harmful change that interferes with the normal structure, function, or appearance of the body or any of its parts. Every disease has characteristic effects on the body, or *symptoms.* Fever, swelling, pain, tiredness, dizziness, nausea, and rashes are some common disease symptoms. Diseases have many causes, but they can be classified into two broad groups: infectious and noninfectious.

INFECTIOUS DISEASES

Infectious diseases are caused by microscopic organisms called *germs,* or *pathogens.* Germs are usually so small that they can be seen only with a microscope. They include a wide variety of viruses, bacteria, and protozoans. A person infected with these organisms that can pass them to another person is said to be *contagious.*

NONINFECTIOUS DISEASES

Diseases not known to be caused by an infectious agent include some of the biggest killers in the United States, such as heart disease and most cancers. *Noninfectious diseases* have many causes. Some are the result of aging, some are inherited, and some are caused by environmental factors or diet.

BODY DEFENSES AGAINST DISEASE

Germs exist all around us, in water, waste products, unwashed clothes and dishes, or the droplets given off when a sick person coughs or sneezes. Germs can even be carried from place to place by other organisms, such as mosquitoes, flies, rats, and ticks. If germs are everywhere, why aren't you sick all the time? You stay healthy because your body has defenses that prevent germs from entering it.

The first line of defense includes the skin, the hairs of the nostril, the linings of the nose and throat, the tonsils, and stomach acid. The tough outermost layer of the skin blocks microorganisms and other foreign materials from entering the body. Nostril hairs act like a sieve, filtering particles that may carry disease from the air we breathe. The sticky mucus produced by the linings of the nose, throat, and tonsils trap microorganisms. Stomach acid kills many microorganisms in food.

Germs that make it past the first line of defense are surrounded by white blood cells. Some white blood cells engulf germs and foreign matter the same way an amoeba engulfs food. Another type produces ***antibodies,*** chemicals that identify and destroy germs that enter the body. The strongest body defense is *immunity,* which is resistance to infection by a particular germ. Humans are *naturally immune* to many diseases because antibodies present in the body at birth kill that type of germ. Humans *acquire immunity* to particular diseases when their body learns how to make the antibodies that kill the germs that cause the disease. Once the body learns how to make a particular antibody, it can make that antibody quickly if the germ invades the body again.

Practice Review Questions

1. A characteristic of *all* known living things is that they

1. are capable of locomotion
2. are made of one or more cells
3. use atmospheric oxygen
4. use carbon dioxide

2. The shape and structure of a cell is most directly related to its

1. age
2. size
3. function
4. growth rate

3. Which is composed of a group of similar cells working together?

1. tissue
2. organ
3. organism
4. organ system

4. Which statement best describes human ligaments?

1. They are made of cartilage and cushion bones at joints.
2. They look striated and are controlled by the nervous system.
3. They are made of tough, elastic tissue and connect the ends of bones at movable joints.
4. They are made of tough, inelastic tissue and connect muscles to bones.

5. Which human body systems interact to control, coordinate, and regulate body functions?

1. circulatory and respiratory
2. digestive and excretory
3. nervous and endocrine
4. skeletal and reproductive

6. The diagram below shows one motion of a human arm.

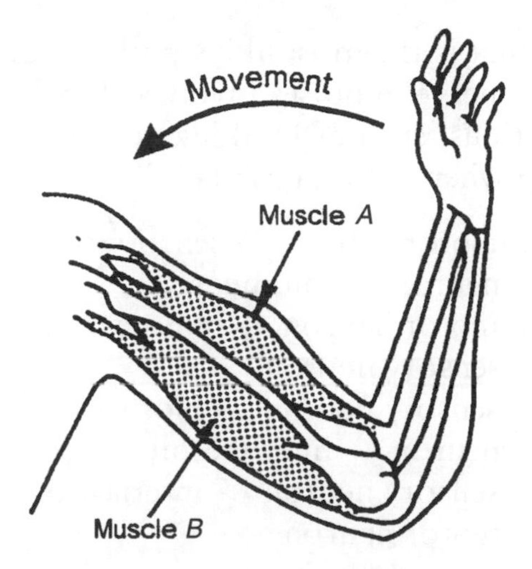

For the arm to move in the direction shown, what must happen to muscles *A* and *B*?

1. Muscle *A* must shorten while muscle *B* lengthens.
2. Muscle *A* must lengthen while muscle *B* shortens.
3. Both muscle *A* and muscle *B* must shorten.
4. Both muscle *A* and muscle *B* must lengthen.

GO ON

7. Which human body system controls production of hormones that regulate body functions?

1. digestive
2. endocrine
3. respiratory
4. skeletal

8.** When a person sees something, what carries the message from the eyes to the brain?

1. arteries
2. glands
3. muscles
4. nerves

9. What is the most likely pathway of the nerve impulses involved in reflex actions, such as blinking and jumping up when sitting on a tack?

1. motor neuron → sensory neuron → interneuron
2. motor neuron → interneuron → sensory neuron
3. sensory neuron → motor neuron → interneuron
4. sensory neuron → interneuron → motor neuron

10. The chemical digestion of carbohydrates begins in which organ?

1. large intestines
2. small intestines
3. esophagus
4. mouth

Base your answers to questions 11–13 on the diagram below.

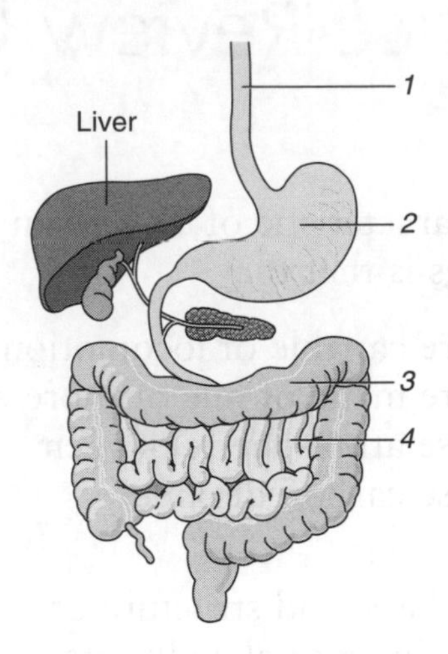

11.** What is the main job of the organ labeled *1*?

1. carrying air
2. carrying food
3. carrying blood
4. carrying messages from the brain

12. Which organ produces and secretes hydrochloric acid?

1. *1*
2. *2*
3. *3*
4. *4*

13. In which organ does absorption of food into the bloodstream mainly occur?

1. *1*
2. *2*
3. *3*
4. *4*

** Reproduced from NAEP Released Items, *National Center for Education Statistics, U.S. Department of Education.*

GO ON →

14.* What is the main function of red blood cells?

1. to fight diseases in the body
2. to carry oxygen to all parts of the body
3. to remove carbon monoxide from all parts of the body
4. to produce materials that cause the blood to clot

Base your answers to questions 15 and 16 on the diagram below, which represents a model of part of the human respiratory system.

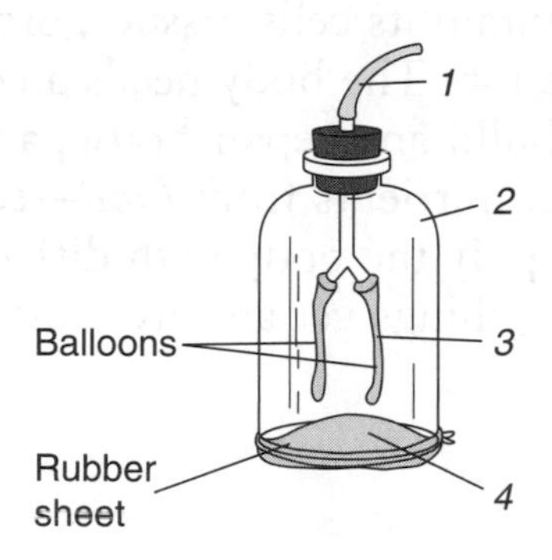

15. The movement of which part demonstrates the contraction and relaxation of the diaphragm during breathing?

1. *1*
2. *2*
3. *3*
4. *4*

16. The rib cage is represented by structure

1. *1*
2. *2*
3. *3*
4. *4*

* Reproduced from IEA TIMSS 2003 Science Released Items for Grade 8, TIMSS International Study Center, Boston College, MA

17. The diagram below shows an air sac surrounded by capillaries in the lung.

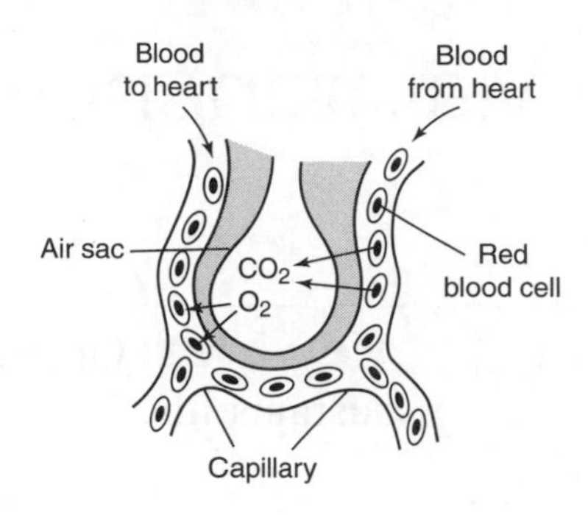

Which two human body systems are working together in the diagram?

1. respiratory and circulatory
2. respiratory and nervous
3. circulatory and digestive
4. circulatory and reproductive

18. Which human body system disposes of liquid and gaseous wastes and aids in the removal of excess heat energy?

1. nervous
2. skeletal
3. digestive
4. excretory

19. Which term best completes the statement below?

Sperm is to testis as egg is to

1. uterus
2. penis
3. ovary
4. mammary gland

Chapter 6

The Maintenance of Life

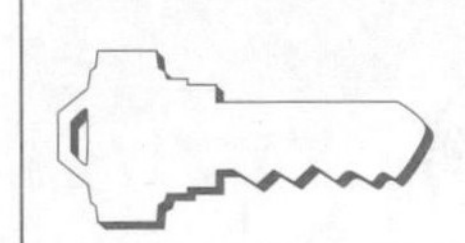

Key Idea: Organisms maintain a dynamic equilibrium that sustains life.

NUTRITION

In order to stay alive, an organism has to support and maintain its cells, tissues, organs, and organ systems. Keeping the body going is not a simple job. The body needs a constant supply of ***nutrients***—substances that can be used to build and repair body parts, provide energy, or control body functions. Living things get nutrients from ***food***—edible materials that contain nutrients. Different kinds of food supply the body with different nutrients. Together, the many different ways in which living things get and use nutrients to satisfy their needs is called ***nutrition.***

TYPES OF NUTRIENTS

Nutrients are complex chemical compounds. There are six basic types of nutrients. See Table 6.1 on page 81.

ENERGY IN FOODS

The energy in a nutrient exists in the chemical bonds of its molecules. In order for the energy to be released, these chemical bonds must be broken. Cells get energy by a process called ***respiration***, in which nutrients from digested food are combined with oxygen to release energy and produce the wastes carbon dioxide and water. Respiration in human cells typically involves combining the simple sugar glucose with oxygen. This process can be written as a formula:

glucose + oxygen → energy + water + carbon dioxide

The energy content in foods is measured in units called Calories. A *food Calorie* is the amount of heat energy needed to raise the temperature of 1 kilogram of water by 1°C. The total caloric value of each type of food varies. Fats and oils tend to have high caloric values because they contain lots of energy stored in chemical bonds.

Table 6.1 **Nutrients**

Nutrient	Uses	Some Food Sources
Water	Cells are mostly water. Life cannot exist without water. Water dissolves many chemicals. Once dissolved in water, chemicals react more easily with one another. Dissolved chemicals are so small that they can go into and out of cells.	Drinking water, beverages, most food contains at least some water
Carbohydrates	Energy source. Compounds of carbon, hydrogen, and oxygen. Three basic types: sugars, starches, and cellulose.	Cereal grains, potatoes, fruits, vegetables, beans, peas, sugarcane, beets, milk, baked goods, pasta
Fats	Energy source, insulate body cushion joints, and protect nerves. Compounds of carbon, hydrogen, and oxygen. Have more chemical bonds than complex carbohydrates and thus contain more energy. In addition to providing energy, fats insulate the body, cushion joints, and protect nerves.	Oils, lard, butter, nuts
Proteins	Used to build and repair body parts. Help control body activities. Basic building blocks of cells. Many structures and substances in the body are built of or contain proteins: e.g., muscles, ligaments, nails and hair, hemoglobin, antibodies, hormones, enzymes, and DNA.	Meat, poultry, fish, eggs, milk, cheese, beans, nuts
Vitamins	Organic, or carbon-based, compounds found in tiny amounts in food. Work together in enzymes in the complex chemical reactions of the body. When lacking in the diet, reactions cannot take place properly and body malfunctions.	Most foods contain small amounts of vitamins. Some foods are rich in certain vitamins, e.g., citrus fruits—vitamin C; fish oils—vitamin A
Minerals	Elements needed to make body chemicals and structures. Calcium and phosphorus build strong bones and teeth. Sodium and potassium are needed for muscles to contract and for nerve cells to carry messages. Iron is needed to form the hemoglobin blood uses to transport oxygen. Other minerals, such as iodine, fluorine, copper, and selenium, are needed by the body only in small amounts.	Calcium: dairy products, dark green vegetables; potassium: citrus fruits, bananas, leafy greens; iron: liver, red meat, eggs; iodine: seafood

The number of Calories someone needs varies from person to person. An organism's need for energy changes constantly depending on its activities. A person running almost constantly for an hour during a soccer game needs more energy than a person surfing the Internet for an hour. Even if you sit perfectly still or are asleep, your body needs energy to keep your heart pumping, to keep your lungs breathing, and to do the thousands of other tasks necessary to sustain life.

TRY THIS

1. Why do athletes eat foods containing carbohydrates before competing in athletic contests?
2. Why is it important for children to include protein-rich foods in their diet?

A BALANCED DIET

In order to stay healthy, your body needs carbohydrates, fats, protein, water, vitamins, and minerals. Eating a ***balanced diet*** means eating a variety of foods each day so that your body gets all of the nutrients it needs. The MyPlate chart shown in Figure 6.1 is based on research by the United States Department of Agriculture. The four sections of the plate and glass represent the five food groups. Each of the food groups in the MyPlate chart provides some, but not all, of the nutrients you need. The size of the food group sections on the plate and glass suggests how much food a person should choose from each group. Not getting the correct amount of each nutrient may result in weight gain, weight loss, or a diseased state. You should also make physical activity a regular part of your day.

The table beneath the MyPlate chart shows a daily amount of each major food group based on a 2,000-calorie diet. The amount that is right for you will depend on how many Calories you need. This, in turn, depends on your age, sex, size, and how active you are. If you follow the recommended proportions for each food group in the MyPlate chart, your body will get the necessary nutrients and at the same time get the right amount of food calories to maintain or improve your weight.

TRY THIS

Based upon the recommendations in the USDA MyPlate food chart, prepare a three-day diet that would supply you with all the nutrients necessary to stay healthy.

TYPES OF NUTRITION

Living things can be divided into two broad groups based on how they get the nutrients they need. These groups are producers and consumers.

PRODUCERS

A ***producer*** is an organism that can make its own food from inorganic substances, getting its energy either from light (photosynthesis) or chemicals (chemosynthesis). Green plants, algae, and certain bacteria are producers.

In a process called *photosynthesis*, green plants use energy from the Sun and the chlorophyll in their cells to change carbon dioxide and water into simple sugars and oxygen. See Figure 6.2. Plant leaves contain special cell structures called *chloroplasts* that contain a green pigment called *chlorophyll*. When sunlight strikes a leaf, chlorophyll captures the light energy. Carbon dioxide enters the leaf through tiny openings called *stomates*. Water comes up through the roots. The chloroplasts use the energy from the Sun to split up water into hydrogen and oxygen. The oxygen from the water is

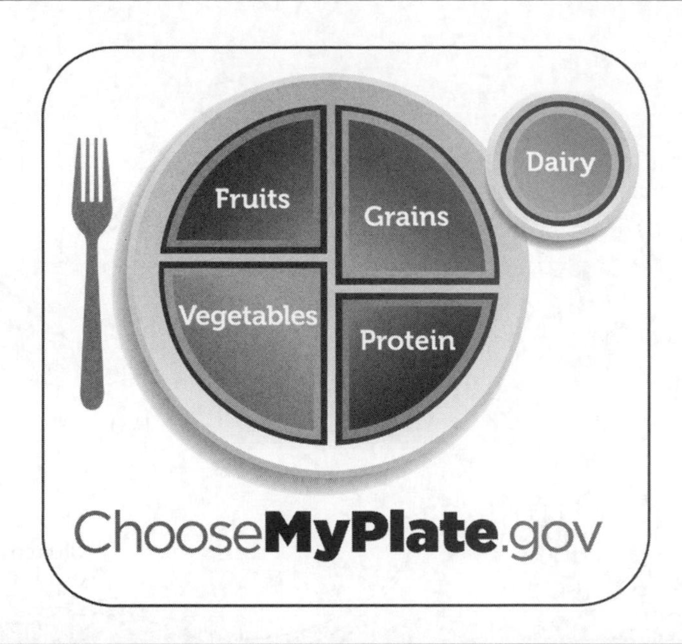

	Grains Group	Vegetable Group	Fruit Group	Dairy Group	Protein Foods Group
	Any food made from wheat, rice, oats, cornmeal, barley, or another cereal grain is a grain product.	Eating vegetables is important because they provide vitamins and minerals and most are low in calories.	Fruits provide nutrients vital for health, such as potassium, dietary fiber, vitamin C, and folate (folic acid). Most fruits are naturally low in fat, sodium, and calories. None have cholesterol.	The Dairy Group includes milk, yogurt, cheese, and fortified soymilk. They provide calcium, vitamin D, potassium, protein, and other nutrients needed for good health throughout life.	Protein foods include both animal (meat, poultry, seafood, and eggs) and plant (beans, peas, soy products, nuts, and seeds) sources.
Recommendations	Make at least half your grains whole.	Vary your veggies.	Focus on fruits.	Get your calcium-rich foods.	Go lean with protein.
Daily Amount (based on a 2,000-calorie diet)	6 oz.	2–3 cups	1½–2 cups	3* cups	Most people, ages 9 and older, should eat 5 to 7 ounces* of protein foods each day.

*What counts as a cup in the Dairy Group? 1 cup of milk or yogurt, 1½ ounces of natural cheese, or 2 ounces of processed cheese.

Figure 6.1
The USDA MyPlate food chart.

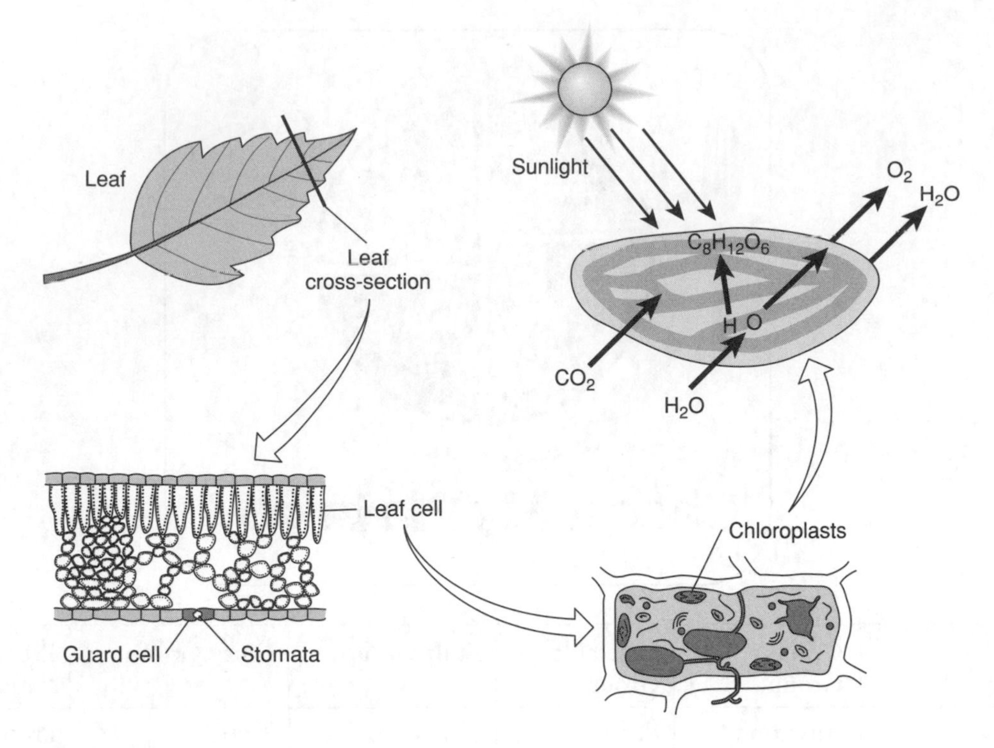

Figure 6.2
Photosynthesis.

given off into the air. The hydrogen from the water is combined with carbon dioxide to form simple sugars. The simple sugars are food for the plant. The formula for photosynthesis can be written as follows:

$$\text{water} + \text{carbon dioxide} + \text{light energy} \xrightarrow{\text{chlorophyll}} \text{glucose} + \text{oxygen} + \text{water}$$

$$12H_2O + 6CO_2 + \text{energy} \xrightarrow{\text{chlorophyll}} C_6H_{12}O_6 + 6O_2 + 6H_2O$$

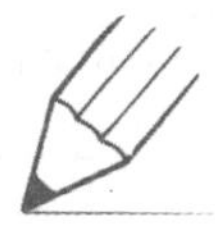

TRY THIS

1. Why are green plants important to all living things?
2. What are the raw materials needed for photosynthesis?

CONSUMERS

A ***consumer*** is an organism that cannot make its own food but depends on other organisms as its source of food. Animals, fungi, protozoans, and most bacteria are consumers. Consumers can be divided into groups based on the organisms they depend upon for food.

Herbivores feed on plants. Herbivores can use the energy stored in plant tissue to form animal tissue. Only herbivores can live on a diet rich in the cellulose of plant tissue. They can use plant tissue as a food source because they have teeth adapted to crushing and grinding tough plant cells as well as complicated stomachs and long intestines containing microorganisms that break down cellulose. A rabbit eating grass and a deer feeding on shrubs are examples of herbivores feeding on plants.

Carnivores feed on other animals. They get the nutrients they need from the tissues of the animals they eat. Animals that capture other animals as their main source of food are called *predators*. The animals predators capture are called *prey*. A hawk feeding on a mouse and a wolf killing and feeding on a deer are examples of carnivores. The hawk and wolf are predators, and the mouse and deer are prey.

Omnivores feed on both plants and animals. Humans are omnivores, as are red foxes, raccoons, and some fish. Not all consumers fit neatly into one category. The red fox feeds on birds, insects, and small rodents but also eats berries and seeds. The white-footed mouse feeds mainly on seeds but also eats insects, small birds, and bird eggs.

Decomposers feed on dead plants and animals. To some degree, all consumers are decomposers since they either break down food by digestion or break it into smaller pieces. *Scavengers* are animals that feed on dead plants and animals. Vultures, gulls, and hyenas are examples of scavengers that feed on animal remains. Earthworms, termites, and various beetles are examples of animals that feed on dead plants. *Saprophytes* are plants that feed on dead plants and animals. Many bacteria and fungi, such as mushrooms, are saprophytes. Since they do not need sunlight as an energy source, saprophytes can live in deep shade and dark caves.

True decomposers feed on dead organic matter, or detritus. Some true decomposers, such as many bacteria and protozoans, are *micro*scopic. Others, such as insect larvae, earthworms, slugs, and crabs, are *macro*scopic. Together, there may be well over one million decomposers in the top 7 to 10 centimeters of forest soil! Decomposers break down organic matter until it finally reaches an inorganic state.

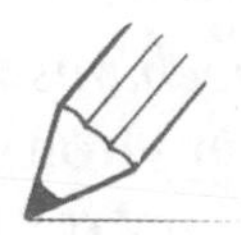

TRY THIS

A table of animal characteristics is shown below.

Animal	Length of Life (in the wild)	Diet	Range
Elephant	About 70 years	Roots, leaves, and grasses	Asia, Africa
Tiger	About 20 years	Antelope and other animals	Asia
Lynx	About 10 years	Deer, rodents, and other small animals	Europe, Asia, North America
Coyote	5 to 6 years	Grasses, grains, and small animals	North America

Based on the information in the table, identify each animal as a carnivore, an omnivore, or an herbivore.

METABOLISM

As mentioned in Chapter 4, the life functions carried on by cells, tissues, organs, and systems are all accomplished through a huge number of chemical reactions, which together are called ***metabolism***. Some metabolic activities build large molecules from small ones or change nutrients into cellular material. Other metabolic activities break large molecules into their smaller building blocks or break them apart to release the energy in their chemical bonds.

Many factors affect metabolism. For example, cells need to keep the chemistry (e.g., levels of water, acid, salt) of the fluids in and around them stable. The chemical reactions of metabolism occur only within a narrow range of temperatures. Therefore, living things must maintain a stable body temperature. Metabolism can also be influenced by hormones, exercise, diet, and aging.

MAINTAINING A BALANCED STATE

All organisms live in a physical environment in which light, temperature, moisture, and available nutrients are constantly changing. To stay alive, organisms must maintain livable conditions inside themselves no matter what their surroundings are like. The maintenance of proper internal conditions is ***homeostasis***. Homeostasis literally means "staying the same"; it is a state of balance between the environment inside and outside of a living thing. For example, the human body must keep its internal temperature within a narrow range around 37°C. An increase or decrease of only a few degrees can be fatal. Likewise, other factors such as water levels, acidity, and salinity must also remain stable. Homeostasis requires a constant exchange of energy and nutrients between the organism and the outside environment.

REGULATION

What keeps the environment inside an organism fairly constant is a *feedback mechanism*. See Figure 6.3. This means some way in which the body senses a change in the environment and responds to it. Feedback mechanisms are how organisms *regulate*, or control, their internal environment. For example, if it gets hot outside and human body temperature rises above 37°C, sensory nerves in the skin send a message to the brain. The brain relays the message to receptors that increase blood flow to the skin. This causes sweating, and evaporation of the sweat cools the body. It may also stimulate behaviors like seeking shade or slowing down.

If the weather gets cold outside, another reaction takes place. This time, blood flow to the skin is reduced (to prevent the loss of heat from the blood), and muscles shiver. The involuntary shivering exercises muscles and produces more heat. Messages from the brain may also stimulate the person to seek shelter or sunlight or to put on heavier clothes.

If outside temperatures become too hot, homeostasis breaks down. When the body cannot get rid of enough heat because the environment is too warm, metabolism speeds up to get blood to the skin even faster. This further raises body temperature until the

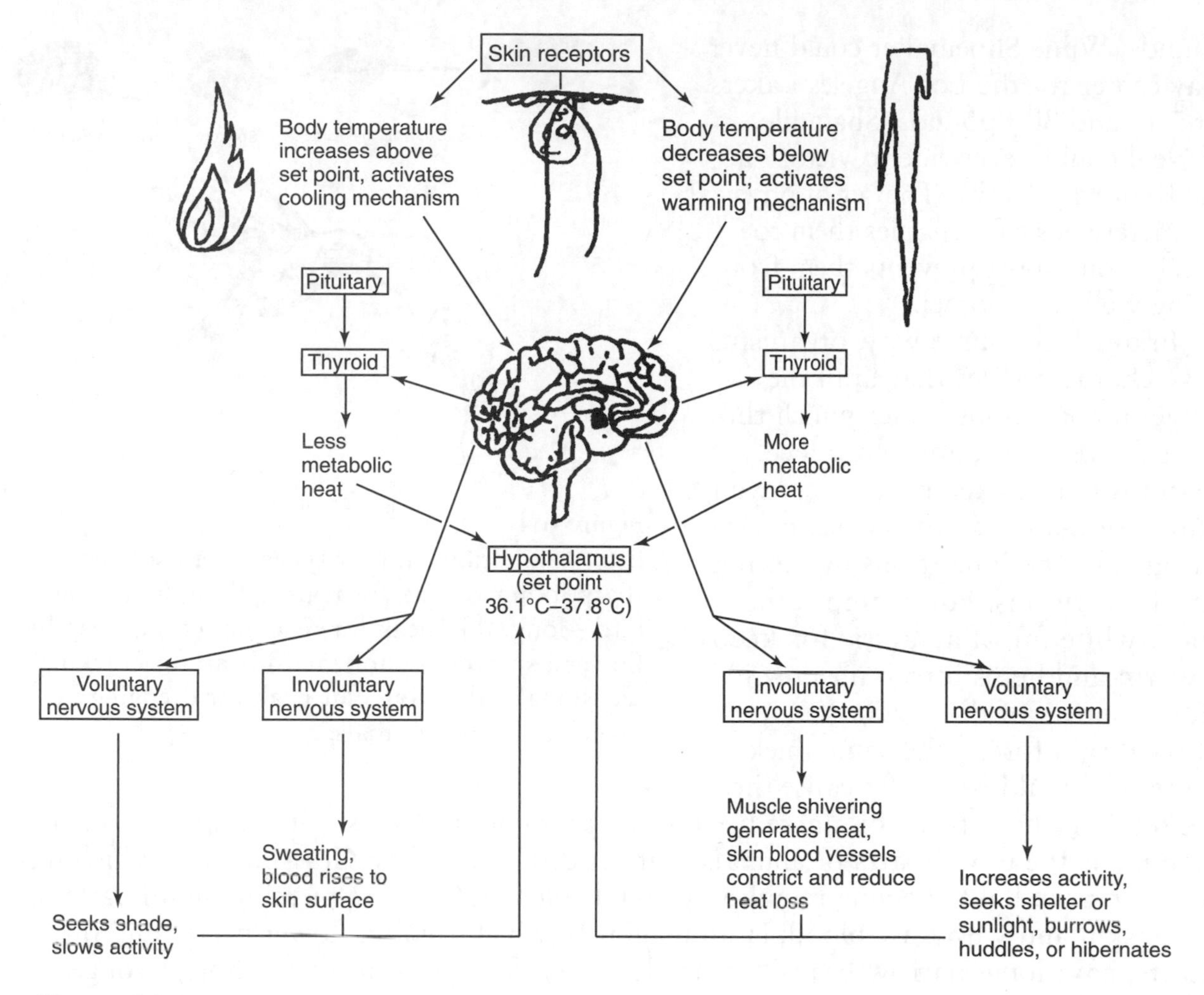

Figure 6.3
Feedback system for temperature regulation in humans.

organism dies of heatstroke. If body temperature gets too low, metabolism slows down, further decreasing body temperature, until death occurs by freezing.

Many feedback mechanisms that regulate body chemistry involve hormones. For example, insulin regulates blood sugar, and adrenaline regulates heart rate (see Chapter 5).

FUNCTIONING IN DIFFERENT ENVIRONMENTS

Animals and plants have a great variety of body plans and internal structures that help them to maintain a balanced condition. For example, birds of the open sea, such as gulls and petrels, can drink salt water because they have special salt-secreting glands. Many desert plants have thick, waxy coatings to prevent water loss or can store water in their cells.

However, characteristics that enable an organism to do well in one environment may be drawbacks in another environment. Willie Shoemaker was probably the best jockey ever to ride a racehorse, and Shaquille O'Neal is one of basketball's great centers. At 4′11″ and 98 pounds, Willie Shoemaker could never play center for the Los Angeles Lakers. At 7′1″ and 300 pounds, Shaquille O'Neal could never ride to victory at the Kentucky Derby. The set of physical characteristics that enables them to excel at one sport prevents them from doing well at the other.

In much the same way, organisms have characteristics that limit the range of conditions under which they can survive, grow, and reproduce. Features that make an organism well suited, or *adapted*, for one set of conditions may be limitations in another set of conditions. For example, the thick, white fur of an arctic fox keeps it warm and helps it to hunt prey in the icy snows of the arctic. Yet, in a tropical rain forest, the same thick, white fur would probably cause the fox to die of heatstroke or starve because its prey could easily see it coming. Another example is the mouthparts of animals that eat different foods. Grasshoppers, which feed on vegetation, have chewing mouthparts, while mosquitoes have piercing mouthparts to pierce skin and withdraw blood. Hummingbirds, which feed on flower nectar and tiny insects, have long, narrow beaks. Cardinals, which feed on seeds, have short, strong, pointed bills. See Figure 6.4.

Figure 6.4
Beak adaptations in four types of birds: falcon—sharp, pointed beak for tearing flesh; hummingbird—long, thin beak for reaching nectar deep in flowers; sparrow—short, hard beak for cracking seeds; mallard—wide, flat beak for dredging up plants and small animals.

The characteristics that enable each organism to feed effectively on one kind of food also prevent it from feeding on another type of food. Together, an organism's overall body plan and its environment determine the way that the organism carries out the life functions.

Practice Review Questions

1. In humans, most of the energy needed for life activities comes from

1. water
2. carbon dioxide
3. food
4. chlorophyll

2. To which group do carbohydrates, proteins, and minerals belong

1. nutrients
2. hormones
3. waste products
4. respiratory gases

3. Oxygen is used to release the energy stored in food by the process of

1. digestion
2. photosynthesis
3. respiration
4. absorption

4. The energy content of food is measured in

1. ounces
2. degrees
3. grams
4. Calories

Base your answers to questions 5 and 6 on the following chart, which shows foods that are a good source of each nutrient listed.

Nutrient	Food Source
Carbohydrates	Whole grain bread
Proteins	Chicken
Fat	Olive oil
Vitamins	Leafy green vegetables

5. Which food listed in the chart is a good source of the nutrient mainly used for growth and repair in the human body?

1. whole grain bread
2. chicken
3. olive oil
4. leafy green vegetables

6. Which nutrient listed in the chart provides a quick source of energy for body cells?

1. protein
2. carbohydrates
3. vitamins
4. fat

GO ON

7. Most of the food and oxygen in the environment is produced by the action of

1. photosynthetic plants
2. decomposition by bacteria
3. respiration by animals
4. saprophytic plants

8. The raw materials used by green plants for photosynthesis are

1. oxygen and glucose
2. carbon dioxide and glucose
3. oxygen and water
4. carbon dioxide and water

9. The equation below shows a summary of a biological process.

carbon dioxide + water → glucose + water + oxygen

This process takes place in

1. cell walls
2. ribosomes
3. chloroplasts
4. mitochondria

10. Dead plants and animal wastes are broken down by

1. producers
2. carnivores
3. decomposers
4. herbivores

11. Which sequence is an example of a producer-herbivore-carnivore relationship?

1. cricket → frog → grass
2. tree → mountain lion → squirrel
3. lion → shrub → rabbit
4. grass → field mouse → owl

12. In consumers, energy for life processes comes from chemical energy stored in the bonds of

1. food molecules
2. mineral molecules
3. water molecules
4. oxygen molecules

13. Short-tailed shrews and ruby-throated hummingbirds have high metabolic rates. As a result, these animals

1. use energy rapidly
2. need very little food
3. have very few predators
4. hibernate in hot weather

14. Which term includes all the activities required to keep an animal alive?

1. metabolism
2. excretion
3. growth
4. nutrition

GO ON →

15. Homeostasis in humans is made possible through coordination of all body systems. This coordination is carried out mainly by the

1. nervous and endocrine systems
2. circulatory and digestive systems
3. respiratory and reproductive systems
4. skeletal and excretory systems

16. Humans breathe more rapidly during exercise than before it because during exercise the blood contains

1. an increased level of oxygen
2. an increased level of carbon dioxide
3. a decreased number of red blood cells
4. a decreased amount of heat energy

17. Which statement best describes an activity that would help an organism maintain homeostasis?

1. A polar bear sheds most of its fur during the coldest winter months.
2. The roots of a willow tree grow away from a region of the soil containing a lot of moisture.
3. A shark swims toward a highly polluted area in the ocean.
4. A desert rattlesnake enters an underground burrow on a hot summer day.

18. When a person exercises hard, small blood vessels near the surface of the skin enlarge. This change allows more heat to leave the blood and the body is cooled. This is an example of

1. locomotion
2. excretion
3. homeostasis
4. synthesis

19. The diagram below shows the actions of two hormones in the human body

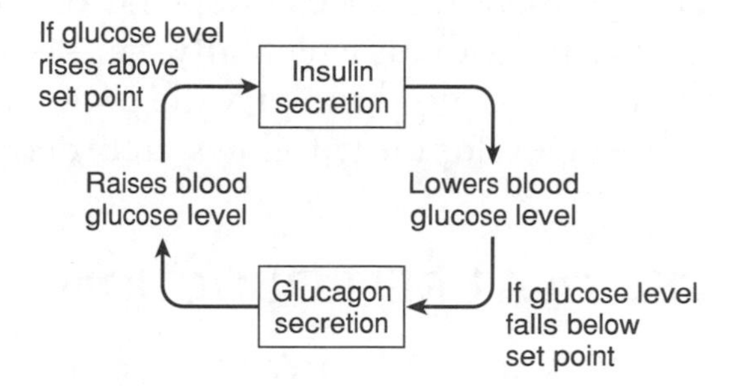

What does this diagram best illustrate?

1. a feedback mechanism
2. cellular respiration
3. mechanical digestion
4. a reflex arc

20. In Yellowstone National Park, some species of algae and bacteria can survive and reproduce in hot springs at temperatures near the boiling point of water. The ability to survive and reproduce at these temperatures is an example of

1. adaptation
2. decomposition
3. photosynthesis
4. respiration

Chapter 7

The Continuity of Life

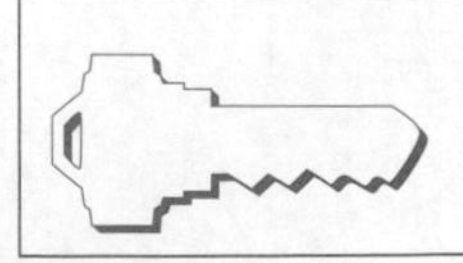

Key Idea: The continuity of life is sustained through reproduction and development.

REPRODUCTION AND DEVELOPMENT

The survival of a species depends on the ability of a living organism to produce offspring. Individuals eventually die. Because of reproduction, though, the species lives on. Living things go through a life cycle involving both reproductive and developmental stages. Development follows an orderly sequence of events.

CELLULAR REPRODUCTION

All cells come from preexisting cells. This occurs when an original cell divides into two cells. One-celled organisms reproduce by dividing into two identical but separate cells, each of which is a separate organism. See Figure 7.1*a*. In multicellular organisms, body cells divide into daughter cells that remain part of the original organism. See Figure 7.1*b*. This type of cell division results in the growth in the size of the organism or the replacement of tissue. In some cases, division of body cells can be a method of reproduction; new cells may separate from the body of the multicellular organism and grow into a new individual.

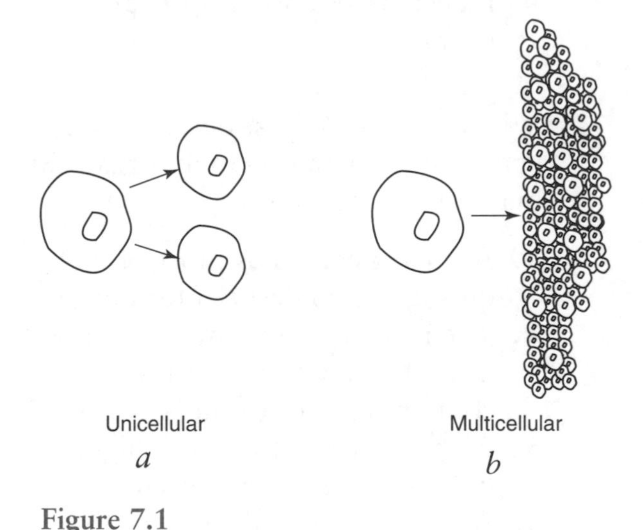

Figure 7.1
(*a*) Cell division in unicellular organisms.
(*b*) Cell division in multicellular organisms.

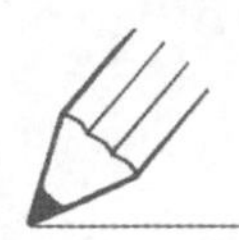

TRY THIS

1. How does cell division in a one-celled organism differ from cell division in a multicellular organism?
2. Identify two different results of cell division in a multicellular organism.

MITOTIC CELL DIVISION

Cell division involves two processes—division of the nucleus (*mitosis*) and division of the cytoplasm (*cytoplasmic division*). Together, the two processes result in division of the entire cell, or *mitotic cell division*. See Figure 7.2.

The nucleus is the control center of the cell. Without the nucleus and the hereditary material it contains, the cell soon dies. Therefore, the first step in making a new cell is copying the nucleus.

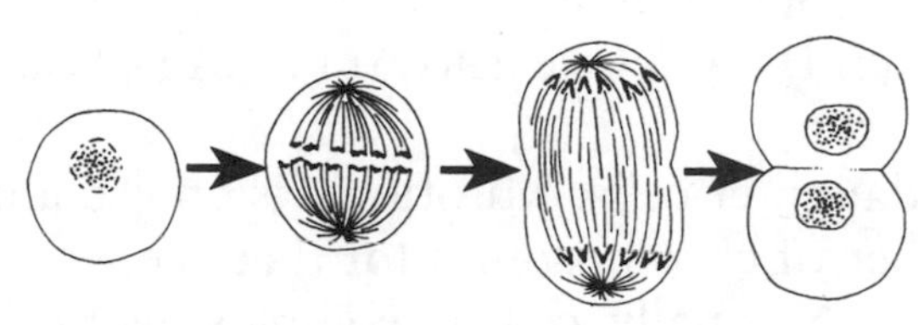

Figure 7.2
Mitotic cell division.

Each and every cell nucleus contains ***chromosomes***, rodlike structures made of ***DNA*** (deoxyribonucleic acid) molecules. DNA molecules are shaped like a spiral ladder—the sides are long, twisted strands of sugar and phosphate, and the rungs are pairs of nitrogen-containing bases. See Figure 7.3*a*. All of the information needed to carry on the life functions and determine the structure and function of the organism as a whole is coded into the DNA molecules. DNA stores information in a pattern, much like a book stores information in a pattern of letters and words. The pattern in DNA is the sequence of nitrogen-containing bases along each strand.

Each species has a set number of chromosomes, and each DNA molecule in a chromosome has a unique pattern. Each body cell of an organism has the same number of chromosomes. Chromosomes occur in pairs that are similar in size and shape. Fruit flies have 8 chromosomes—4 similar pairs; humans have 46 chromosomes—23 similar pairs; and potatoes have 48 chromosomes—24 similar pairs. Since each chromosome has only part of the total information needed to direct the cell, a cell must have a complete set of chromosomes to function properly.

An important characteristic of DNA is that it can ***replicate***, or duplicate itself. Each DNA molecule has the ability to split, or come apart at the rungs. New nitrogen-containing bases, sugars, and phosphates then join each side of the ladder until an exact copy of the old ladder is made. See Figure 7.3*b*. Then there are two new, identical DNA strands. For a short time, just before a cell divides, it has two complete sets of chromosomes—a double of each pair of chromosomes.

After the chromosomes have been duplicated, they line up at the center of the cell. Soon, each double pair of chromosomes separates—one pair moves to one side of the cell, and the other pair moves to the opposite side. Then the cell membrane pinches in between the two sets of chromosomes, separating the original cell into two identical

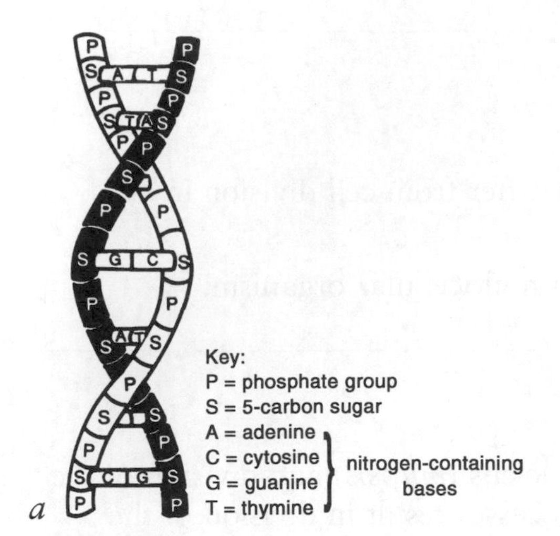

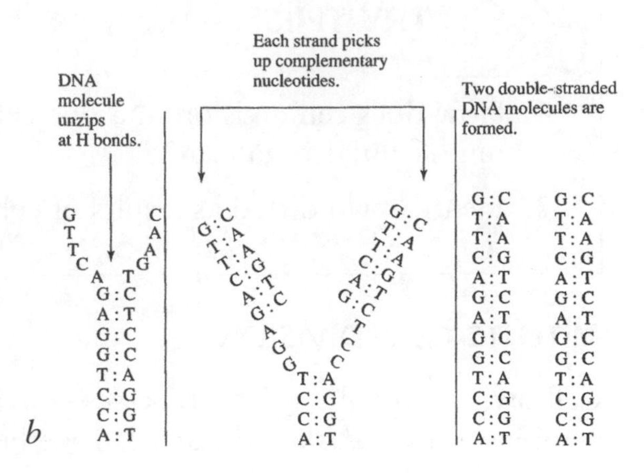

Figure 7.3
(*a*) DNA double helix. (*b*) DNA replication.

daughter cells. Mitotic division is complete, and each daughter cell has the correct number of chromosomes for that species.

Normally, cells in complex multicellular organisms divide only as needed for growth or repair of tissue. Sometimes, however, a group of cells may begin to divide in an uncontrolled way. The wild growth may invade surrounding tissue and interfere with body functions. Harmful, uncontrolled cell growth is called ***cancer***. Research into what causes cells to divide and what controls cell division would be of great help in fighting cancer.

METHODS OF REPRODUCTION

There are two basic methods of reproduction—sexual and asexual. ***Sexual reproduction*** involves two parents and the merging of special sex cells to begin development of a new individual. The prefix *a* in the word asexual means *not*, so the word asexual means *not sexual*. In ***asexual reproduction***, only one parent and no special sex cells or organs are involved. The new organism is a separated part of the parent organism.

Some organisms reproduce sexually. Other organisms reproduce asexually. Some organisms can reproduce sexually or asexually. Reproduction is not necessary for the survival of an individual, but it is necessary for the survival of a species.

ASEXUAL REPRODUCTION

One-celled organisms, many simple animals, and many plants reproduce asexually. There are many methods of asexual reproduction. These include mitotic division of a cell into two cells and the separation of part of an animal or plant from the parent, resulting in the growth of another individual. Figure 7.4 shows some of the different methods of asexual reproduction.

The main disadvantage of asexual reproduction is its lack of diversity. All of the offspring are exact copies of the original organism. If the environment changes, all of the

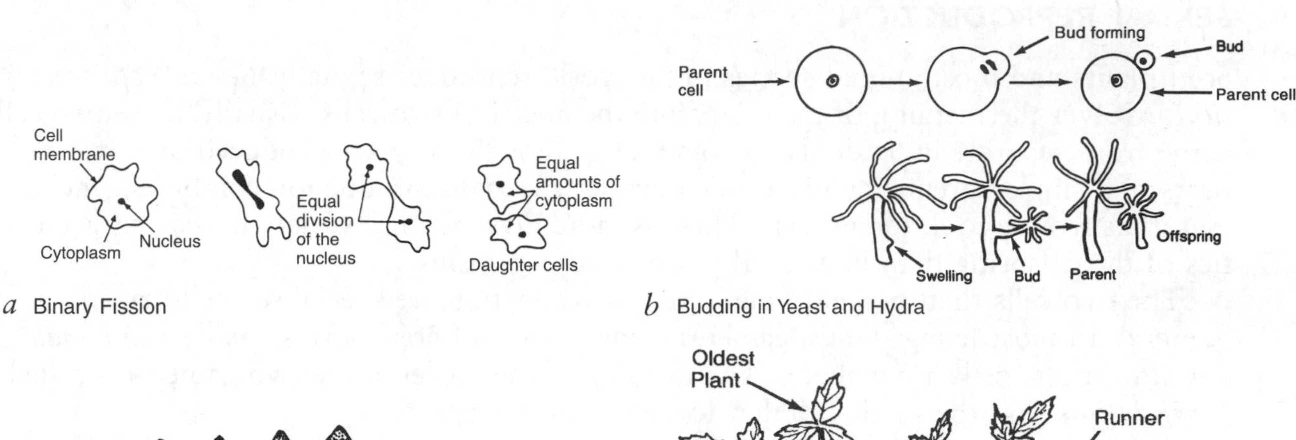

a Binary Fission

b Budding in Yeast and Hydra

c Regeneration in Planaria Worm

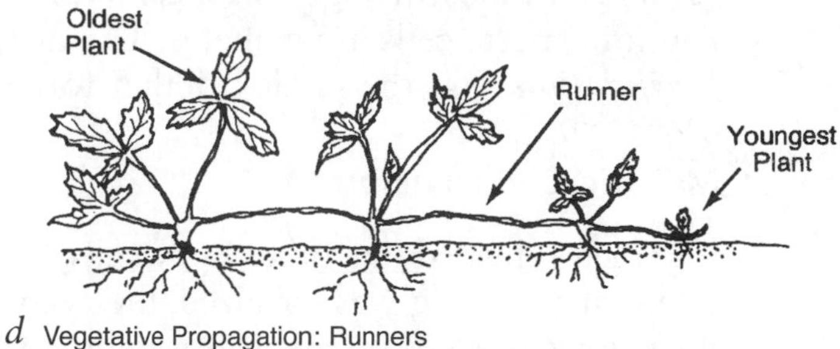

d Vegetative Propagation: Runners

Figure 7.4
Some methods of asexual reproduction.

offspring will respond in the same way. For example, if the environment becomes cold enough to kill one of the organisms, it will probably be cold enough to kill them all. Asexual reproduction does have one advantage—speed. Bacteria and other one-celled organisms can reproduce every 20–30 minutes. At that rate, in just one day, a single bacterium could produce more than 281 *trillion* offspring!

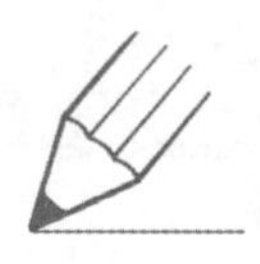

TRY THIS

The diagram on the right shows a series of divisions of a one-celled organism.

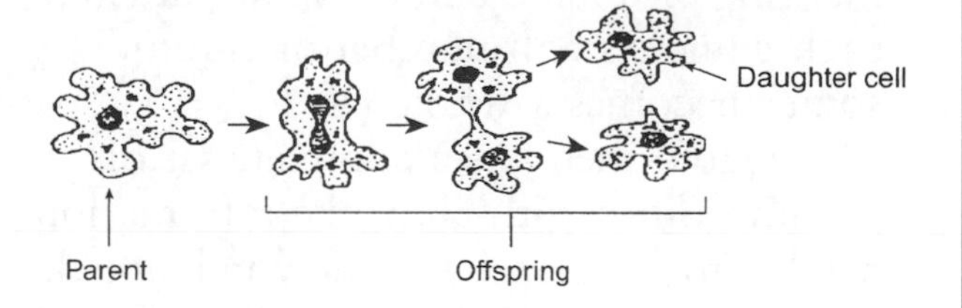

1. Which type of reproduction is shown in this diagram?
2. How does the genetic material of the daughter cell compare to the genetic material of the parent cell?

SEXUAL REPRODUCTION

Seed plants and most animal species in the world reproduce sexually. *Sexual reproduction* involves the merging of material from the nuclei of two cells. Usually, these two cells come from separate individuals. In some cases, though, they are produced by different parts of a single parent organism. In either case, sexual reproduction combines genetic materials from two different cells. There is more diversity and variety in the characteristics of the offspring than in asexually produced organisms.

The two cells that merge during sexual reproduction are special sex cells called ***gametes***. In most living things, gametes come in two different forms—*male* and *female*. Egg and sperm cells are gametes. The merging of the nuclei of the two gametes is called ***fertilization***, and the single cell that forms is called a ***zygote***.

Meiotic Cell Division

If gametes divided like other body cells, each would have a full set of chromosomes. Then when an egg was fertilized, the zygote would contain *twice as many chromosomes* as the species needed. By the next generation, a zygote would have even more chromosomes. Therefore, this does not happen. Doubling of chromosomes in gametes is prevented by ***reduction division***, a special kind of cell division that occurs only in sex cells. Reduction division occurs during ***meiosis***, the formation of specialized sex cells such as eggs or sperm.

In meiosis, the chromosomes replicate once and then the cell divides twice. See Figure 7.5. As in mitosis, the first step is replicating the chromosomes. Then the cell divides in the usual way, forming two daughter cells. During the second division, however, the two daughter cells divide *before* replicating their chromosomes. Four daughter cells called gametes are formed. Each gamete receives one member of each pair of chromosomes in the parent cell. Thus, each gamete ends up with only one-half the number of chromosomes present in body cells.

When a zygote is formed by the merging of a male and a female gamete, each gamete's half number of chromosomes becomes a whole ($^1/_2 + ^1/_2 = 1$). The zygote then has a complete set of chromosomes with all of the information needed to develop into a new individual.

One of the advantages of sexual reproduction is that it produces diversity. Since the zygote is a mix of chromosomes from two parents, it is a completely different organism. Organisms resulting from sexual reproduction are one of a kind! This mix of traits creates diversity and helps the species survive changes in the environment.

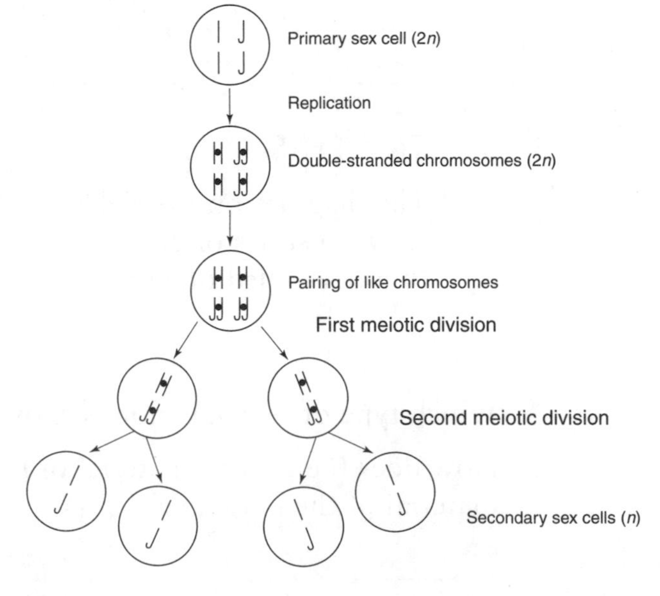

Figure 7.5
Meiosis.

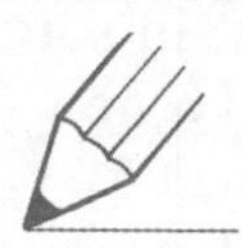

TRY THIS

1. Which process is shown in the diagram on the right?
   1. metamorphosis
   2. regulation
   3. fertilization
   4. respiration

A Sperm + B Egg → C

2. What happens to the structure shown at *C* after *A* and *B* combine?
3. A male chimpanzee has 48 chromosomes in each of his regular body cells. How many chromosomes would be found in each of his sperm cells?
   1. 96
   2. 48
   3. 24
   4. 12

Fertilization in Animals

Egg cells and sperm cells are single cells that contain only half the normal number of chromosomes. Neither can live by itself for more than a few days. Egg cells are large, round, and move slowly, if at all. Sperm cells are very small compared with egg cells and have tiny tails that whip around to propel the sperm forward. Many species produce millions of sperm at a time. All the sperm move toward the egg cell, but only one sperm can fertilize the egg, or unite with it, to form a zygote. The successful sperm forces itself through the egg's cell membrane. The pierced membrane quickly seals the break and, almost immediately, becomes so tough that no other sperm can get through. Once inside the egg, the sperm continues to move until it meets the nucleus of the egg cell and its chromosomes. Then the sperm's nucleus breaks open, and the chromosomes from the two cells unite. The egg is now fertilized. The fertilized egg, or zygote, starts to develop into a new organism. The process of fertilization may occur inside or outside of the body.

External Fertilization: In ***external fertilization***, eggs are fertilized outside the body of the female parent. For example, a female frog may lay a mass of thousands of tiny eggs onto the stem of an underwater plant. A male frog swims by and releases sperm into the water over the eggs. The sperm swim through the water toward the eggs. Under the right conditions, the sperm may fertilize the eggs. External fertilization commonly occurs in fish, amphibians, and other animals that live in or near water.

Internal Fertilization: Some fish and all reptiles, birds, and mammals use internal fertilization. In ***internal fertilization***, egg cells are fertilized inside the body of the female parent. The male puts sperm inside the female's body, where fertilization occurs. In some species, such as humans, the fertilized egg develops inside the female parent and a new organism is born. In other species, the fertilized egg is deposited outside the female's body and the new organism is hatched. For example, a male grasshopper may deposit sperm in a pouch inside the female. The sperm is stored until the eggs are fertilized. Then the female digs a hole in the ground into which she lays the fertilized eggs. The fertilized eggs remain in the hole until the new grasshoppers hatch.

Fertilization in Plants

The flower is a special part of the plant's body that makes both egg and sperm cells. Flowering plants have many different kinds of flowers, but they all contain the same reproductive parts. See Figure 7.6 The ***stamens*** are the male organs of reproduction, and the ***pistil*** is the female organ. The egg cells are encased in a structure inside the pistil called the *ovary*, and the sperm cells are found in ***pollen*** grains produced by the stamens. For fertilization to take place, the pollen must be transported to the ovary. The first step is transferring the pollen to a sticky pad at the top of the pistil—a process called ***pollination***. Insects, wind, or rain can transport pollen from the stamen to the pistil. Pollination that takes place within the flower is called ***self-pollination***. If pollen from one plant pollinates the flower of another plant, it is called ***cross-pollination***.

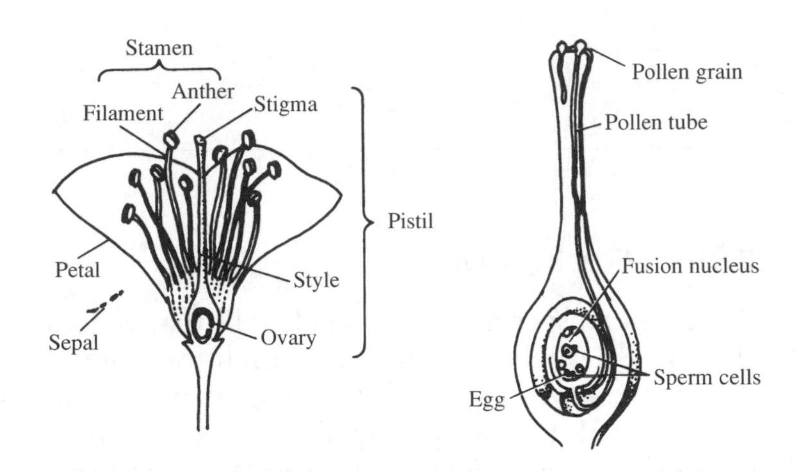

Figure 7.6
Parts of a flower.

After pollination, the pollen grain grows a ***pollen tube*** long enough to reach down to the eggs in the ovary. When the pollen tube reaches the eggs, fertilization takes place.

GROWTH AND DEVELOPMENT

After fertilization, many-celled living things go through complex stages of development until they become adults. The fertilized egg goes through many cell divisions that result in a multicellular organism in which each cell has an identical set of chromosomes.

Although a zygote is a single cell, it has a full set of chromosomes. Many things must happen before the zygote becomes an adult. First, the zygote divides into two identical but smaller cells. Then this process is repeated many times. The two cells divide to form four, the four divide to form eight, and so on. In a short time, there are millions of cells. The zygote has become an ***embryo***, an organism in its early stages of development. Gradually, signals from the DNA cause the new cells to take on different shapes and functions, a process called ***development***. The cells at one end of the embryo begin to form the head, with eyes, mouth, and other organs. Cells along the side become nerve tissue that eventually forms the spinal cord. Inside, other cells develop into organs such as the heart or intestines. Eventually, a complete organism is formed, and not just any organism, but one of the same species as its parents. See Figure 7.7.

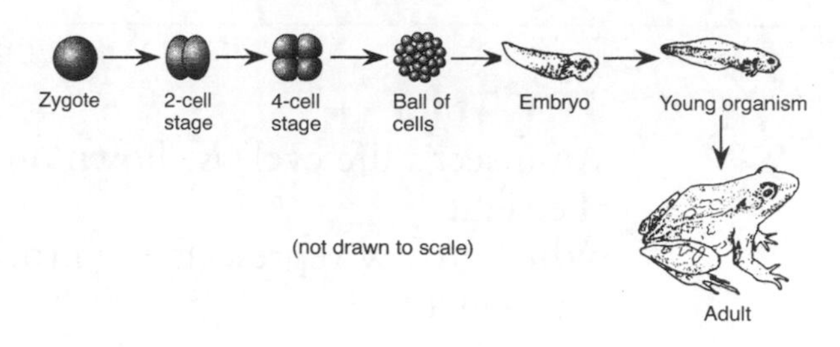

Figure 7.7
Stages of development of an embryo.

As an individual organism ages, various body structures and functions change. The patterns of development vary among living things.

PATTERNS OF DEVELOPMENT IN ANIMALS

In some animals, the young resemble the adult. In others, they do not. Some insects and amphibians go through a series of changes called ***metamorphosis*** as they mature.

Insects

Insects reproduce sexually, and fertilization takes place inside the female. However, insects go through a series of changes as they develop called metamorphosis, which takes place in two ways.

Incomplete Metamorphosis: The first stage is the fertilized egg. The insect as it hatches from the egg is a ***nymph***. The nymph looks like a tiny adult but lacks wings and reproductive structures. As the nymph grows, it will develop wings and reproductive structures and then become an adult. See Figure 7.8.

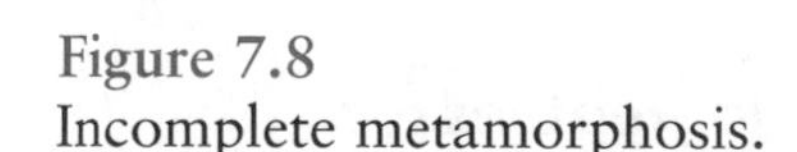

Figure 7.8
Incomplete metamorphosis.

Complete Metamorphosis: Most insects go through a complete metamorphosis, a series of four stages in their life cycle. See Figure 7.9. Again, the first stage is a fertilized egg. In the second stage, the newly hatched insect, or ***larva***, has a wormlike form. The larva eats large amounts of food and grows quickly. When full grown, the larva spins a case, or forms a shell, around its body and changes into a ***pupa***. While inside the case, the pupa develops into an adult. When fully developed, the adult breaks out of the case.

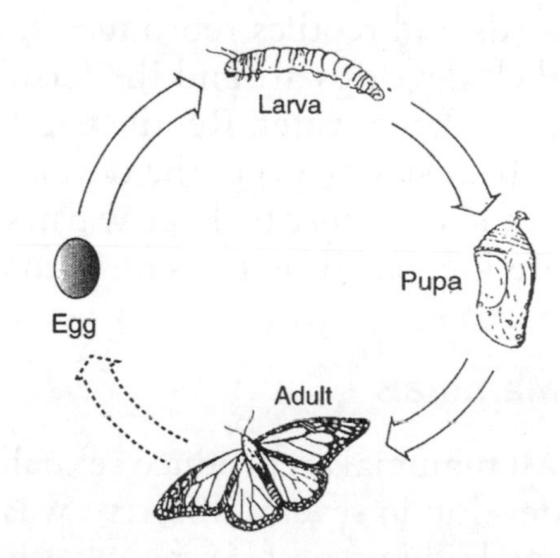

Figure 7.9
Complete metamorphosis of a butterfly.

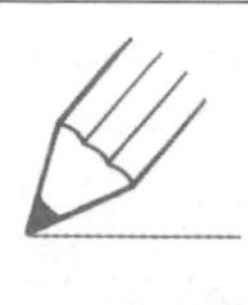

TRY THIS

An insect's life cycle is shown on the right.
Which arrow represents the process of reproduction?

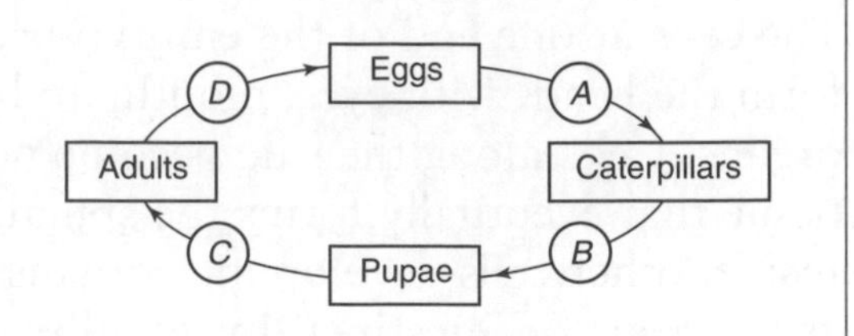

Fish

Fish reproduce sexually. Fertilization usually takes place outside of the female. Fish eggs have no shell, and the embryo develops inside a jellylike egg. When the embryo is fully developed, it looks like a tiny fish.

Amphibians

Like fish, amphibians reproduce sexually, and fertilization takes place outside of the female's body. However, since amphibians spend part of their life in the water and part on land, they develop in a different way. Like insects, some amphibians undergo a complete metamorphosis as they grow. For example, frog eggs hatch into tiny tadpoles, which look like tiny fish. Tadpoles live in water and have a tail for swimming and gills for taking oxygen from water. As it grows, the tadpole's body goes through a series of changes. Gradually, it grows limbs, and its tail gets shorter until it disappears. At the same time, the body develops lungs to take oxygen from air. The young frog leaves the water and spends the rest of its life on land. However, it returns to water to lay its eggs.

Birds and Reptiles

Birds and reptiles reproduce sexually. Fertilization takes place inside the female, and a shell develops around the fertilized egg. The shell allows birds and reptiles to reproduce away from water. Reptile eggshells are tough and leathery. Bird eggshells are hard and stiff. Inside the egg, the developing embryo has water and food. To develop, the embryo in the egg must be kept warm, or incubated. Reptiles lay their eggs in places where they are incubated by the sun. Bird eggs are incubated by their parents.

Mammals

All mammals reproduce sexually, and fertilization takes place inside the female. Mammals develop in several different ways. A very few mammals, like the platypus, develop in an egg laid in a nest. Some, like the kangaroo, start developing inside the mother, but after it is born, it finishes developing inside the mother's pouch. Other mammals, like humans, develop completely inside the mother. The time for mammals to develop, from fertilization until birth, is called ***gestation***. Gestation time varies widely among mammals, from as little as 13 days for an opossum to as much as 2 years for an elephant.

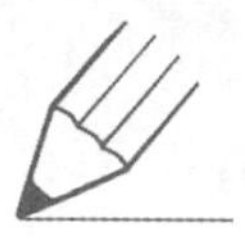

TRY THIS
Compare, by means of a chart, external fertilization and internal fertilization. Be sure to include advantages and disadvantages of each method of fertilization.

PATTERNS OF DEVELOPMENT IN PLANTS

Patterns of development vary among plants. One-celled plants reproduce asexually by cell division. Plants such as mosses and ferns can reproduce asexually or sexually by forming spores. Seed-bearing plants reproduce sexually and form seeds that contain stored food for future development. The pattern of growth by which seeds develop into adult plants varies from species to species.

One-Celled Plants

One-celled plantlike organisms, such as simple green algae, reproduce by dividing into two cells. Each cell becomes a new organism.

Mosses and Ferns

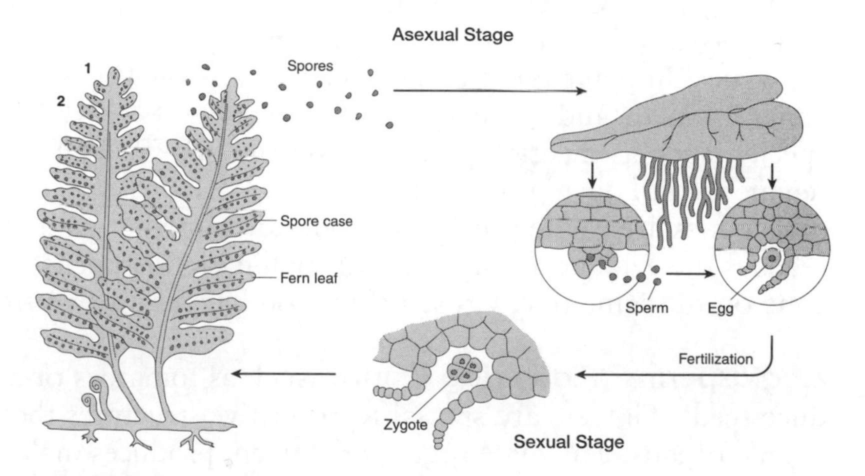

Figure 7.10
The Life Cycle of a Fern consists of a sexual stage and an asexual stage.

Mosses and ferns have a life cycle in which reproduction has a sexual stage and an asexual stage. In the sexual stage, mosses and ferns develop male parts that produce sperm and female parts that produce eggs. When the plant gets wet, the sperm swim to the female branch, fertilize the eggs, and form zygotes. The zygotes then grow into a plant structure that forms spores inside a case. *Spores* are reproductive cells that contain a complete set of chromosomes and can grow into a new plant. The spore is the end product of the sexual stage. In the asexual stage, the spores are released. If they fall into moist soil, they grow into new plants with male and female parts and the cycle begins again. See Figure 7.10.

Seed-Bearing Plants

Seed-bearing plants can reproduce asexually by vegetative propagation (see Figure 7.4*d*) or by growing a new plant from part of a parent plant. That part may be a root, stem, or leaf. For example, potatoes are underground roots. The eyes on a potato are buds.

Each one can grow into a new plant. In water or moist soil, stems or leaves cut from a plant, such as an African violet, may grow into a new plant.

Seed-bearing plants reproduce sexually by producing seeds. Each seed is the product of two sex cells. A ***seed*** is a young plant with stored food in a cover. See Figure 7.11. There are two types of seed-bearing plants. ***Gymnosperms*** do not bear flowers but produce seeds that are exposed, that is, the seed is not inside a fruit. ***Angiosperms*** bear flowers and produce seeds that are contained in a fruit.

Figure 7.11
A seed is a young plant with stored food surrounded by a protective covering.

Gymnosperms (Nonflowering Plants) such as pine trees, produce seeds but do not bear flowers. Most gymnosperms grow reproductive organs called ***cones*** on their branches. Cones are made of scales, and the seeds develop on the surface of the scales. In plants with cones, some cones are male and some are female. Some species, such as the ginkgo, have male and female trees. Pollen grains released by male cones are carried to female cones by rain and wind. The pollen grows a tube that reaches the egg and fertilizes it. Self-pollination happens when pollen reaches a female cone on the same tree. Cross-pollination happens if two trees are involved.

Angiosperms (Flowering Plants) such as tomatoes or apples, bear flowers and produce seeds. Flowers are special reproductive structures that contain both male and female organs. The male organ, or stamen, produces male sex cells called pollen. The female organ, or pistil, produces female sex cells called eggs, inside in a structure called the ovary. Pollen from the stamen is carried to the sticky surface of the pistil by insects, wind, or rain. The pollen grows a tube down to the eggs, and fertilization occurs. After fertilization, the ovary grows and becomes a fruit with seeds inside. The fruit houses and protects the seeds. Seeds contain tiny plant embryos. The seeds remain dormant until environmental conditions are right for further development. Then the embryos begin to grow and, under the right conditions, become adult plants.

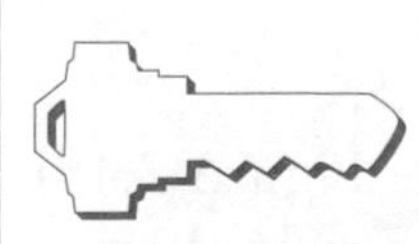

Key Idea: Organisms inherit genetic information in a variety of ways that result in continuity of structure and function between parents and offspring.

PATTERNS OF INHERITANCE

GENETICS

Have you ever been told that you look like someone in your family? Have you ever seen people in a particular family that looked alike? Family members often look alike because they share certain ***traits***, or characteristics. All organisms pass their traits on to their children, or ***offspring***. Bulldogs look like other bulldogs. Calico cats have litters of calico kittens. The passing on of traits from parents to their children is called ***heredity***.

Heredity alone does not determine what an organism will be like. All organisms are also affected by their environment. A watermelon seed may have inherited the ability to grow into the biggest, sweetest watermelon in the world. However, if the environment in which the seed grows does not have enough sunshine and water or the proper soil and temperature, the seed will not become as good a watermelon plant as it could have been.

GREGOR MENDEL: FATHER OF MODERN GENETICS

More than 100 years ago, an Austrian monk named ***Gregor Mendel*** did scientific experiments with pea plants that later became the basis for the modern study of heredity, or ***genetics***. Genetics deals with the ways in which traits are passed on from one generation to the next.

Mendel wanted to understand how traits were passed on from parents to offspring. Mendel had planted seeds from tall pea plants and observed that although most of the plants that grew from "tall" seeds were tall, some were short. He wanted to find out why the short pea plants were unlike their parents.

First, Mendel grew plants that were pure for the trait of tallness, that is, plants that always produced tall plants from their seeds. He did this by a process called self-pollination. Mendel took pollen from the *anther* (male sex organ) of a tall plant and put it onto the *pistil* (female sex organ) of the same tall plant. This fertilized the *ovary*, and seeds were produced. Mendel then used these seeds to grow new tall pea plants. After doing this for several generations, Mendel was sure that he had pure tall plants. A ***generation*** is made up of all offspring that are at the same level of descent from a common ancestor. For example, your grandparents are one generation, your parents are another generation, and you are another generation. By using the same process of self-pollination, Mendel also bred plants that were pure for the trait of shortness. Once Mendel had pure tall and pure short plants, he was ready to carry out his experiment.

Mendel called the pure tall and pure short plants the P_1 generation. P_1 stood for the first, or parent, generation of pure plants. Mendel wanted to find out what kind of pea plant would grow if it had one pure tall parent and one pure short parent. To do this, he used a process called *cross-pollination*. Mendel took the pollen from the anther of a pure tall plant and placed it onto the pistil of a pure short plant. He also did the reverse, placing pollen from the anther of a pure short plant onto the pistil of a pure tall plant. In this way, he produced seeds that had two different parents. Mendel called the plants that grew from these seeds the F_1 generation (*F* stands for filial, or son).

Mendel expected that when he planted the seeds from the crossed plants, they would grow into medium-sized offspring. To his surprise, however, all of the seeds produced tall offspring. Mendel crossed thousands of pure tall and pure short pea plants. The results were always the same—all of the plants in the F_1 generation were tall. Mendel was puzzled. What had happened to the trait for shortness? Why had it disappeared?

To find out, Mendel grew another generation of pea plants by self-pollinating the F_1 generation and planting their seeds. He called this third set of plants the F_2 generation. One can only admire Mendel's patience. For three generations, from P_1 to F_1 to F_2, he carefully tended the pea plants in his monastery garden and kept detailed records. When the F_2 plants were fully grown, Mendel discovered that the trait for shortness had reappeared! In the F_2 generation, $^3/_4$ of the plants were tall, but $^1/_4$ were short. Mendel repeated this experiment many times and always got the same result in the F_2 generation.

Mendel also experimented with other traits of pea plants, such as seed color, pod color, and seed texture. Each of these traits followed the same pattern as tallness. For example, when pure yellow-seed plants were crossed with pure green-seed plants, the F_1 generation all had yellow seeds. When the F_1 generation was crossed, the F_2 generation was $^3/_4$ yellow and $^1/_4$ green.

Mendel concluded that pea plants must contain some factor that causes them to have traits like tallness or seed color. He also reasoned that this factor was passed along in units—one from the mother and one from the father. We now know that Mendel's factors are ***genes***, a DNA sequence at a specific location on a chromosome that contains the genetic code for a particular characteristic of an organism.

PUNNETT SQUARES

One of Mendel's experiments was to cross pure tall and pure short plants. In this Punnett square, a capital T stands for the tall trait, and a small t stands for the short trait. Thus, TT stands for a pure tall plant, and tt stands for pure short plant. To set up a Punnett square, the genotype of one parent is placed across the top and that of the other parent down the left side. It does not matter which parent is on the side or which is on the top of the Punnett square. Note that only one letter goes into each box for the parents. Next, you fill in the boxes by copying the row and column letters across or down into the empty boxes. This tells you the possible combinations of genes that offspring can have every time reproduction occurs. It also shows you the percentage of offspring that will have a particular combination of traits. In this example, 100% of the offspring had one tall T gene and one short t gene.

Pure Short Parent

Pure Tall Parent	F_1	t	t
	T	**Tt** hybrid tall	**Tt** hybrid tall
	T	**Tt** hybrid tall	**Tt** hybrid tall

Figure 7.12
Punnett square.

As you can see, when Mendel crossed pure tall and pure short plants, the short trait did not disappear. However, all of the plants in this F_1 generation were tall. Mendel reasoned that this was because the tall trait had a stronger influence than the short trait. Mendel called the tall trait that showed up ***dominant*** and the short trait that was hidden ***recessive***. Plants that have more than one gene for the same trait are called ***hybrids***. Even though the hybrids in Mendel's F_1 generation had both a tall and a short gene, they all grew tall because their growth was more strongly influenced by the dominant tall gene. The genetic makeup of an organism is its ***genotype***. The expression of a trait, that is, the way an organism looks, is its ***phenotype***. Pure tall and hybrid tall plants have the same phenotype but different genotypes.

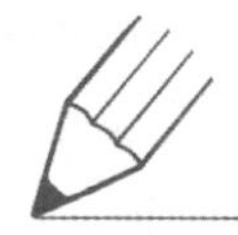

TRY THIS

1. Make a Punnett square for the F_2 generation of Mendel's pea plants. In the F_2 generation, each parent plant is a hybrid tall pea plant with both a tall and a short gene (**Tt**).
2. What percentage of the F_2 generation are hybrids?
3. What percentage are pure tall or pure short?
4. What is the genotype of a short plant?
5. What two genotypes will produce a tall phenotype?

Some traits are inherited by mechanisms other than dominance and recessiveness. For example, flower color in garden peas is either red or white, with red flowers dominant. However, when Japanese four o'clock plants with pure red and pure white flowers are crossed, the offspring is pink. The two traits for color blend to produce a third form of that trait. Another example is hair color in shorthorn cattle. When pure red-haired shorthorn cattle are crossed with pure white-haired shorthorn cattle, the offspring have *both* red and white hairs, giving them a roan color. Yet another example is how sex is determined in humans. The chromosomes that determine sex come in two forms—X and Y. Females have two X chromosomes, males have one X and one Y chromosome. Sex is determined at the moment the egg cell and sperm cell unite. Females have only X chromosomes, so egg cells can contain only X chromosomes. Males have both X and Y chromosomes, so sperm cells may contain either an X or a Y chromosome. If the egg unites with a sperm cell containing an X chromosome, the offspring will be female. If the egg unites with a sperm containing a Y chromosome, the offspring will be male.

Sex chromosomes carry genes for traits other than sex. Traits carried on the X chromosome but not on the Y chromosome are called ***sex-linked traits***. Hemophilia, a disease in which blood does not clot properly, is a sex-linked trait. The gene for hemophilia is carried only on the X chromosome. Thus, a male needs only one recessive gene to get hemophilia, while a female needs two recessive genes.

Another model that shows patterns of inheritance is a pedigree chart. A pedigree is a diagram that shows the pattern of inheritance in a family. A pedigree chart shows many more individuals than a Punnett square. On a pedigree chart, you can see which individuals are carriers of a recessive gene such as the one for hemophilia. A ***carrier*** shows the dominant trait but carries a recessive gene. See Figure 7.13.

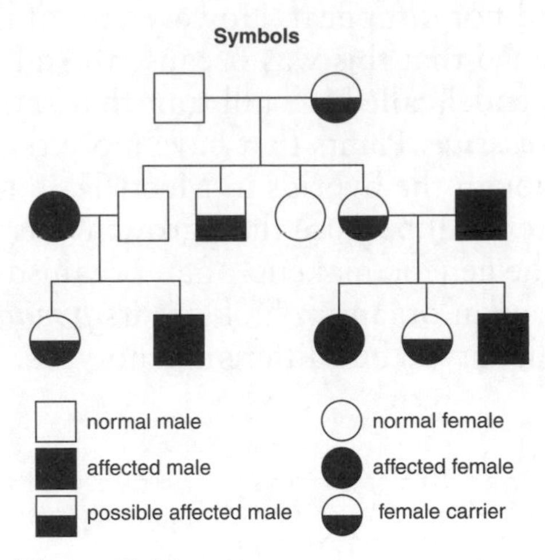

Figure 7.13
Pedigree chart for hemophilia in a family.

TRY THIS

The pedigree chart below represents the inheritance of sickle-cell anemia through three generations.

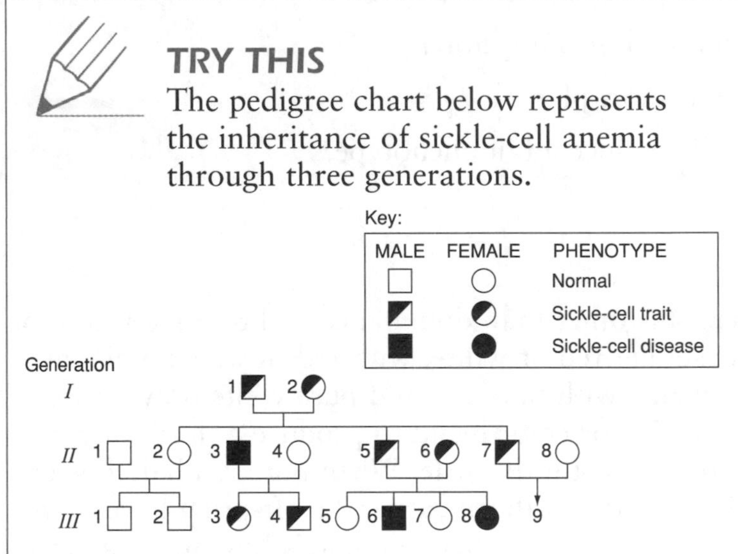

1. Which possible symbols could be used to represent person 9 in generation III?

   1. ○ and ●
   2. ■ and ◪
   3. ■ and ●
   4. ◐ and □

2. Female 3 in generation III marries a man who is a hemophiliac. What percentage of their children can be expected to have hemophilia?

Practice Review Questions

1. The type of molecule shown below is found in the cells of all organisms.

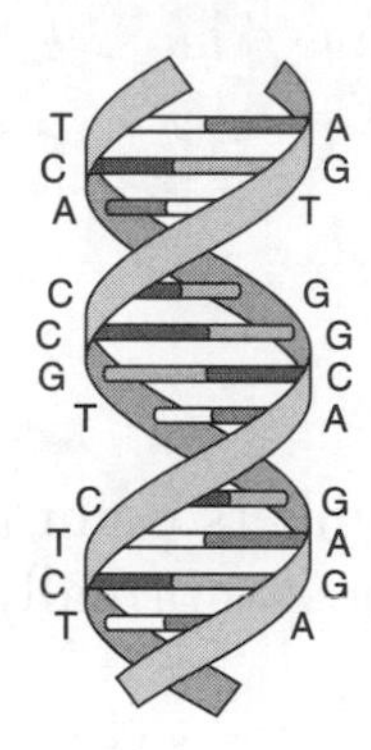

The function of this type of molecule in a cell is to

1. control the synthesis of starch
2. regulate the movement of water into and out of the cell
3. determine the characteristics that will be inherited
4. provide energy for life activities

2. The uncontrolled division of abnormal cells may result in

1. cancer
2. hemophilia
3. Down syndrome
4. diabetes

3. The diagrams below show various processes related to reproduction.

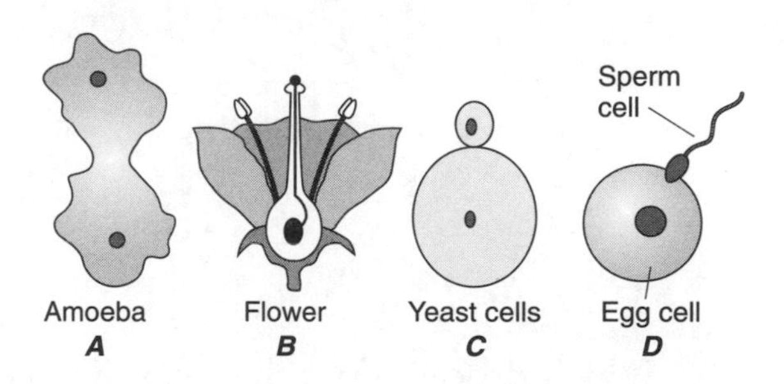

Asexual reproduction is represented by

1. *A* and *C*
2. *B* and *D*
3. *B* only
4. *A* only

Base your answers to questions 4 and 5 on the diagram below

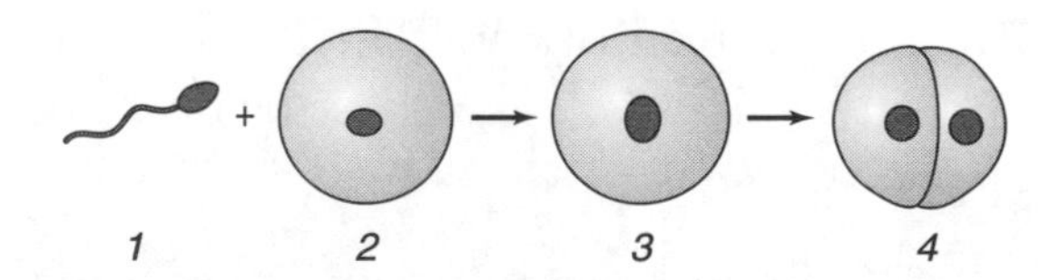

4. Cell *1* is the result of the process of

1. mitosis
2. mutation
3. meiosis
4. fertilization

5. If cell 2 has 30 chromosomes, cell 3 should have

1. 45 chromosomes
2. 60 chromosomes
3. 15 chromosomes
4. 30 chromosomes

GO ON

Base your answers to questions 6 and 7 on the diagram below of the life cycle of a frog.

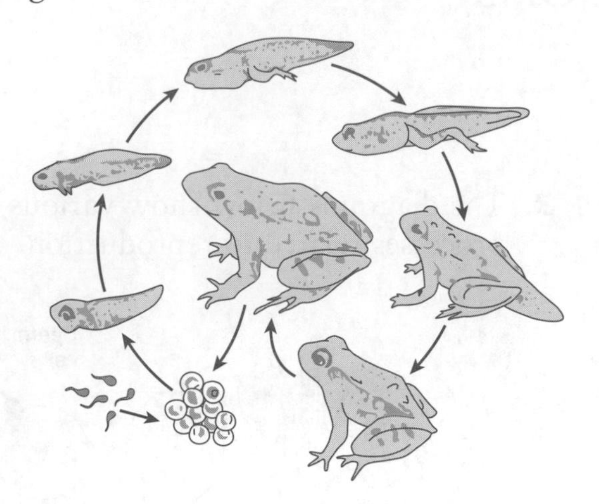

6. Which process does the frog undergo as it changes from a fertilized egg to a tadpole to an adult frog?

1. natural selection
2. mutation
3. metamorphosis
4. photosynthesis

7. The method of reproduction involved in this life cycle can best be described as

1. asexual reproduction by external fertilization
2. sexual reproduction by external fertilization
3. asexual reproduction by internal fertilization
4. sexual reproduction by internal fertilization

Base your answers to questions 8–10 on the diagram of a flower below.

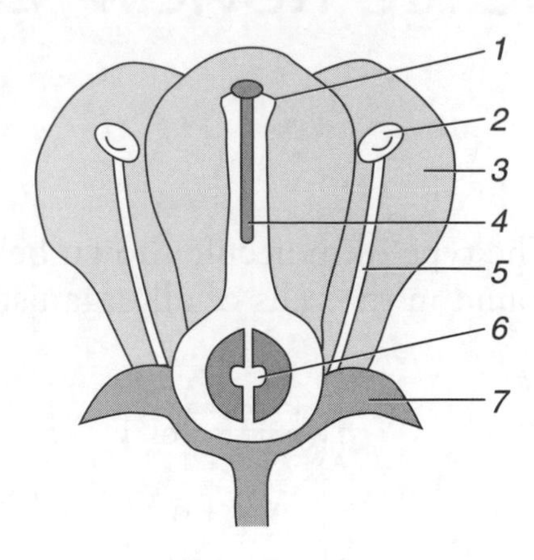

8. Which numbers point to structures where meiosis normally occurs?

1. *2* and *6*
2. *3* and *7*
3. *4* and *5*
4. *1* and *3*

9. In which structure would fertilization take place?

1. *1*
2. *2*
3. *4*
4. *6*

10. During pollination, pollen is transferred from

1. *6* to *1*
2. *2* to *1*
3. *5* to *3*
4. *3* to *7*

GO ON

Base your answers to questions 11 and 12 on the pedigree chart below, which shows the inheritance of handedness in humans over three generations. The gene for right-handedness (*R*) is dominant over the gene for left-handedness (*r*).

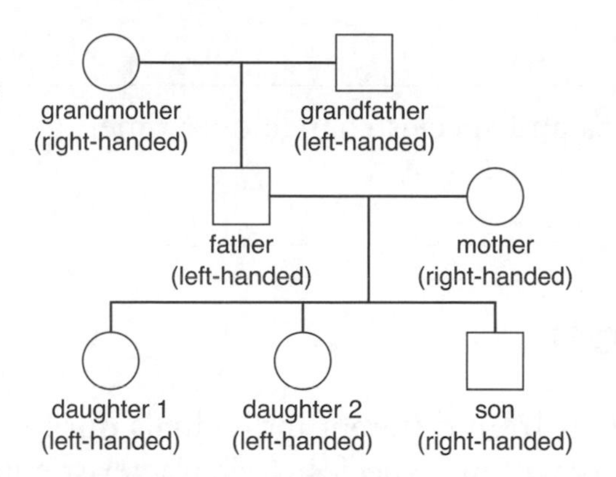

11. For which individual is *Rr* the most probable genotype?

1. grandmother
2. grandfather
3. daughter *1*
4. father

12. Which two individuals have identical genotypes for handedness?

1. mother and daughter *1*
2. mother and father
3. grandmother and grandfather
4. mother and son

CONSTRUCTED-RESPONSE QUESTIONS

Base your answers to questions 13–16 on the information and Punnett square below, which shows a cross between two squash plants.

	D	*d*
D	*DD*	*Dd*
D	*DD*	*Dd*

Key:
D = disc-shaped squash (dominant)
d = round squash (recessive)

13. State the process represented by the Punnett square.

14. According to the Punnett square, what percentage of the offspring squash will be disk shaped?

15. Use the Punnett square below to show the results of crossing two Dd parents.

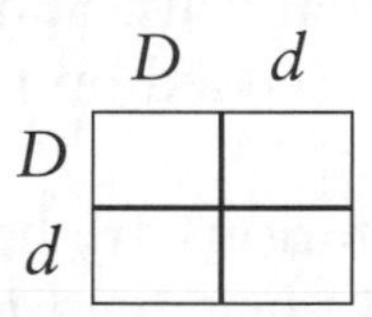

If 100 offspring were produced from the Punnett square above, how many would have a round shape?

Chapter 8

The Evolution and Diversity of Life

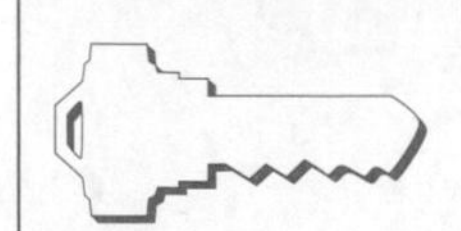

Key Idea: Individual organisms and species change over time.

VARIATION

Over many generations, the mixing of chromosomes from different individuals during sexual reproduction gives rise to a mixture of traits within a species. A characteristic in an individual that is different from other individuals of the same species is called a ***variation***. The millions of distinctly different people that exist are all the result of the different combinations of genes in the chromosomes.

MUTATION

Sometimes an organism is born with a trait that none of its ancestors has. Such a trait is known as a ***mutation***, or change in the genes that cause an organism to have a particular characteristic. Some mutations happen by chance, as when a chromosome does not replicate perfectly. Others occur when radiation, such as X rays, change part of a DNA molecule and that change is passed on when the chromosome replicates. Some chemicals, such as certain poisons that kill insect pests, cause mutations by damaging the chromosomes of normal cells. Some cockroaches had a mutation that made them resistant to the poison. It was passed on to their offspring, and now ridding infested homes of these pests is harder.

Many mutations are harmful. For example, a fly born without wings cannot fly to obtain food. Other mutations seem to increase an organism's ability to survive. For example, a tomato plant that is resistant to a fungus may survive while others infected with the fungus die.

By using genetics, scientists are able to breed plants and animals that have desired traits. For example, the seedless orange is the result of a mutation. In nature, this mutation would be fatal since oranges that do not produce seeds cannot produce offspring from seeds. However, people like to eat oranges without having to spit out seeds. So growers graft branches of seedless oranges onto trees of oranges that make seeds.

The process of breeding plants and animals for particular genetic traits is called ***selective breeding***. For thousands of years, people have mated individuals with desirable traits to produce offspring with those traits. Over time, these have resulted in new varieties of cultivated plants and domestic animals. For example, Greyhounds have been bred to hunt by outrunning their prey. Dog owners did this by mating only the quickest and most intelligent dogs.

TRY THIS

Tomato seeds were taken into space during a shuttle flight, where they were exposed to cosmic rays. When brought back to Earth, the seeds were planted. Many did not germinate, and one seed produced a plant that had no chlorophyll in its stem. What is the most likely cause of this result?

ADAPTATIONS

Both the mixing of genes and mutation result in populations made up of individuals with a range of differences instead of all being identical. These variations give the population a much better chance of survival in the long run. Should the environment change or a disaster strike, there is a much greater chance that some individuals will have traits that allow them to survive. Those individuals will then reproduce and carry on the species. Individuals without the traits needed to survive in the new environment will die and not be able to reproduce. Over time, the makeup of the population changes. Eventually, it consists mostly of individuals that have traits that make them better suited, or ***adapted***, to survive in the environment. See Figure 8.1. Successful adaptations help individuals survive and continue the species.

The number of different kinds of adaptations we find in living things is remarkable. Some organisms, such as the zebra, are colored in a way that helps them confuse predators. When zebras in a herd run in different directions, their striped bodies make it hard for a lion to focus on just one of them during the chase. Other organisms have protective coloration that helps them blend in with the landscape. Baby deer have light brown fur with white spots that helps them hide on leafy forest floors dappled with sunlight. Flounder and chameleons have special skin cells that allow them to change color to match their background.

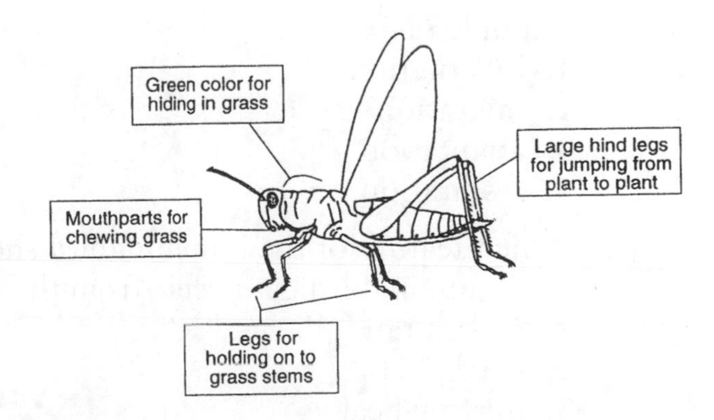

Figure 8.1
Adaptations in a grasshopper.

Some organisms have a body shape or color that looks like another living or nonliving thing, an adaptation called ***mimicry***. Walking-stick insects look like tree twigs.

Pipefish and many eels look like sea grass. Some organisms have coloration that imitates the warning colors of poisonous species. For example, the viceroy butterfly looks almost exactly like the monarch butterfly. The monarch butterfly makes birds sick because monarchs feed on milkweed, a mildly poisonous plant. On the other hand, birds can safely eat viceroy butterflies. However, once a predator, such as a bird, eats a monarch and gets sick, it avoids the monarch *and* the look-alike viceroy butterfly. See Figure 8.2.

Figure 8.2
Mimicry. Monarch (*a*) and viceroy (*b*) butterflies.

Some organisms have body parts that are adapted to eating a certain type of food. Eagles and hawks have sharp, pointy beaks for tearing prey. Hummingbirds have long, thin beaks to reach nectar deep in a flower. Sparrows have strong, stubby beaks that help them crack open seeds.

However remarkable adaptations may seem, remember that they are literally the result of a process of elimination. For example, as predators pick off viceroy butterflies that do *not* resemble monarchs, the genetic makeup of the population gradually changes, or evolves. Over time, the traits that make a viceroy look like a monarch become more common while those that make it look different disappear.

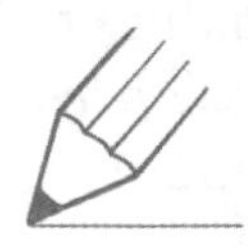

TRY THIS

1. Just before winter, the hair on horses grows longer and thicker. This is an example of
   1. hibernation
   2. migration
   3. adaptation
   4. respiration
2. Which feature of the hummingbird shown below is an adaptation that helps the hummingbird get nectar from the flower?
   1. forked tail
   2. small feet
   3. narrow beak
   4. keen eyesight

EVOLUTION

Over time, small differences between parents and offspring can accumulate. Eventually, the descendants may be very different from their ancestors. Changes in a species over time is called ***evolution***. Evolution is the idea that existing life-forms have developed from earlier, different life-forms. Much of the history of life on Earth has been pieced together from fossils found in rocks. ***Fossils*** (from the Latin word *fossilis* meaning "dug up") are the remains or traces of extinct organisms. Fossils provide information about the structure of animals and plants that lived in the past.

Many thousands of layers of sedimentary rock contain fossils. Fossils in layers of rock that are next to each other are more alike than fossils in layers of rock that are far apart. The earliest traces of life on Earth are fossils of simple, one-celled organisms in 3.5-billion-year-old rocks. Since then, fossils of many new life-forms have appeared, and most old life-forms have disappeared.

TRY THIS

The diagram below represents a process that happens over a long period of time.

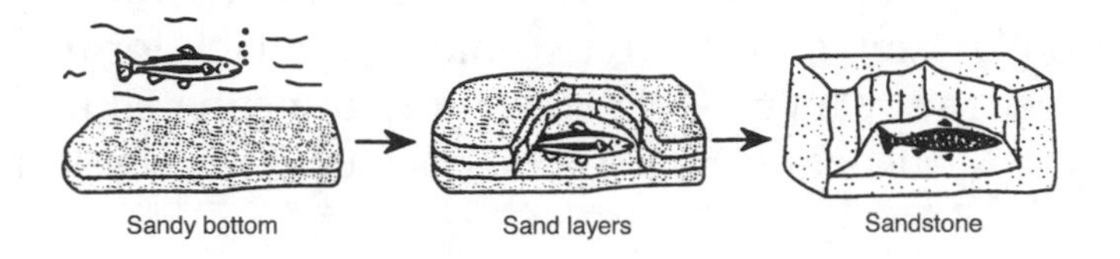

Which process is represented in this diagram?

1. condensation
2. fossilization
3. photosynthesis
4. reproduction

The fossil record shows that life-forms have changed gradually over time, so that modern life-forms are very different from ancient ones. This supports the theory that changes have occurred in living things over time. Scientists have traced the gradual changes in the body forms of many different living things over time. They have observed the steady buildup of differences from one generation to the next. They are convinced that this is what has led to species as different from one another as algae are from whales.

NATURAL SELECTION

The fossil record shows that over time, species change, species become extinct, and new species appear. Charles Darwin offered an explanation for how these changes could occur. Darwin's key idea was that conditions in *nature* have an influence on which individuals in a population will survive and reproduce. Known as ***natural selection***, Darwin's idea was that organisms with favorable variations are better able to survive and reproduce than organisms without them. Natural selection proposes that evolution happens in the following way:

1. Individuals of the same species have different characteristics, and sometimes these differences give one organism an advantage in surviving and reproducing.
2. Offspring that inherit the advantage are more likely to survive and reproduce.
3. Over time, the number of individuals with the advantageous characteristic will increase.

In this way, the process of natural selection causes a species to evolve. This leads to organisms that are well suited for their environment.

Changes in the environment can affect the survival value of some inherited characteristics. Before the middle of the 1800s, England's peppered moth, *Biston betularia*, was always white with black speckling on its wings and body. In 1850, a mutant black form was caught for the first time near the industrial city of Manchester. After the rise of factories, grime and soot caused blackening of the tree trunks. The black form, *Biston carbonaria*, increased steadily over the years until 95% or more of peppered moths found were of this type. Why do you think this happened? If you guessed natural selection, you are right. Birds eat peppered moths. Before factories polluted the woods, the black form stood out on the pale-colored tree trunks and was picked off by birds. Then the environment changed. Factory pollution caused the trees to darken in color. Now the white moths showed up on the dark background and were more easily spotted and eaten by birds. The environment favored the black moths. Black moths were more likely to survive and reproduce. They became the dominant moth in the area. See Figure 8.3.

Figure 8.3
The peppered moth. Both photos contain a black form and a white form of the peppered moth. Can you find the second moth?

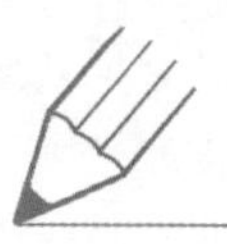

TRY THIS

Certain insects resemble the twigs of trees on which they live. The most probable explanation for this resemblance is that

1. the trees caused a mutation to occur
2. no mutations have taken place
3. natural selection has favored this trait
4. the insects needed to camouflage themselves

As the environment changes, the characteristics that will be advantageous for survival might change. Therefore, changes in the environment lead to changes in organisms. Small differences between parents and offspring can accumulate over many generations so that descendants may become very different from their ancestors.

Evolution by natural selection does not imply long-term progress toward a goal or that organisms have a set direction. Nor does evolution always occur gradually. Recent research shows that evolution often occurs in spurts brought on by sudden, large-scale changes in the environment. Thus, a good analogy for evolution is the growth of a hedge. Some branches exist from the beginning of the hedge's life with little or no change. Other branches grow and then die out. Still others grow a little and then branch apart repeatedly. The end result is a complex network of large and small branches. See Figure 8.4.

One important reason to preserve life on Earth is that evolution builds on what exists. The more variety there is, the more variety can exist in the future. Human behavior that results in the extinction of organisms decreases the variety of life-forms on Earth and may have far-reaching effects in the future.

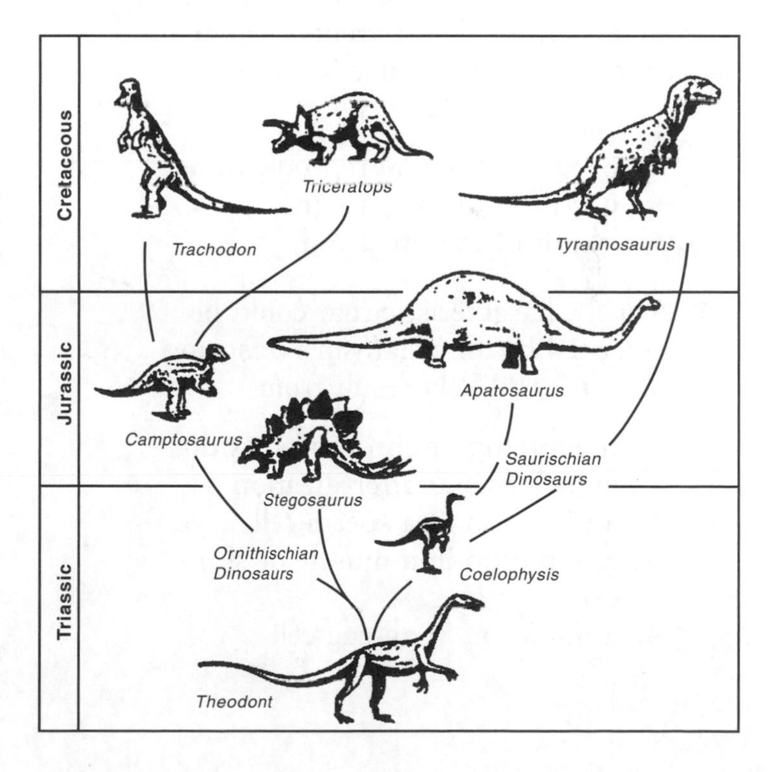

Figure 8.4
Branching out in evolution.

Practice Review Questions

1. The diagram below shows two guinea pig parents and their six young offspring.

Some of the offspring are colored differently from their parents. The main cause of this difference is the

1. age of the offspring
2. process of sexual reproduction
3. nutrition of the parents
4. health of the offspring

2. A *new* characteristic that could be inherited by an organism's offspring would *least* likely result from

1. the mixing of chromosomes during meiosis and fertilization
2. a mutation in a sperm cell
3. a mutation in a muscle or skin cell
4. a mutation in an egg cell

3. One characteristic of mutations is that they usually

1. change the genetic pattern of an organism's chromosomes
2. do not happen at random
3. happen because an organism needs a particular trait
4. benefit only the organism

4. Certain strains of bacteria that were killed by penicillin in the past have become resistant to penicillin. The probable explanation for this is that

1. the penicillin decided not to attack the bacteria any more
2. a mutation enabled some bacteria to survive penicillin and they reproduced, creating a more resistant population
3. the bacteria that survived exposure to penicillin learned to avoid it
4. the resistant bacteria came from populations that had never been exposed to penicillin

5. In the spring and summer, the Arctic fox has dark fur that matches the brown dirt in its environment. In the fall and winter, its fur turns white to match the surrounding snow. This change is an example of

1. adaptation
2. metamorphosis
3. homeostasis
4. cellular respiration

GO ON 

6. In an area in Africa, temporary pools form along rivers during the rainy months. Some fish have developed the ability to use their lower fins as "feet" to travel over land from one temporary pool to another. Other fish in these pools die when the pools dry up. What can be expected to happen in this area after many years?

1. All the varieties of fish will survive and produce many offspring.
2. The fish using lower fins as "feet" will be present in increasing numbers.
3. All fish species will develop "feet" in the form of lower fins.
4. The fish using lower fins as "feet" will develop real feet.

7. Scientists studying a moth population in a forest in New York State kept track of wing color. The graph below shows the distribution of moth wing color. The majority of the trees in the forest had brown bark.

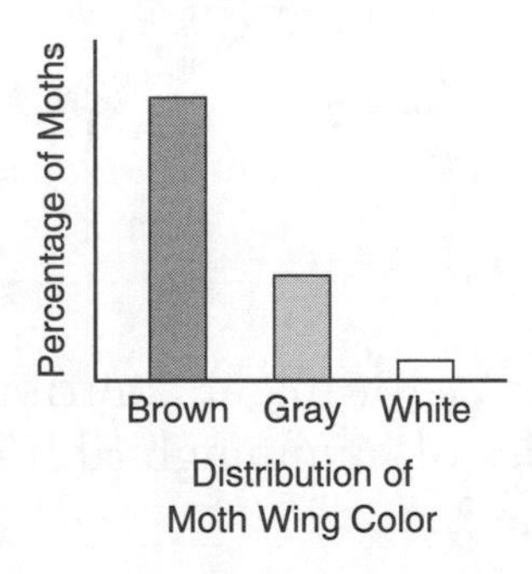

A fungus infection affected nearly all the trees in the forest. It did not kill the trees, but it changed their bark to a gray-white color. Which graph shows what the distribution of wing color in the moth population would be after a long period of time?

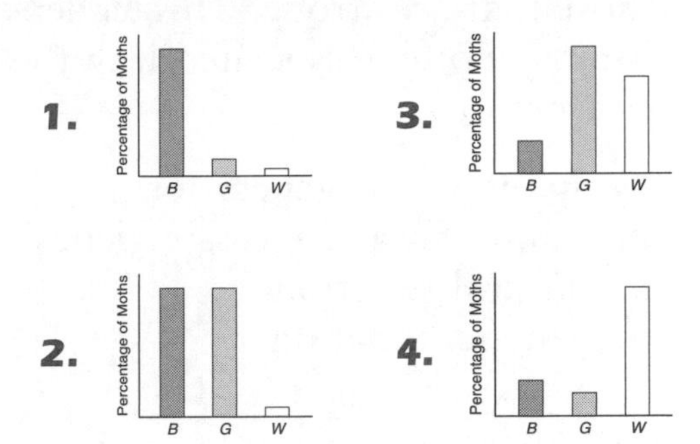

8. Evolution can best be described as

1. a process of growth in an organism
2. a change in size of body structures through use or disuse
3. the formation of fossils
4. the process of change through time

9. The fossil record shows that most plants and animals that lived on Earth in the past

1. lived on land
2. are still living today
3. appeared about 60 million years ago
4. have changed over time

10. The diagram below shows the changes in foot structure in a bird population over many generations.

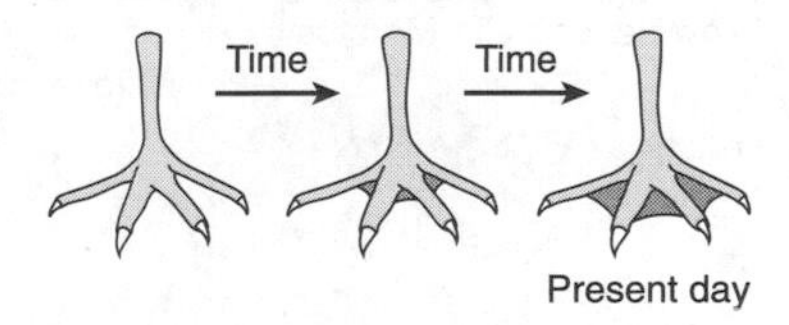

These changes are best explained by the concept of

1. evolution by natural selection
2. extinction
3. stimulus and response
4. climate change by global warming

GO ON

11. A wolf in Alaska tends to attack and kill animals that are weak rather than those that are strong. This tendency is most closely associated with the concept of

1. spontaneous generation
2. dominant and recessive genes
3. natural selection
4. genetic mutation

Base your answers to questions 12 and 13 on the diagram below, which shows the evolutionary relationships between animals in a possible canine family tree.

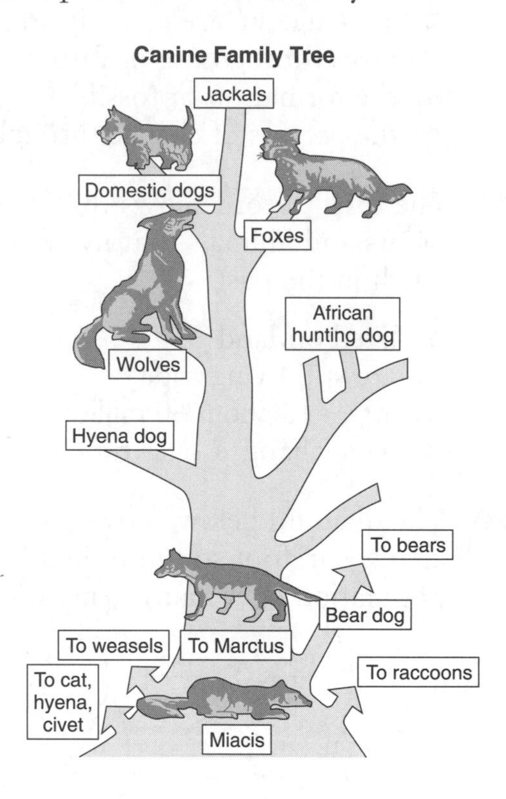

12. According to the diagram, weasels, foxes, and domestic dogs all most likely originated from the

1. wolf
2. bear dog
3. *Miacis*
4. *Marctus*

13. According to the diagram, which group of organisms has the most closely related members?

1. cats, weasels, wolves
2. African hunting dogs, hyena dogs, and domestic dogs
3. jackals, foxes, and domestic dogs
4. bears, raccoons, and hyena dogs

14. In which type of organism would changes due to natural selection be seen most rapidly?

1. bacteria
2. oak trees
3. dogs
4. humans

15. The drawing below shows a hummingbird using its long beak to obtain nectar from a flower.

What factor might contribute to this species of hummingbird becoming extinct?

1. a new source of nectar in the environment
2. a mutation that produces more flowers per plant
3. an increase in nectar production by the plant
4. the widespread use of pesticides on flowers it feeds from

Chapter 9

Ecology

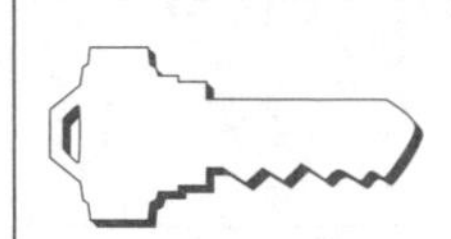

Key Idea: Plants and animals depend on each other and their physical environment.

THE NATURE OF ECOLOGY

Ecology is the scientific study of the relationship between living things and their environments. All the living and nonliving things that surround an organism make up its ***environment***. An organism's relationship with the environment includes interactions with the physical world and with other organisms. When things *interact*, they act on each other. The environment affects living things, and living things affect the environment.

TRY THIS

Figure 9.1 shows an organism (a goldfish) in an environment. List some of the living things and nonliving things in its environment.

Living Things	Nonliving Things

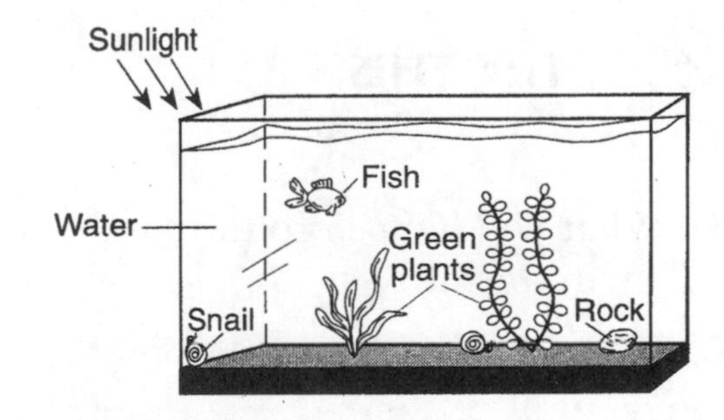

Figure 9.1
Goldfish in an aquarium.

THE PHYSICAL ENVIRONMENT

All living things get the materials they need to carry out their life functions from their environments. Organisms interact with the nonliving things around them, or the *physical environment.* Light, water, air, climate, and Earth's surfaces are some of the things that make up the physical environment. Green plants get minerals and water from the soil. They get carbon dioxide from the air. They use the energy in sunlight to turn the carbon dioxide and water into food and release oxygen into the air. Animals get oxygen from the air and release carbon dioxide. Animals burrow into the ground. People build houses and roads that cover the ground.

TRY THIS
What are some materials living things get from their environment?

THE BIOLOGICAL ENVIRONMENT

Organisms also interact with other living things around them, or their *biological environment.* Animals get nutrients and energy by eating other living things in their environment. Some animals eat plants, some eat other animals, and some eat both. Rabbits eat grasses. Wastes from rabbits enrich the soil and help the grasses grow better. The rabbits, grasses, and soil have an effect on one another. An interaction also takes place between rabbits and foxes. Foxes eat rabbits. If a lot of foxes are hunting the same rabbits, the number of rabbits will go down. With fewer rabbits, some foxes may lack food, and the number of foxes will go down.

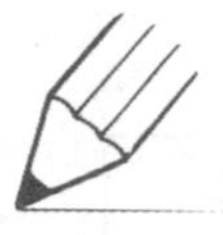

TRY THIS
Hawks eat mice.

1. What will happen to the number of mice if more hawks move into their environment?
2. What will happen to the number of mice if the number of hawks decreases?

THE NATURE OF AN ECOSYSTEM

An *ecosystem* is made up of the living and nonliving things in a particular place that interact with one another. The *eco-* part of the word refers to the environment. The *-system* part of the word means that an ecosystem is a system. A system is a set of parts that work together as a single unit. Think of a stereo system. It has many parts, such as speakers, a CD player, a tape player, and a radio. Each of those parts is also a system. A CD player has a laser, a motor that spins the CD, and circuit boards. They all work together, or interact, to play music.

Now think of a forest. It has nonliving parts, such as the air, soil, water, and climate. It has living parts, such as trees, birds, flowers, and insects. The living and nonliving parts of a forest interact. The trees capture sunlight and use it to make food. In doing so, they change the environment beneath them, making it darker and cooler. Birds eat insects and worms on the forest floor. Then there are fewer worms and insects. This changes the environment for other animals that eat worms and insects. A forest is an ecosystem. It is made of parts that work together as a whole. Earth has many ecosystems that differ in size and makeup. A squirrel in an upstate New York forest interacts with different living and nonliving things than a fish in a coral reef off the coast of Florida. Together, all of Earth's ecosystems form the ***biosphere***, that portion of Earth in which living things exist.

THE ORGANIZATION OF AN ECOSYSTEM

INDIVIDUAL ORGANISMS

The smallest unit of an ecosystem is an organism. An organism is a complete living thing. You are an organism. You interact with your environment when you inhale oxygen and exhale carbon dioxide. You interact with plants when you eat them and breathe in the oxygen they give off. Every organism interacts with the other living things that share the Earth with it.

Habitat

The place where a plant or animal lives is its ***habitat***. The habitat has the resources an organism needs to live. An organism's habitat also provides it with shelter and a place to reproduce. A habitat may be large or small. A habitat may be on the land or in the water. A rotting log can be a habitat for termites. An entire ocean is the habitat for a salmon. The forest is the habitat for a squirrel.

Niche

Many kinds of organisms may share the same habitat. A pond may contain bass, minnows, aquatic plants, water beetles and other aquatic insects, snails, algae, and bacteria. Though they share the same habitat, different kinds of organisms have different needs and play different roles in the habitat. The role of each organism in a habitat is its ***niche***. The algae capture sunlight and make food. Minnows eat algae and water plants. Minnows are eaten by bass. Water beetles eat other aquatic insects, snails, small fish, and even earthworms. Bacteria break down dead plants and animals.

POPULATIONS

A ***population*** is a group of organisms of the same species that are found together at a given place and time. Since they are of the same species and live together, members of a population can interbreed. The black bears living in the Catskill Forest Preserve are one population. So are the hemlock trees. The bass in Lake George form a population of interbreeding fish. All of the people in New York State make up its human population.

TRY THIS

Figure 9.2 shows a backyard garden with three rabbits eating lettuce. The population of rabbits in the garden is three. What is the population of lettuce plants? Of humans?

Figure 9.2
Backyard garden.

A population that has enough resources, and no diseases or predators, will tend to increase. As a population grows, it uses more resources and its activities change the environment. The growing population may deplete food or nesting sites. It may lose more individuals to predators, parasites, or disease. If a disaster, such as a flood or fire, occurs, habitat may be lost. Such factors limit the growth of certain populations in an ecosystem. Over time, ecosystems reach a balance as a result of interactions between their populations and their environment.

COMMUNITIES

The populations in an ecosystem interact with one another. One population may be a source of food for another population. Populations may compete with one another for food, water, or space. Two populations can benefit each other, with each doing better because the other is present. All the populations living together in an ecosystem are called a ***community***. A community includes all the different kinds of living things that live together in a place. A community consists only of living things. Figure 9.3 shows some of the populations in a pond community.

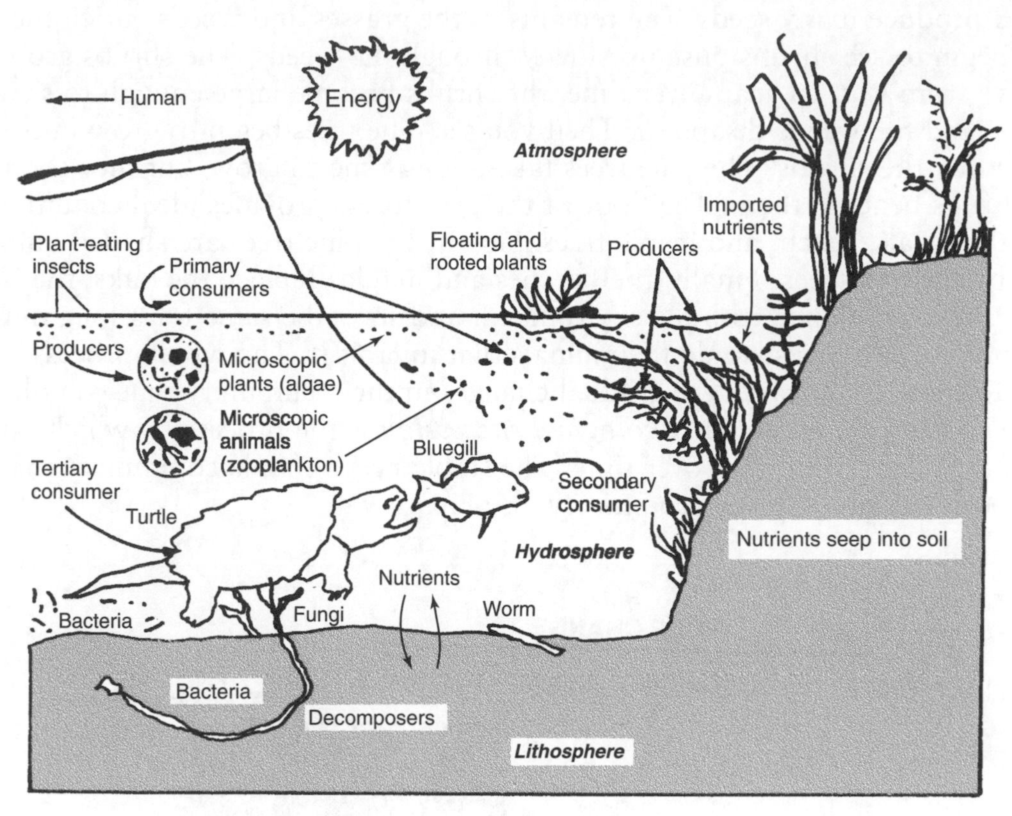

Figure 9.3
Populations in a pond community.

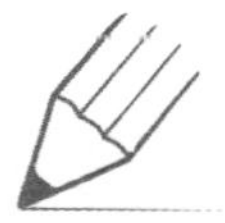

TRY THIS

1. Why is a pond community not permanent?
2. Describe several habitats that exist in a pond.
3. Why do many species live only in certain habitats?

ECOLOGICAL SUCCESSION

Over time, the community in any one place changes. Suppose you stand on an empty piece of ground near a forest and look around. All you see is bare soil and rocks. Then, every summer you come back to the same place and observe the area as time passes. Where first you saw only bare soil and rock, you begin to see lichens, mosses, and ferns that can grow in poor soil and get their start from spores carried by the air. Then grasses and weeds arrive. They grow quickly in the open sunlight with limited water and produce many seeds. The remains of the grasses and weeds enrich the soil, and you begin to see shrubs push up slowly through the weeds. The shrubs grow slowly but survive from year to year. Over time, the shrubs become large enough to shade out the weeds, and the weeds disappear. Then you see pine trees begin to grow in the rich soil shaded by the shrubs. The pine trees take a long time to grow, but they soon shade out the shrubs beneath them. The floor of the pine forest provides ideal conditions for the growth of oak, beech, and maple trees. Slowly, the pine trees are shaded out and replaced by the oak trees. Finally, the beeches and maples replace the oaks; the forest reaches a state of balance with the environment and stays the same for a long time. As the plant community changes, so do the animals that inhabit it. The whole process has taken between 150 and 200 years. The gradual changes in the community that you observed are called ecological succession. ***Ecological succession*** is the process by which one community is replaced by another over time. The stable beech-maple community that eventually formed is called a ***climax community***. See Figure 9.4.

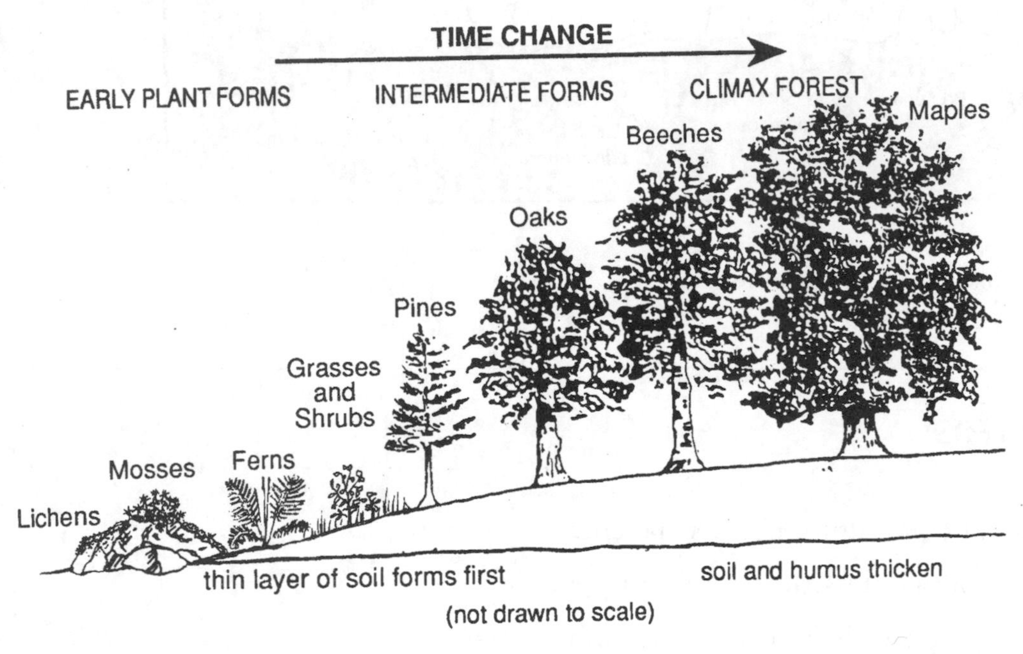

Figure 9.4
Succession from bare field to climax forest.

THE ECOSYSTEM

When communities and their physical environment work together as a system, they form ecosystems. See Figure 9.5. Ecosystems may be as large as a prairie or as small as a pile of rotting leaves. Rivers, ponds, forests, and deserts can be ecosystems. Even a goldfish bowl can be an ecosystem.

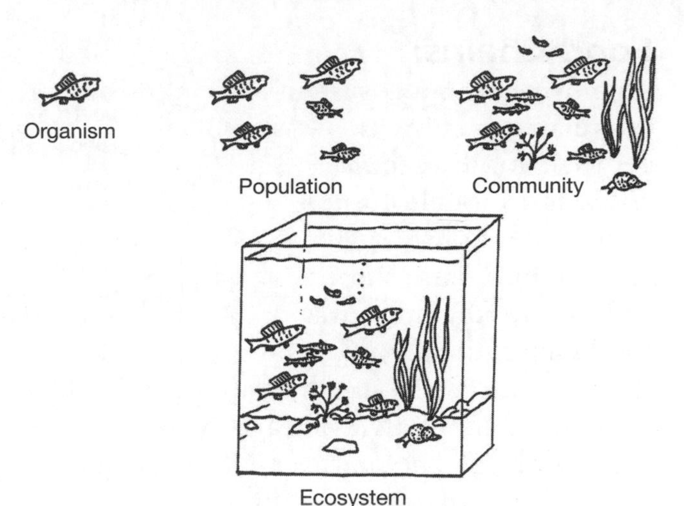

Figure 9.5
Organism, population, community, ecosystem.

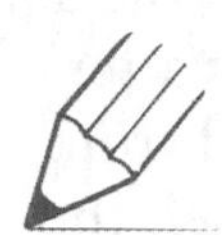

TRY THIS

1. What is the difference between the biosphere and an ecosystem?
2. What is the difference between a population and a community?

INTERACTIONS IN AN ECOSYSTEM

Like many complex systems, ecosystems tend to change until they become stable. Then they remain that way unless their surroundings change. Ecosystems can be reasonably stable over hundreds or thousands of years. Ecosystems become stable when things are happening that exactly counterbalance one another.

ENERGY FLOW IN ECOSYSTEMS

The Sun is the source of energy in most ecosystems. Plants capture the Sun's energy and make their own food. Energy can change from one form to another in living things. Organisms get energy from oxidizing their food, releasing some of its energy as heat. Organisms also convert the chemical energy in food into motion, sound, light, and even electricity.

Food Chains, Food Webs, and Energy Pyramids

The flow of energy in an ecosystem can be shown with models such as food chains, food webs, and energy pyramids. Energy flows through ecosystems in one direction, usually from the Sun, through producers, to consumers, and then to decomposers.

Food Chains: The organisms in an ecosystem are related by how they get their food. A ***food chain*** is a model of the feeding relationships in an ecosystem. Figure 9.6 shows a food chain. In a food chain, the arrows show the direction in which nutrients and energy move along the chain.

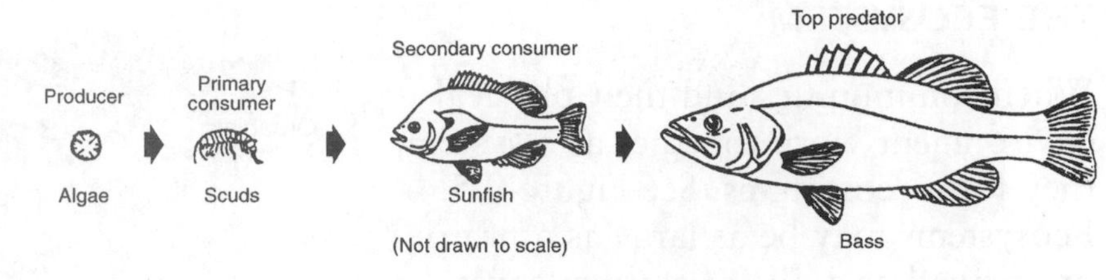

Figure 9.6
Food chain in a freshwater pond ecosystem.

A food chain always starts with a ***producer***, a green plant that makes its own food, such as algae. The matter and energy in the producer then pass to a ***primary consumer***, an organism that eats producers. In this food chain, the scud—a primary consumer—eats the algae. The energy and nutrients in the algae move to the scud. There may be several levels of consumers in a food chain. ***Secondary consumers*** eat primary consumers. In this food chain, a sunfish eats the scud and is therefore the secondary consumer. The energy and nutrients that moved from the algae to the scud now move from the scud to the sunfish. ***Tertiary consumers*** eat secondary consumers. In this food chain, the bass is a tertiary consumer, the top predator in the pond community. The energy that moved from the algae to the scud to the sunfish now moves from the sunfish to the bass. When the bass dies, its body is broken down by ***decomposers***, organisms that break down the remains or wastes of other organisms. Many decomposers are bacteria or fungi.
The nutrients and energy in the bass then move to the decomposers or are released back into the environment.

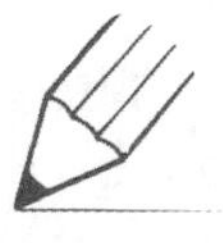

TRY THIS
Construct two food chains based on the diagram of a pond community in Figure 9.3.

Energy Pyramids: The *energy pyramid* in Figure 9.7 is a model showing how energy moves through a food chain. Notice that the producers are at the base of the pyramid. This is because producers far outnumber consumers in an ecosystem. The producer level also contains the most energy and nutrients. Not every producer is eaten by a consumer. Even when eaten by a consumer, the consumer does not use all of a producer's matter and energy. Some is given off as waste. Some is used as fuel and does not become part of the consumer's body. In general, *only 10%* of the matter and energy in an organism moves from one level of the pyramid to the next. So, as you move up the pyramid, each level contains less energy, fewer nutrients, and fewer organisms. Therefore, each level can support fewer consumers at the next level. In most ecosystems, so little energy gets to the top level that large populations of tertiary consumers cannot be supported.

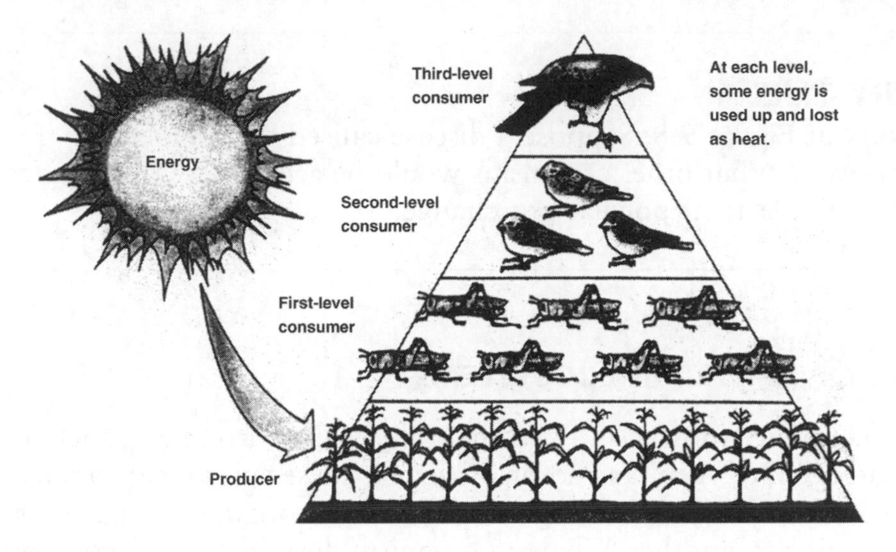

Figure 9.7
Energy pyramid (trophic levels).

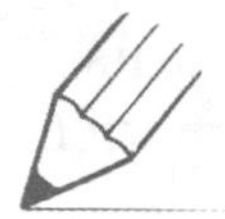

TRY THIS

1. Why are there always fewer consumers than producers in an ecosystem?
2. Which could survive better alone—producers or consumers?
3. How is a food chain like a pyramid?

Food Webs: Many possible food chains can exist in an ecosystem. Mice do not eat *only* seeds. Owls do not eat *only* mice. A food web shows how a number of food chains are related. See Figure 9.8. A food web shows that producers, consumers, and decomposers in an ecosystem are all interrelated. It shows how changing the population of one organism in an ecosystem may affect other organisms in the ecosystem.

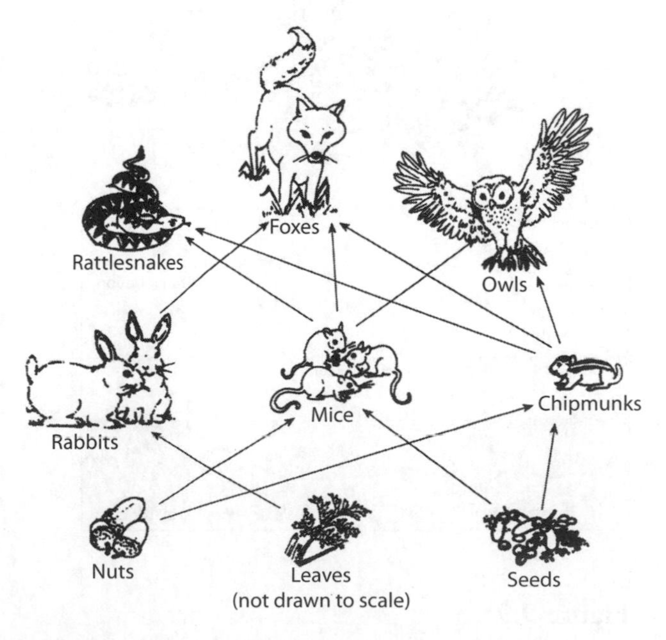

Figure 9.8
Food web.

TRY THIS

Look at Figure 9.8. Suppose a disease caused the population of mice to decrease. What other organisms would be affected by this change? How might their populations change?

THE CYCLING OF MATTER IN ECOSYSTEMS

Living things take many substances from their physical environment, such as water, carbon dioxide, nitrogen, oxygen, and so on. They use these substances to make more complex chemicals in their body. When organisms eat other organisms, matter is transferred from one living thing to another. When an organism dies, its body decomposes, and matter is released back into the environment. Then another organism may use this matter again. Over a long time, matter is transferred from one organism to another repeatedly and between organisms and their physical environment. In an ecosystem, matter is recycled, or used over and over again. The total amount of matter stays the same, but its form and location change. Figure 9.9 describes some of the processes by which matter flows in and through ecosystems.

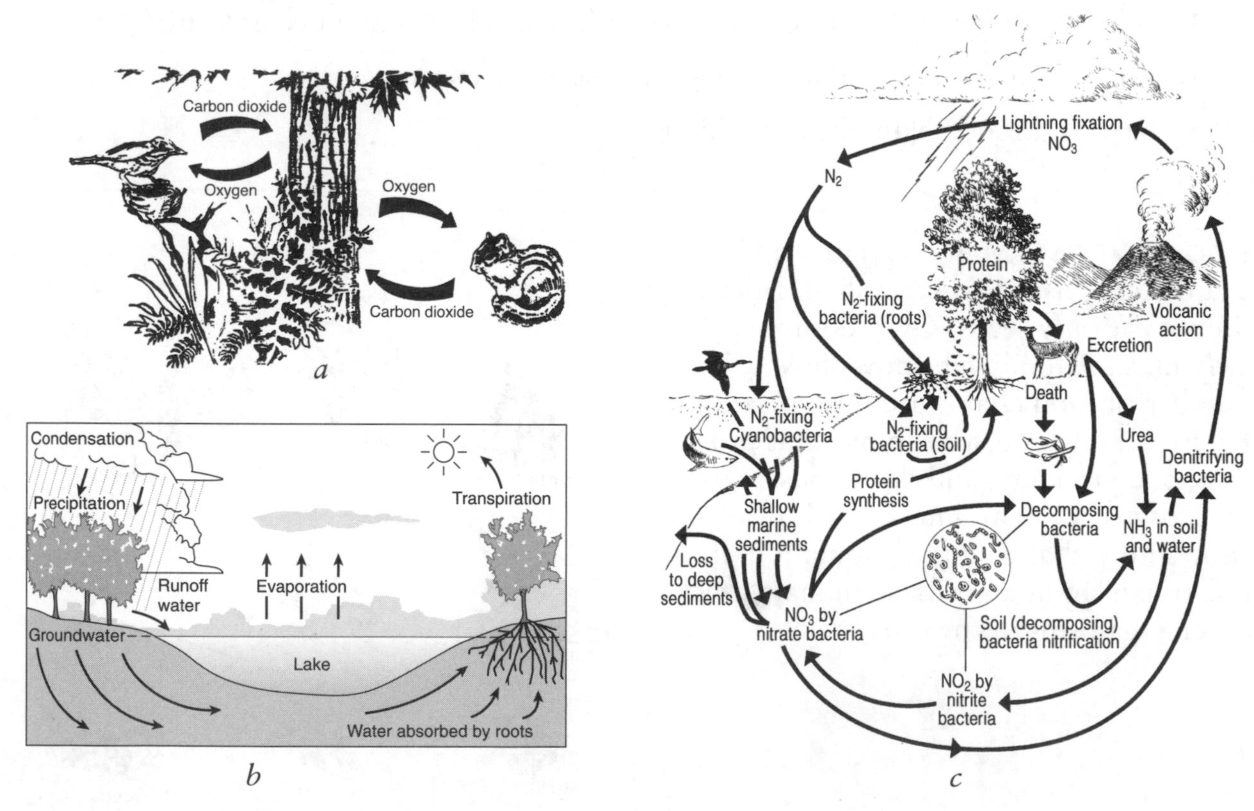

Figure 9.9
Some ways in which matter cycles through an ecosystem. (*a*) The carbon dioxide–oxygen cycle. (*b*) The water cycle. (*c*) The nitrogen cycle.

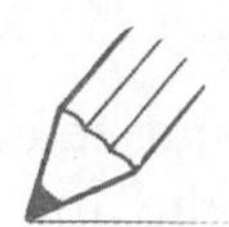

TRY THIS

1. Explain why we do not run out of water, oxygen, or nitrogen in nature.
2. Explain why the biosphere needs a constant supply of energy.

RELATIONSHIPS BETWEEN SPECIES

In relationships between species, each may be affected in a positive (+, or beneficial), negative (–, or harmful), or neutral (0, neither benefits nor harms) way. The effects of a relationship between two species can be shown with a pair of symbols, one for each species.

Predator-Prey (+, –)

In a predator-prey relationship, the predator population benefits, and the prey population is harmed. When a lion preys on a zebra, the lion benefits, and the zebra is harmed.

Parasitism (+, –)

Parasitism is another relationship in which one species benefits and the other is harmed. In parasitism, the parasite benefits, and the host is harmed. When a dog (a host) has a tick (a parasite), the tick benefits, but the dog is harmed.

Mutualism (+, +)

Mutualism is a relationship in which two species benefit from one another's presence. A good example is trees and fungi. The fungi infect and penetrate the cells of the tree roots. They form a network of tiny fibers that act as extended roots for the tree but do not change the shape or structure of the tree roots. They draw in minerals from soil beyond the reach of the tree roots. In return, the fungi get carbohydrates (a source of energy and carbon) from the tree. Fungi are nongreen plants and cannot make their own food.

Some species have adapted to depend on each other so much that one cannot survive without the other. One example is coral and a tiny plant with a big name—zooxanthellae (zo-zan-thell-ay). *Zooxanthellae* are tiny, one-celled plants that live in coral tissues and make food by photosynthesis. They help the coral make its rocky skeleton and make food for the coral. In return, the zooxanthellae get a steady supply of nutrients such as nitrogen and phosphorus from the coral and a place to live. Coral and zooxanthellae depend on each other so much that neither one can survive very well without the other.

Commensalism (+, 0)

Commensalism is a relationship in which one species benefits and the other is unaffected. For example, orchids grow on tree trunks or limbs and have roots that hang in the air. The trunk or limb of the tree provides a place for the orchid to grow. High in the tree, the orchid gets sunlight, nutrients from the air, and moisture through its aerial roots. The tree is unaffected.

TRY THIS

Questions 1–3 refer to the following list showing some relationships between species.

A. Barnacles on whales (+, 0)

B. Nitrogen-fixing bacteria in the roots of legumes (+, +)

C. Athlete's foot fungus on humans (+, –)

D. Protozoa in termite digestive tracts (+, +)

E. Orchids on tropical trees (+, 0)

F. Tapeworms in dogs (+, –)

1. Which relationships are examples of mutualism?
2. Which relationships are examples of commensalism?
3. Which two organisms are parasites?

For each of the following examples, show the way the two species relate to one another using the symbols +, –, and 0.

4. Lamprey eels attach to the skin of lake trout and absorb nutrients from the body of the trout. (____, ____)
5. A squirrel eats some acorns from an oak tree and buries others. The buried ones later germinate and begin to grow. (____, ____)
6. An owl eats a field mouse. (____, ____)

Practice Review Questions

1. The science of ecology is best defined as the study of

1. the classification of plants and animals
2. weather and its effects on food production in the ocean
3. the interactions of living organisms and their environment
4. technology and its effects on society

Questions 2 and 3 refer to the following diagram of an aquarium setup.

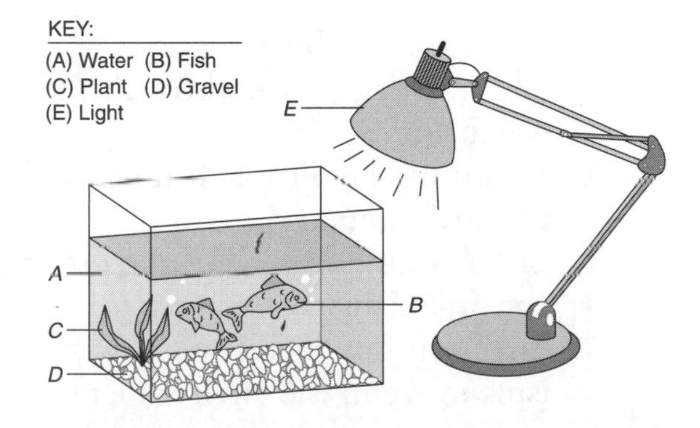

2. In the diagram of the aquarium setup, which letters indicate parts of the physical environment?

1. *A*, *B*, *C*, *D*, and *E*
2. *A*, *D*, and *E*, only
3. *A*, *B*, *D*, and *E*, only
4. *A*, *B*, and *C*, only

3. Which term includes all the interactions between the various organisms in this aquarium and their physical environment?

1. competition
2. ecosystem
3. population
4. habitat

4. In a freshwater pond, a carp eats decaying material from around the bases of underwater plants, while a snail scrapes algae from the leaves and stems of the same plants. They can survive at the same time because they occupy

1. the same habitat but different niches
2. the same niche but different habitats
3. different habitats and niches
4. the same habitat and the same niche

5. An example of a population is

1. animals in a barn and their food
2. animals in a barn and their surroundings
3. field mice and owls living in a barn
4. field mice living in a barn

GO ON

6. When a partially rotted log was turned over, fungi, termites, pill bugs, ants, slugs, and earthworms were found to be living in and around it. Together, these organisms represent a

1. community
2. population
3. habitat
4. niche

7. Which sequence shows the levels of ecological organization from simplest to most complex?

1. community, ecosystem, biosphere
2. biosphere, ecosystem, community
3. ecosystem, biosphere, community
4. biosphere, ecosystem, community

8. The diagram below shows changes that might occur over time after a forest fire in an area.

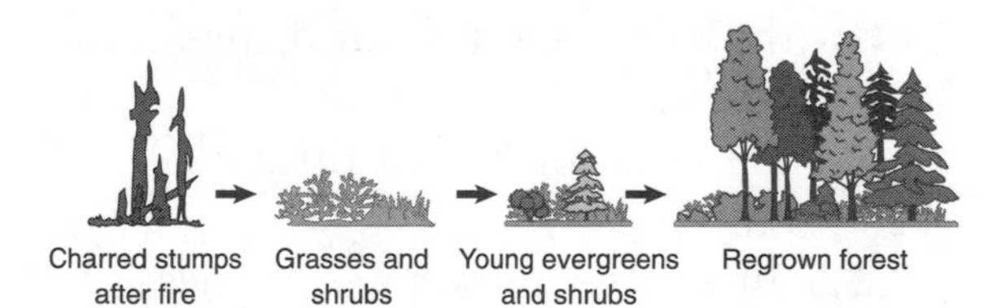

Which statement best explains the events shown in the diagram?

1. Once a stable ecosystem is destroyed, it never recovers.
2. After a sudden change, an ecosystem returns immediately to its original state.
3. After a natural disaster, an ecosystem gradually returns to its original state by succession.
4. After being damaged by the fire, the forest ecosystem evolves into a new ecosystem totally different from the original.

9. The first organism in a food chain must be

1. a producer
2. a consumer
3. a herbivore
4. a decomposer

10. The diagram below shows a food chain.

The arrows in the diagram show the

1. direction in which energy and nutrients move through the organisms
2. return of chemical substances to the environment
3. order of importance of the different organisms
4. direction in which these organisms move in the environment

11. Orchids grow on large tropical trees. The orchids depend on the support given by the trees but do not harm the trees. This type of relationship is known as

1. mutualism
2. commensalism
3. parasitism
4. predator–prey

GO ON

Questions 12–15 refer to the following diagram, which shows a food web.

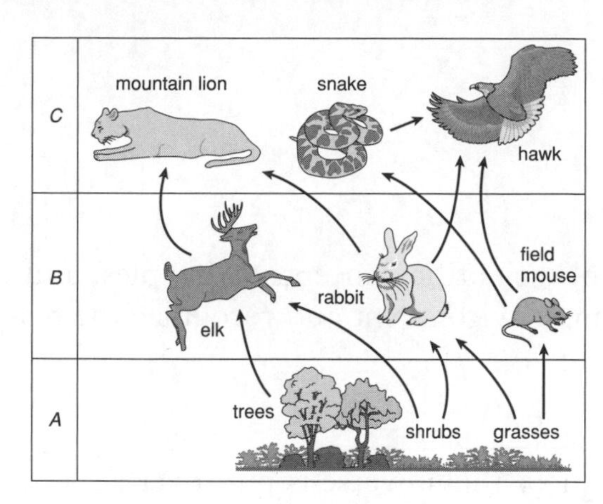

12. Which organisms in the diagram are parts of the same food chain?

1. trees, rabbit, elk, and shrubs
2. grasses, mouse, snake, and hawk
3. trees, mountain lion, snake, and hawk
4. grasses, mouse, rabbit, and elk

13. What is the original source of energy for this food web?

1. the Sun
2. chemical energy in sugars
3. chemical reactions in bacteria
4. gravity

14. What parts of the food web are indicated by letters *A*, *B*, and *C*?

1. *A*—producers; *B*—primary consumers; *C*—secondary consumers
2. *A*—primary consumers; *B*—producers; *C*—secondary consumers
3. *A*—primary consumers; *B*—secondary consumers; *C*—producers
4. *A*—producers; *B*—secondary consumers; *C*—decomposers

15. What group of organisms is missing from this food web?

1. herbivores
2. carnivores
3. producers
4. decomposers

16. In an ecosystem, which part is not recycled?

1. energy
2. water
3. oxygen
4. carbon

Unit 3

The Physical Setting

Standard 4: Students will understand and apply scientific concepts, principles, and theories pertaining to the physical setting and living environment and recognize the historical development of ideas in science.

People have always been interested in finding out how the universe is put together, how it works, and their place in the universe. What they have figured out so far is that all natural objects, events, and processes are interconnected and can be explained by a few underlying concepts.

In this unit, you will learn about the composition of the universe, the physical principles upon which it seems to operate, and how our modern view of the physical universe came into existence. You will also learn about the nature of matter and energy, the many forms in which both exist, and how energy and matter interact.

The following chapters will discuss the major areas of science that deal with the physical setting—astronomy, geology, meteorology, chemistry, and physics.

Chapter 10

Astronomy

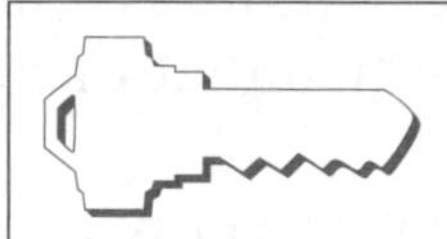

Key Idea: The Earth and celestial phenomena can be described by principles of relative motion and perspective.

PERSPECTIVE AND RELATIVE MOTION

If you kept a list of things observed in the sky, it might include birds, smoke, clouds, rainbows, halos, lightning, stars, the Moon, the Sun, and comets. One of the first ideas that might occur to you is that the sky has depth. Some things in the sky appear closer to you, and some appear farther away. Why? Perspective! ***Perspective*** is how things appear to you from a certain point of view.

From everyday experiences, we know that closer objects block our view of more distant objects. If a bird flying by blocks your view of a cloud, you conclude that the bird is closer to you than the cloud. If you see a solar eclipse, you conclude that the Moon is closer to you than the Sun. By using the idea of perspective, you could put all of the objects on your list in order from nearest to farthest. Perspective is one of the ways in which early astronomers tried to make sense of the objects they saw in the sky.

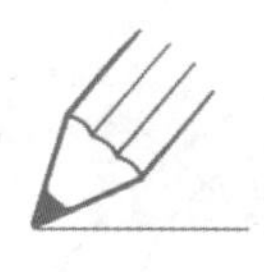

TRY THIS

By using the idea of perspective, decide which objects are closer to the observer than the bird and which are farther away.

CELESTIAL OBJECTS

Objects that can be seen in the sky and are not connected with Earth or its atmosphere are known as ***celestial objects***. Stars, the Sun, the Moon, the planets, and comets are examples of celestial objects. Clouds, rainbows, halos, and other things seen in the sky that are part of or occur in Earth's atmosphere are *not* considered celestial objects. By using the idea of perspective, early astronomers worked out that the Moon was the closest celestial object to Earth and that the stars were very, very far away.

CELESTIAL MOTION

If you observe the sky for even a short while, it is clear that celestial objects change position in the sky over time. You have probably seen the Sun in different places in the sky at different times of the day. If you observe the Moon and stars carefully, you find that they are also seen in different places in the sky at different times of the night.

If you keep track of all this motion, you will discover that with very few exceptions, every single one of the thousands of objects seen in the sky appears to *move in the same general direction at the same speed!* All celestial objects appear to move across the sky from east to west along a path that is an arc, or part of a circle, at a steady 15° per hour. This equals one complete circle every day (24 hours/day × 15°/hour = 360°/day). Therefore, this motion is called *apparent daily motion*. In the Northern Hemisphere, the arcs of circles along which celestial objects move are all centered very near the star Polaris. See Figure 10.1.

TRY THIS

In order to make the photograph in Figure 10.1*a*, the camera was aimed at the star Polaris (due north).

Explain why the stars seen by an observer on a clear night appear to move from east to west at 15° per hour.

a

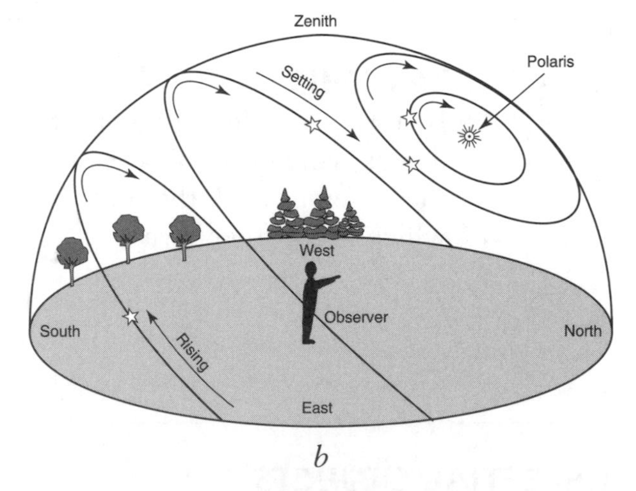

b

Figure 10.1
(*a*) A time exposure taken with a camera aimed at Polaris over Mauna Kea Observatory, latitude 20°. Note the circular star trails.

(*b*) The apparent motion of celestial objects to an observer in New York State.

RELATIVE MOTION

We have used the word *appear* to describe motion so far because there are several possible reasons why an object may look like it is moving to an observer. One possibility is that the observer is standing still and the *object* is moving. Another is that the object is standing still and the *observer* is moving. Yet another is that *both* the observer and the object are moving, but one is moving faster, slower, or in a different direction than the other. The motion of an object is always judged relative to some other object or point. Therefore, *how* motion is perceived depends on your point of view. That is why a change in the position of two objects in relation to one another is called ***relative motion***. See Figure 10.2.

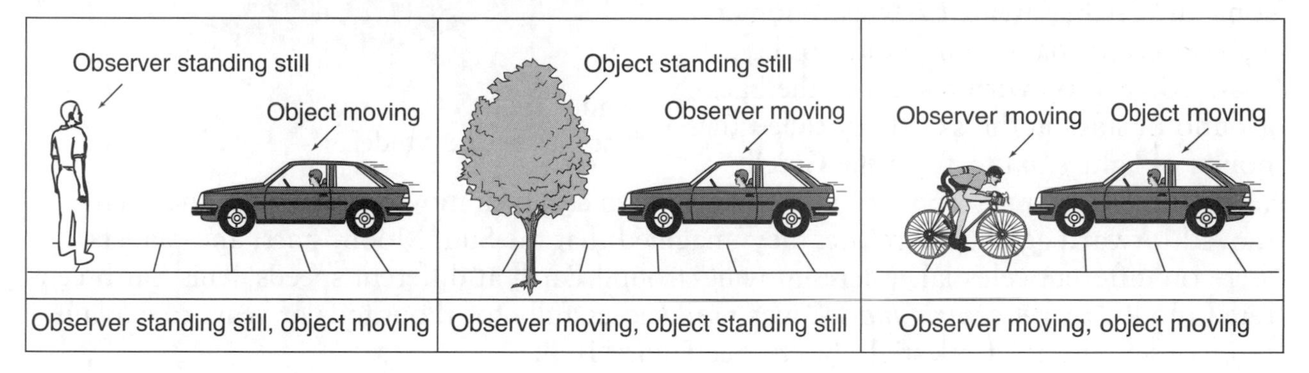

Figure 10.2
Relative motion. The appearance of motion in three possible scenarios.

TRY THIS

Imagine you are looking around while driving along in a car with your family. Give an example of each of the following situations:

1. An object that is not actually moving but appears to you to be moving.
2. An object that is moving but appears to you to be standing still.
3. An object that appears to be moving faster or slower than it is actually moving because you are moving.

MODELS OF THE UNIVERSE

The problem of determining which is moving, the object or the observer, is not always easy to solve. Over time, people have proposed a number of different models of the universe to explain the motions of celestial objects.

THE GEOCENTRIC MODEL

Early observers of the sky believed that they were standing still because their senses gave them no signs that they were moving. They *felt* as if they were standing still. Therefore, when they looked at the sky, they figured that the changing positions of celestial objects meant that the *celestial objects* were moving. One effect of apparent daily motion is that the sky appears to move as if it were a single object. So early observers imagined that the sky was a single object—a huge, celestial sphere slowly spinning around the motionless Earth.

This model had some problems, however. Celestial objects cannot all be on the same sphere because some are closer and some are farther away. Early astronomers also observed that certain points of light changed position with respect to the background of stars in the sky. They called these points of light *planets,* from the Greek word for "wanderer." Some of these planets even stopped, moved backward, and then moved forward again. Therefore, they imagined that the Sun, Moon, stars, and planets were on different celestial spheres moving around Earth at different speeds. This Earth-centered model, or ***geocentric model,*** was used successfully for thousands of years to explain most observations of celestial objects. See Figure 10.3.

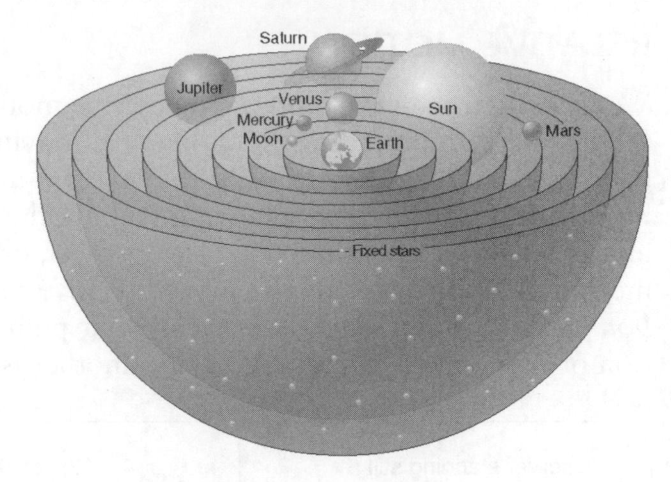

Figure 10.3
The Geocentric Model.

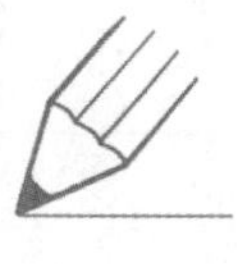

TRY THIS

Explain how the geocentric model could explain the star trails photographed in Figure 10.1.

THE HELIOCENTRIC MODEL

Suppose you stand in the middle of a room and spin around or rotate. The fixed objects in the room seem to move around you in circles. In 1543, Nicolaus Copernicus used this idea to suggest a model that explained the motion of celestial objects in a different way. He suggested that celestial objects seem to circle Earth because Earth rotates once a day. He also suggested that the stars and planets (including Earth) revolved around the Sun once a year. This Sun-centered model, or ***heliocentric model,*** was rejected by almost everyone because the idea that Earth was moving did not agree with what they felt with their senses and went against their belief that Earth was the center of the universe. Over time, though, the evidence for it mounted. The heliocentric model was eventually accepted. Our modern view of planetary motions in the solar system is based upon the heliocentric model. See Figure 10.4.

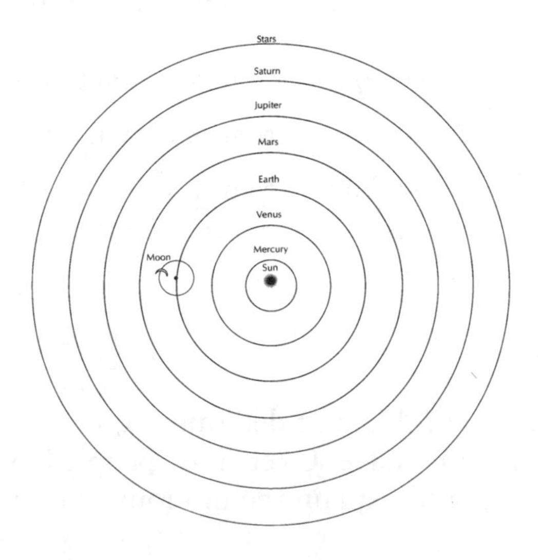

Figure 10.4
Copernicus's heliocentric model.

OUR CURRENT MODEL OF THE UNIVERSE

We now know that *everything* in the universe is moving. We have also discovered that the universe contains a lot more than we can see with the unaided eye. Telescopes have revealed a universe filled with an almost uncountable number of stars, dust clouds, chunks of rock, and radiation.

STARS

To many people, a star is just a dot of light in the night sky. However, a ***star*** is actually a huge sphere of hot, glowing gases. The Sun is just an average-sized star. It looks so large and bright because it is so close to Earth. Other stars are like the Sun but are so far away from Earth that they just look like tiny points of light.

Stars form when gravity causes clouds of dust and gas in the universe to contract. As the dust and gas contract they heat up until nuclear fusion of light elements into heavier ones occurs. Nuclear fusion in stars releases huge amounts of energy over millions of years.

GALAXIES

Stars that are too dim to be seen with the unaided eye can be seen with a telescope. Telescopes gather more light than the human eye. The more light a telescope gathers, the larger, brighter, and more detailed is the image it forms. Patches of sky that look empty to the naked eye are filled with stars when viewed through a telescope. As ever-larger telescopes have been built, ever-dimmer stars have been detected. The number of observable stars has grown from thousands to hundreds of billions.

When viewed through a telescope, some objects that look like dim smudges of light to the naked eye show up as clusters of billions of stars called ***galaxies.*** From what we can see, the universe contains many billions of galaxies of all sizes and shapes. See Figure 10.5*a*. Our Sun is just one of the billions of stars that make up the Milky Way Galaxy. The Sun and our solar system are located in one of the spiral arms of the galaxy about two-thirds of the way out from the center. See Figure 10.5*b*.

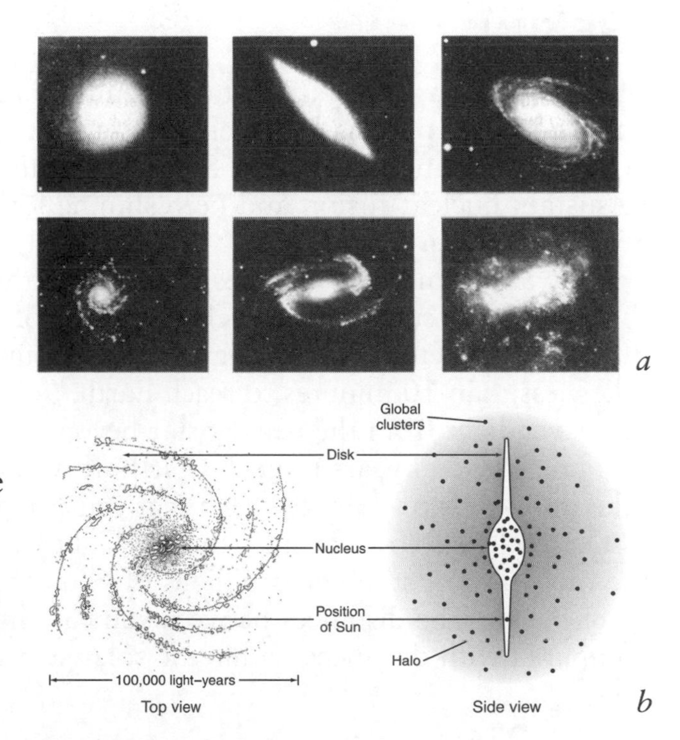

Figure 10.5
(*a*) Galaxies are systems containing billions of stars. The universe is filled with galaxies of all shapes and sizes. (*b*) The Milky Way showing the Sun's position.

NEBULAE

Nebulae are clouds of gas and dust. Some nebulae can be seen because the gas and dust is so hot it glows or is lit up by a group of stars within the cloud. Sometimes the dust cloud is so thick that it blocks visible light, forming a black region against the background of stars.

TRY THIS

What is the difference between a star and a nebula? A star and a galaxy?

THE SOLAR SYSTEM

The *solar system* is made up of the Sun, the planets that orbit the Sun, the satellites of those planets, and other smaller bodies that orbit the Sun, such as asteroids and comets. See Figure 10.6. The Sun contains 99.9% of all the mass of the solar system, so the solar system pretty much is the Sun.

THE SUN

The ***Sun*** is an average-sized star, meaning some stars in the universe are larger than the Sun and some are smaller. Even our average-sized Sun is still more than a million times greater in volume than Earth. Scientists estimate that the Sun contains enough matter to sustain nuclear fusion and keep shining for 15 billion years.

The Sun is millions of times closer to Earth than any other star. Light travels at a speed of 300,000 kilometers/second. Light from the Sun takes less than 10 minutes to reach Earth. However, light from the nearest star beyond the Sun takes several years to get to Earth. It would take thousands of years to reach that star using our fastest rockets. Now consider that the light from the most distant stars takes billions of years to reach Earth. The distances between stars are huge compared with distances within the solar system.

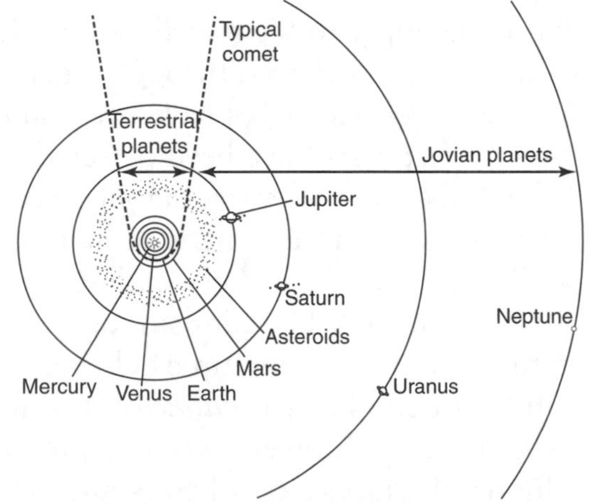

Figure 10.6
The Solar System. Note that in addition to the planets and their moons, the solar system includes the asteroid belt, dwarf planets such as Pluto, and comets that circle the Sun in highly eccentric orbits.

THE PLANETS

Planets are objects that are large enough that their gravity has pulled them into a round shape. Their gravity is also strong enough that they have swept up all nearby objects and debris and now orbit in a clear path around the Sun. *Moons* are satellites, or solid bodies that orbit planets. Beginning at the center, the major planets of the solar system are Mercury, Venus, Earth, Mars, Jupiter, Saturn, Uranus, and Neptune. A good way to remember the names of the eight major planets is the sentence **M**y **V**ery **E**ducated **M**other **J**ust **S**erved **U**s **N**achos. Table 10.1 summarizes some basic information about the bodies in the solar system. Planets orbit the Sun in elliptical orbits. Earth's elliptical orbit is nearly circular. The major planets of the solar system are divided into two groups based upon their size and composition: the four innermost, small, dense *terrestrial planets* and the four large, much less dense, outermost *Jovian planets.*

Table 10.1 Solar System Data

Celestial Object	Mean Distance from Sun (million km)	Period of Revolution (d = days) (y = years)	Period of Rotation at Equator	Eccentricity of Orbit	Equatorial Diameter (km)	Mass (Earth = 1)	Density (g/cm³)
Sun	—	—	27 d	—	1,392,000	333,000.00	1.4
Mercury	57.9	88 d	59 d	0.206	4,879	0.06	5.4
Venus	108.2	224.7 d	243 d	0.007	12,104	0.82	5.2
Earth	149.6	365.26 d	23 h 56 min 4 s	0.017	12,756	1.00	5.5
Mars	227.9	687 d	24 h 37 min 23 s	0.093	6,794	0.11	3.9
Jupiter	778.4	11.9 y	9 h 50 min 30 s	0.048	142,984	317.83	1.3
Saturn	1,426.7	29.5 y	10 h 14 min	0.054	120,536	95.16	0.7
Uranus	2,871.0	84.0 y	17 h 14 min	0.047	51,118	14.54	1.3
Neptune	4,498.3	164.8 y	16 h	0.009	49,528	17.15	1.8
Earth's Moon	149.6 (0.386 from Earth)	27.3 d	27.3 d	0.055	3,476	0.01	3.3

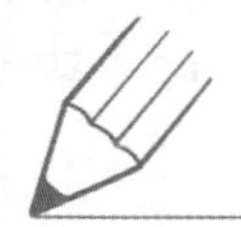

TRY THIS

Answer the following questions using the data in Table 10.1.

1. On which planet would the length of a day be nearest that of a day on Earth?
2. Which planet spins on its axis most rapidly?
3. Name two pairs of planets that are nearly the same size.
4. How does the density of planets near the Sun compare with the density of planets far from the Sun?

The four planets closest to the Sun are Mercury, Venus, Earth, and Mars. These terrestrial, or Earthlike, planets resemble Earth in size and rocky composition. They are also about the same density as Earth.

The more distant planets—Jupiter, Saturn, Uranus, and Neptune—are called Jovian, or Jupiterlike, because like Jupiter, they are all gas giants. Gas giants have thick atmospheres made mostly of gases such as water (H_2O), methane (CH_4) and ammonia (NH_4) surrounding a small rocky or liquid core. Although they have a lot more mass than the terrestrial planets, they are less dense, so they take up much more volume. Despite their large size, the gas giants rotate very rapidly, which causes a distinct equatorial bulge.

In addition to the eight major planets, several smaller dwarf planets orbit the Sun. ***Dwarf planets*** have enough gravity that they are round or nearly round, but they have not swept up everything near their path. Therefore, they may orbit the Sun in a zone that still has many other objects in it. Currently, there are three known dwarf planets—Ceres, Pluto, and Eris. All are less than half the size of the planet Mercury and at best have only a trace of an atmosphere.

OTHER SOLAR SYSTEM BODIES

In addition to the eight planets and three dwarf planets, many smaller objects have been found orbiting the Sun and are therefore also considered part of the solar system. All objects that orbit the Sun but do not fit the definition of a planet or dwarf planet are called ***small solar system bodies***.

Asteroids are irregularly shaped rocky objects orbiting the Sun. They have no atmosphere. Most orbit in the so-called asteroid belt between the planets Mars and Jupiter. However, thousands orbit farther out, like the 950 km diameter asteroid Chiron that orbits between Saturn and Uranus. In addition to the asteroids, the orbits of smaller chunks of matter cross those of the planets. When one of these chunks of matter hits Earth's atmosphere at very high speed, it vaporizes because of friction with the air. As it streaks through the air and vaporizes, it gives off light. To an observer on Earth, this appears as a streak of light. The chunk of matter is a ***meteoroid***, and the streak of light seen in the sky is called a ***meteor***. If any of the matter survives to strike the ground, it is called a ***meteorite***.

Comets are small, icy masses that orbit the Sun like planets, but their orbital ellipses are highly elongated. When a comet approaches the Sun, the ices heat up and turn into a cloud of gases around the comet, called a ***coma***. As the comet gets closer to the Sun and warms up, more gas is given off and the coma gets larger. Particles streaming out of the Sun push gases out of the coma, forming a tail that points away from the Sun.

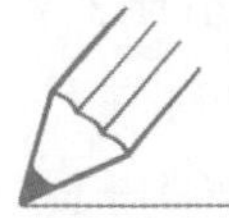

TRY THIS

1. What is the difference between a meteoroid and a meteorite?
2. Why does the tail of a comet always point away from the Sun?

MOTION IN THE SOLAR SYSTEM

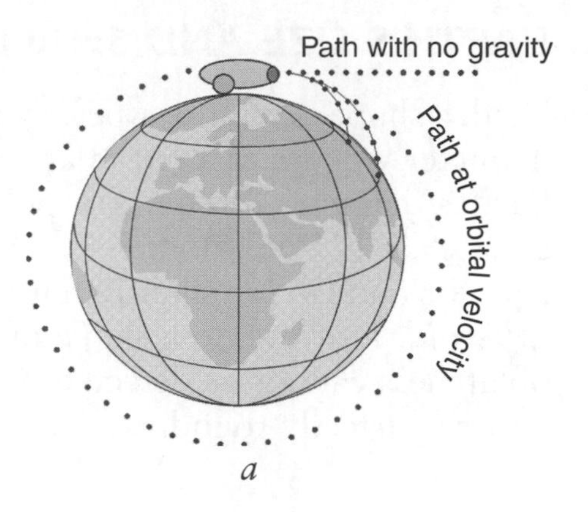

The planets all move around the Sun in nearly circular orbits. They also all orbit the Sun in the same direction and in roughly the same plane of the Sun's equator. All of the planets, except Venus and Uranus, spin in the same direction. Most of the moons also spin in the same direction as their planets and orbit close to the plane of their planet's equators.

GRAVITY AND MOTION

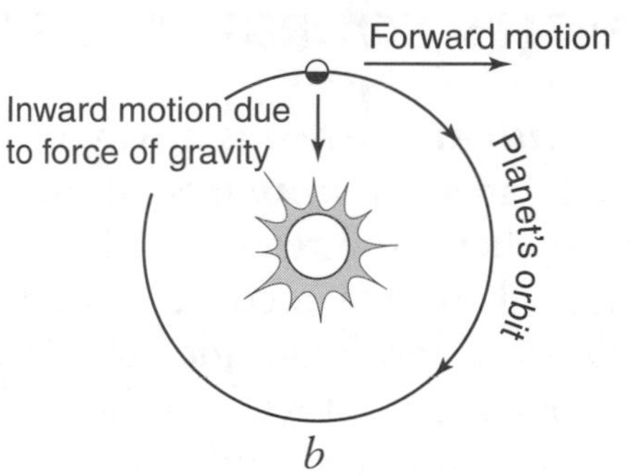

Figure 10.7
(*a*) Orbital velocity. (*b*) The orbit of a planet as the result of two motions.

The force that keeps planets in orbit around the Sun and moons in orbit around the planets is gravity. ***Gravity*** is a force of attraction that exists between all matter. How does gravity keep satellites moving in a curved orbit? Imagine that a cannonball is shot out of a cannon aimed horizontally. If there were no gravity, the cannonball would tend to fly off horizontally until some force stopped it. With gravity pulling the cannonball downward toward Earth's center as the cannonball is flying horizontally, the path of the cannonball would curve downward. Eventually, the cannonball would strike Earth's surface. If a more powerful charge were used in the cannon, the cannonball would travel farther before it struck Earth. If you used a powerful enough charge, the cannonball would travel far enough that as its path curved downward due to gravity, Earth's surface would curve away due to its spherical shape. The ball would never strike Earth's surface. Instead, gravity would cause the cannonball to fall downward at the same rate that Earth's surface curves away from it. The cannonball would fall unendingly in a circular path around Earth—it would be in orbit. See Figure 10.7*a*.

Similarly, the combination of two forces keeps the planets moving in their curved paths around the Sun. The combination of a planet's forward motion and its motion toward the Sun due to gravity results in circular motion—the planet's orbit around the Sun. See Figure 10.7*b*.

THE EARTH-MOON-SUN SYSTEM

Many of the occurrences observed on Earth are due to the relative motion of Earth, Moon, and Sun. These include day and night, seasons, eclipses, tides, meteor showers, and comets.

EARTH'S SIZE AND SHAPE

Earth's shape is almost a perfect sphere. A perfect sphere would have exactly the same diameter when measured in any direction. Earth's actual measurements differ slightly. Earth's diameter through the poles is 12,714 kilometers. Earth's diameter through the equator is 12,756 kilometers. Thus, the Earth's spherical shape bulges very slightly at the equator and is very slightly flattened at the poles. This is called an ***oblate (flattened) spheroid***. However, the shape of Earth is so close to being a perfect sphere that its oblateness cannot be detected by the eye. If you were to view Earth from space, it would appear perfectly round.

EARTH'S COORDINATE SYSTEM

A ***coordinate system*** is a way of locating points by labeling them with numbers called coordinates. ***Coordinates*** are numbers measured with respect to a system of lines or some other fixed reference. In order to describe the position of any point on Earth's spherical surface, a coordinate system has been set up that uses two coordinates known as latitude and longitude. The latitude-longitude system is made up of two sets of lines that cross each other at right angles. ***Latitude lines*** (called parallels) run in an east-west direction, and ***longitude lines*** (called meridians) run in a north-south direction.

Because Earth is a sphere, east-west lines on its surface form circles. Latitude lines are called parallels If you drew a series of east-west lines, the circles formed would all be parallel to one another. The fixed reference line for latitude is the ***equator,*** an east-west line midway between the North and South Poles. ***Latitude*** is a parallel's angular distance north or south of the equator as measured from the center of Earth. The latitude of the equator is 0°, and the latitude of the poles is 90°. See Figure 10.8.

North-south longitude lines on Earth's

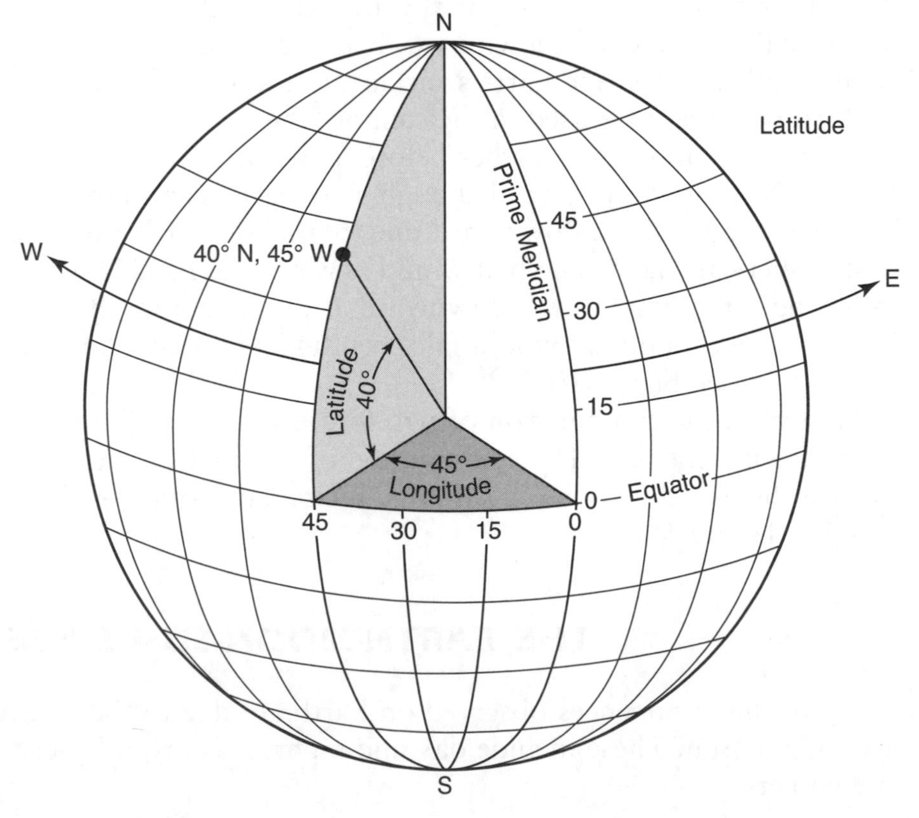

Figure 10.8
Latitude and longitude.

surface also form circles. Any circle that passes through both of Earth's poles is called a ***meridian.*** In 1884, an international conference agreed that the fixed reference line for longitude, or ***prime meridian***, would be the meridian that runs through the Royal Observatory in Greenwich, England. The longitude of the prime meridian is 0°. ***Longitude*** is a meridian's angular distance along the equator east or west of the prime meridian as measured from the center of Earth. See Figure 10.8. Longitude east or west of the prime meridian can be up to 180°.

Any meridian will cross the equator and all other parallels at right angles. So once reference lines are chosen, a system of meridians and parallels forms a grid of lines that intersect at right angles—a pretty neat coordinate system! See Figure 10.9

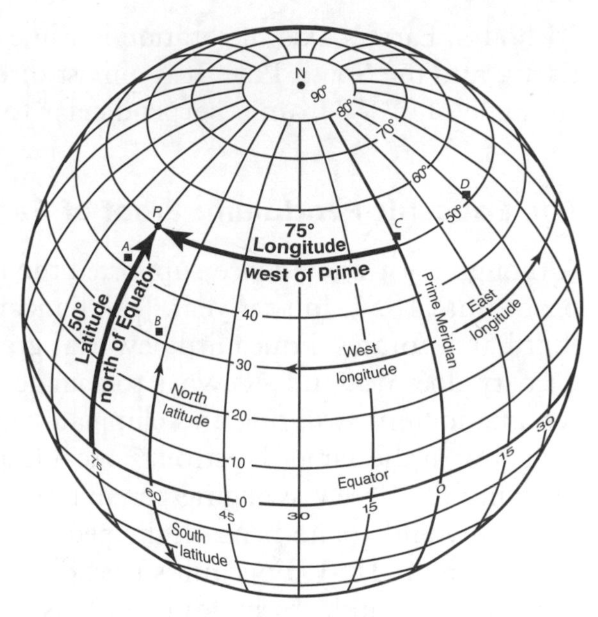

Figure 10.9
The geographic grid of parallels and meridians that allows any point on Earth's surface to be located with a set of coordinates. Point *P* has a latitude of 50°N and a longitude of 75°W.

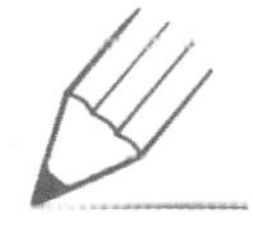

TRY THIS

Match the following locations with the corresponding letters *A–D* on the grid shown in Figure 10.9.

1. Niagara Falls 43°N, 79°W
2. London, United Kingdom 52°N, 0°
3. Bermuda, United Kingdom 32°N, 65°W
4. Kiev, Ukraine 51°N, 30°E

EARTH'S MOTIONS

Earth, like most objects in the solar system, moves in a regular and predictable way. It rotates on its axis and revolves around the Sun.

ROTATION

Rotation is a motion in which every part of an object spins around a central line called the axis of rotation. Earth spins at a rate of 15° per hour, or one complete rotation every

24 hours. Earth's axis of rotation is a line passing through the North Pole, the center of the Earth, and the South Pole. It is almost directly lined up with the star Polaris and is tilted at an angle of $23^1/_2°$ from a perpendicular to the plane of the Sun. See Figure 10.10.

The Foucault Pendulum: Proof of Earth's Rotation

Although a lot of evidence supported the idea that Earth rotated, there was no scientific proof until 1851. In that year, Jean Foucault, a French physicist, using a pendulum, *proved* that Earth rotates. Foucault knew that gravity pulls a pendulum only toward Earth's center. Gravity does not pull sideways to change the direction of the pendulum's swing. So, if you set a pendulum swinging and eliminate any sideways forces, the pendulum should keep swinging in the same direction. Foucault argued that if Earth were motionless, the direction of swing of a freely swinging pendulum would not change. However, if Earth rotated, the ground would rotate beneath the pendulum and change position relative to the swinging pendulum. In 1851, Foucault suspended a freely swinging pendulum from the inside of the dome of the Pantheon building in Paris. Through windows, observers were able to watch the pendulum slowly change its direction of swing—proving Earth rotates. In 24 hours, the direction of swing of the pendulum appeared to make one rotation. See Figure 10.11.

REVOLUTION

Revolution is the motion of one body around another body along a path called an ***orbit.*** Earth revolves around the Sun in a slightly oval orbit in which the Sun is slightly off center. Since Earth's orbit is slightly oval, its distance from the Sun varies during its orbit. It is closest on January 4 and farthest away on July 4.

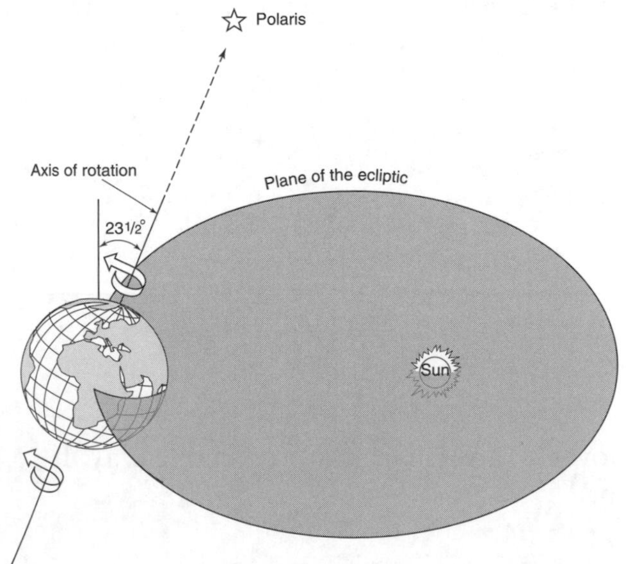

Figure 10.10
Earth rotation diagram showing plane of the ecliptic, Polaris, and circular motion around the axis of rotation.

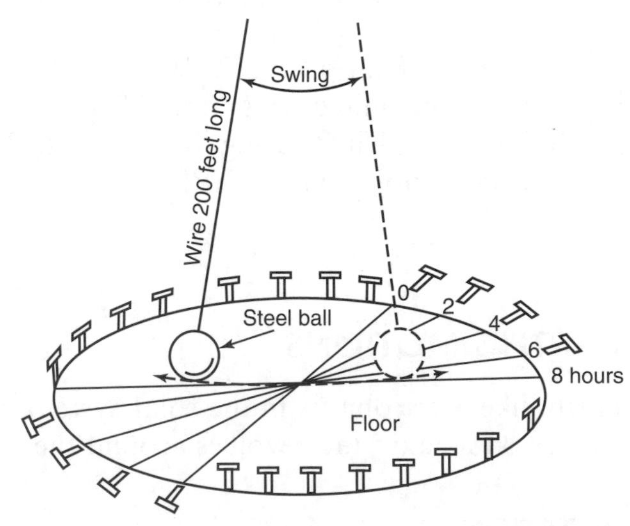

Figure 10.11
The Foucault pendulum.

As Earth revolves around the Sun, its spin keeps the axis of rotation pointing in the same direction. In other words, Earth's axes of rotation at any two points in its orbit are parallel. As you can see in Figure 10.12, this results in the Northern Hemisphere tipping toward the Sun in June and away from the Sun in December.

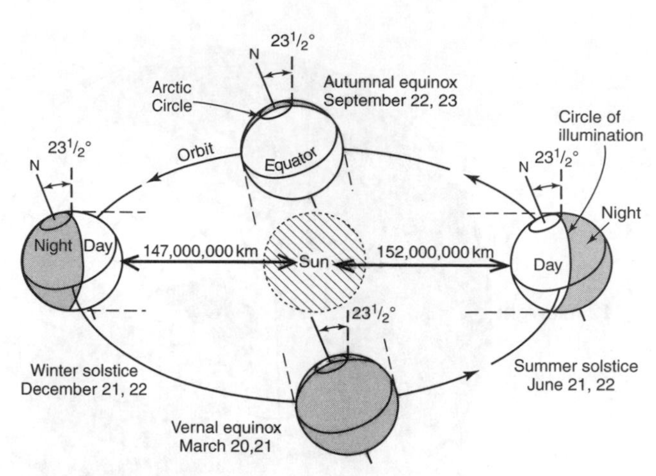

Figure 10.12
The parallelism of Earth's axis of rotation as it changes position in relation to the Sun.

THE EFFECTS OF EARTH'S MOTIONS

An Earth that rotates on an axis tilted at $23^1/_2°$ as it revolves around the Sun explains a wide range of observations. To understand the effects of Earth's motions, we will use a simple model.

THE SKY HEMISPHERE: A SIMPLE MODEL

Even though we now know that the motion of celestial objects is due to Earth's rotation, thinking of them as part of a celestial sphere surrounding Earth is still sometimes useful. The most you would see at any one time would be half of this sphere, or a *hemisphere*. This hemisphere represents the sky as an observer on Earth's surface would see it. The circle formed by the intersection of the sky and the ground is called the ***horizon***. The point in the sky that is right over an observer's head at any given time is called the ***zenith***. An imaginary line that passes through the north and south points on the horizon and through the observer's zenith is called the *celestial meridian*. See Figure 10.13.

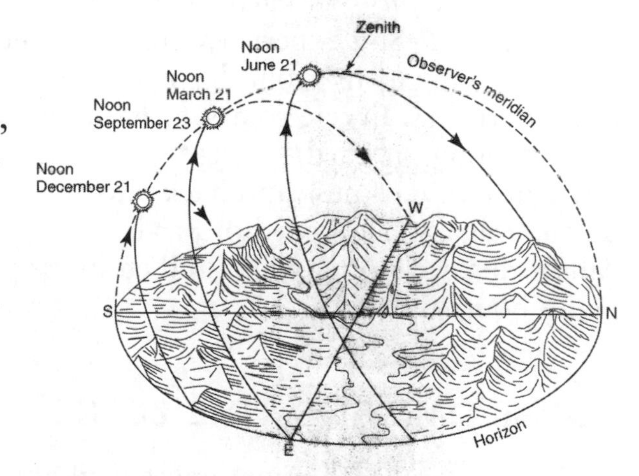

Figure 10.13
The hemisphere model.

APPARENT DAILY MOTION

Each day, the Sun moves along an imaginary path on the sky hemisphere due to Earth's rotation. As Earth (and any observer on Earth's surface) rotates from west to east, the Sun appears to move across the sky hemisphere from east to west.

The Sun is not the only celestial object that appears to move this way. Earth's rotation also causes the Moon and the stars to appear to rise in the eastern sky, arc across the sky hemisphere, and set in the western sky.

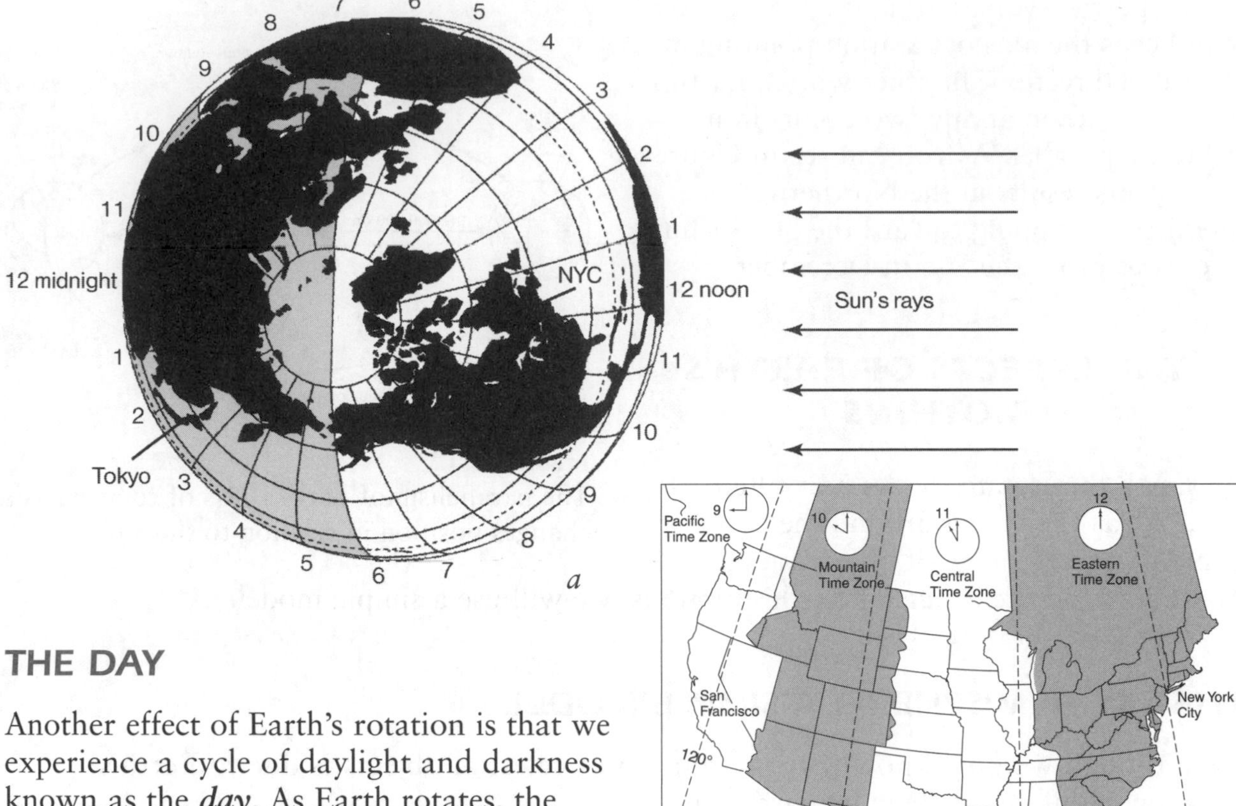

Figure 10.14
(*a*) Day and night at different locations.
(*b*) Time Zones of the United States.

THE DAY

Another effect of Earth's rotation is that we experience a cycle of daylight and darkness known as the ***day***. As Earth rotates, the side facing the Sun experiences daylight and the side facing away from the Sun experiences darkness. In one rotation, an observer carried along on Earth's surface moves from daylight into darkness and back into daylight again. The speed of Earth's rotation causes this cycle of daylight and darkness to be 24 hours in length.

LOCAL TIME AND TIME ZONES

Each place on Earth experiences daylight and darkness at a different time. A careful look at Figure 10.14*a* should make it clear that when it is noon in New York City, it is the middle of the night in Tokyo. In the past, every place set its clocks according to when it was solar noon in that place. Clocks in New York City would be set at a different time from clocks in Buffalo or Syracuse. When people began to travel long distances quickly by train or steamboat, this caused problems with arrival and departure timetables.

In the early 1880s, the countries of the world decided to set up a series of time zones. In a time zone, the same time is used by all places within that zone. Since Earth rotates at 15° per hour, zones were set up at 15° intervals around the globe. To avoid problems, the boundaries of time zones were shifted so that small countries, or whole states, fall within the same time zone. See Figure 10.14*b*.

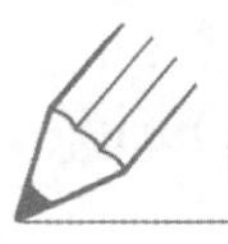

TRY THIS
The map in Figure 10.14*b* shows the boundaries of four time zones. The clocks show the time in each of the zones at the same instant.

1. The time of day in New York City is always later than the time in San Francisco because Earth
   1. has seasons
   2. orbits the Sun
   3. rotates in an easterly direction
   4. is tilted on its axis
2. When it is 5 P.M. in New York City, what time is it in San Francisco?
   1. noon 2. 2 P.M. 3. 8 P.M. 4. 5 A.M.

UNEQUAL DAY AND NIGHT

Except at the equator, the length of the daylight and darkness periods varies throughout the year. This happens because as Earth revolves, its position in relation to the Sun changes. Look at Figure 10.15. Note that on June 21, Earth's axis is tilted toward the Sun. As Earth rotates, a person standing at the North Pole (90° N latitude) moves in a circle that never enters darkness. At the same time, a person in New York City (41° N latitude) is carried along a path that travels mostly through daylight (about 15 hours) and enters darkness for only a short time (about 9 hours). As you travel south of the equator on this date, the daylight period gets shorter and shorter, until at the South Pole, a person never enters daylight and experiences 24 hours of darkness.

On March 21, the situation is quite different. Earth is tilted neither toward nor away from the Sun. As a result, all places on Earth spend exactly one-half of a rotation (12 hours) in daylight and one-half of a rotation (12 hours) in darkness.

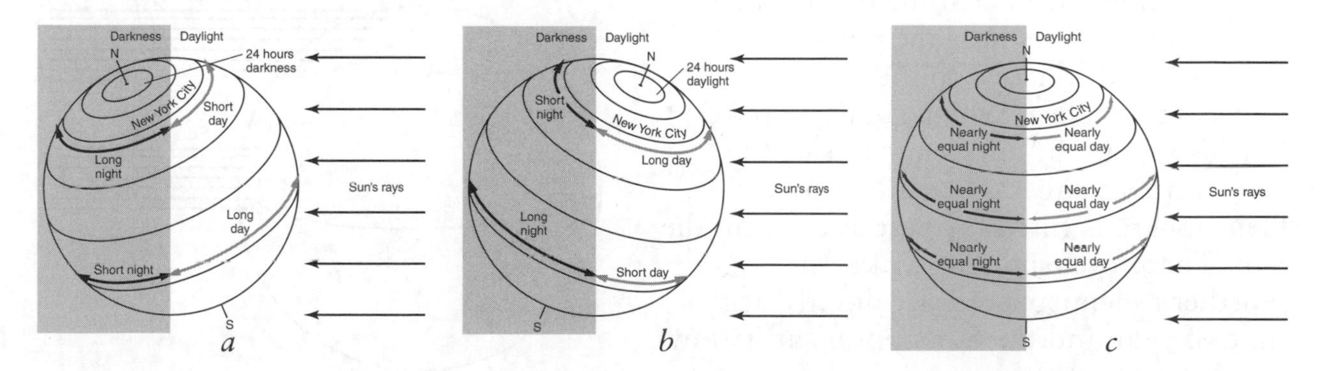

Figure 10.15
Daylight and darkness at the *(a)* winter solstice (Dec. 21), when Earth's axis of rotation is tilted away from the Sun, *(b)* summer solstice (June 21), when Earth's axis of rotation is tilted toward the Sun, and *(c)* equinoxes (Mar. 21 and Sept. 21), when Earth's axis of rotation is not tilted toward or away from the Sun.

THE YEAR

A *year* is the time Earth takes to complete one 360° revolution around the Sun. During that time, Earth completes $365^1/_4$ rotations, so a year is $365^1/_4$ days long. Earth's revolution around the Sun causes several phenomena.

THE ANNUAL TRAVERSE OF THE CONSTELLATIONS

As Earth revolves around the Sun, the side of Earth facing the Sun experiences day and the side facing away experiences night. Since the stars are visible only at night, the portion of the universe whose stars are visible to an observer on Earth varies cyclically as Earth revolves around the Sun. See Figure 10.16.

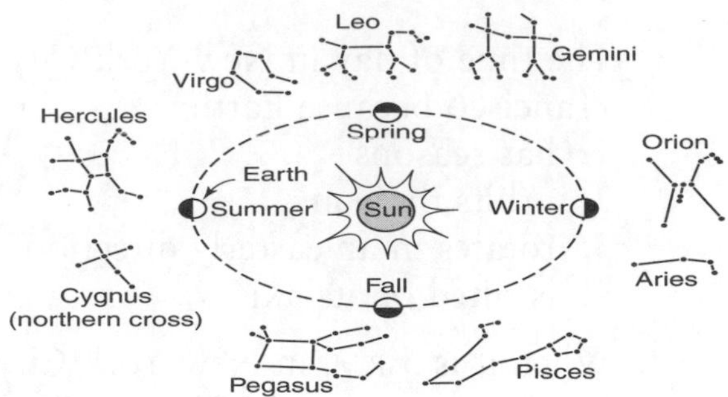

Figure 10.16
The cyclic change in constellations seen throughout the year. In spring, Leo is visible at night, but Pisces is not.

SEASONS

The warming effect of sunlight is related to the angle at which it strikes Earth's surface. Because Earth is a sphere, sunlight does not strike all points on Earth's surface at the same angle on any given day. As the Sun's ***altitude***, or angle above the horizon, increases, the angle at which sunlight strikes Earth's surface increases. The two beams of sunlight approaching Earth in Figure 10.17 are identical. Notice that the sunlight near the equator strikes Earth's surface more directly (nearer 90°) and the energy of the sunlight is concentrated in a smaller area. Near the poles, the sunlight strikes Earth's surface less directly (at a lower angle) and the sunlight's energy is spread out over a larger area. The more concentrated the energy of the sunlight, the greater its warming effect. The difference in the concentration of the incoming sunlight causes higher temperatures near the equator and lower temperatures near the poles.

Earth's tilt relative to the Sun also affects the angle at which sunlight strikes Earth's surface at different times of the year. On December 21, the Northern Hemisphere is tilted farthest away from the Sun. Therefore, sunlight strikes the Northern Hemisphere least directly at this time of year and the Sun appears at its lowest altitude of the year at solar noon. Remember that this is also the time when locations in the Northern Hemisphere experience the fewest hours of daylight. The

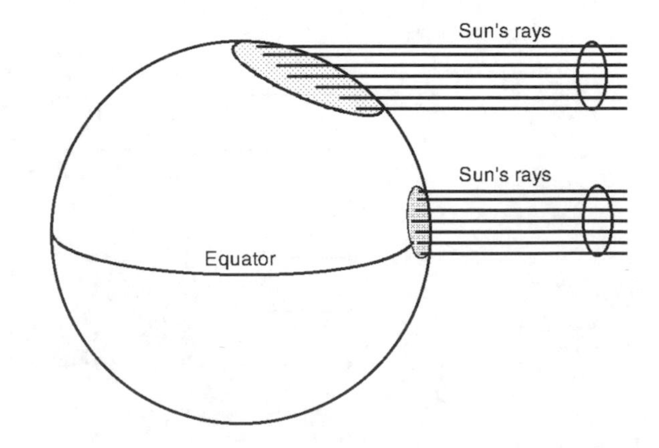

Figure 10.17
Sunlight is more direct near the equator and less direct near the poles.

combination of a short daylight period and less direct sunlight causes the low temperatures of the winter season.

On June 21, the Northern Hemisphere is tilted farthest toward the Sun. Therefore, sunlight strikes the Northern Hemisphere most directly and the Sun appears at its highest altitude of the year at solar noon. It is also the time when the daylight period is longest. The combination of a long daylight period and more direct sunlight results in high temperatures during the summer season.

At the equinoxes on March 21 and September 22, Earth is tilted neither toward nor away from the Sun, resulting in nearly equal periods of daylight and darkness. The sunlight striking Earth's surface is halfway between its most and least direct. The length of day and directness of sunlight are between the two extremes. Therefore, the temperatures are moderate in the spring and fall seasons.

From December to June, the daylight period becomes longer each day and the altitude of the Sun at noon increases. On June 21, the altitude of the Sun at noon stops increasing. Therefore, this date is called the ***summer solstice***, meaning summer "Sun stop." From June to December, the daylight period gets shorter each day and the altitude of the Sun at noon decreases. On December 21, the altitude of the Sun at noon stops decreasing. Therefore, this date is called the ***winter solstice***, meaning winter "Sun stop." March 21 and September 22, when the daylight and darkness periods are nearly equal, are called, respectively, the *spring* ***equinox*** and *fall* ***equinox***, meaning spring and fall "equal night." This pattern of change repeats in an annual cycle.

MOTIONS OF THE MOON

As Earth revolves around the Sun, the Moon revolves around Earth. The Moon's orbit is tilted at an angle of about 5° from the plane of Earth's orbit around the Sun. See Figure 10.18. As the Moon orbits Earth, it also rotates on its axis. In the time the Moon takes to complete one orbit, it also completes one rotation. Thus, the same side of the Moon always faces Earth.

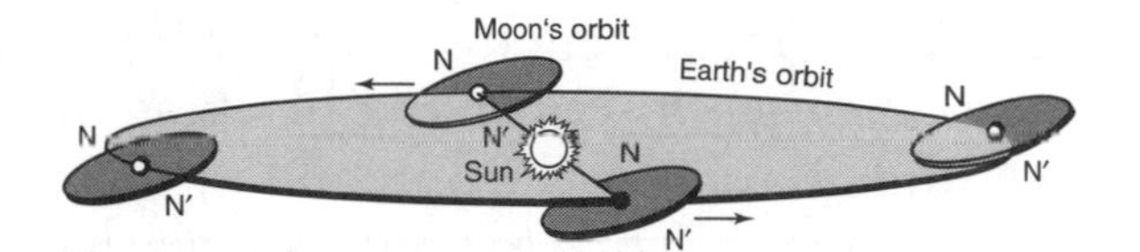

Figure 10.18
The Moon's orbit around Earth is tilted at an angle of 5° to Earth's orbit around the Sun.

PHASES OF THE MOON

The Moon does not produce visible light of its own. The Moon is visible only because of light from the Sun that is reflected from its surface. An observer on Earth is able to see only that part of the Moon that is illuminated by the Sun. As the Moon moves around Earth, different parts of the side of the Moon facing Earth are illuminated by sunlight and the Moon passes through a cycle of visible shapes called ***phases of the moon.*** See Figure 10.19.

The Moon takes 29.5 days to go through a complete cycle of phases. The word *month* has its origin in "moon-th," which refers to this 29.5-day cycle of phases.

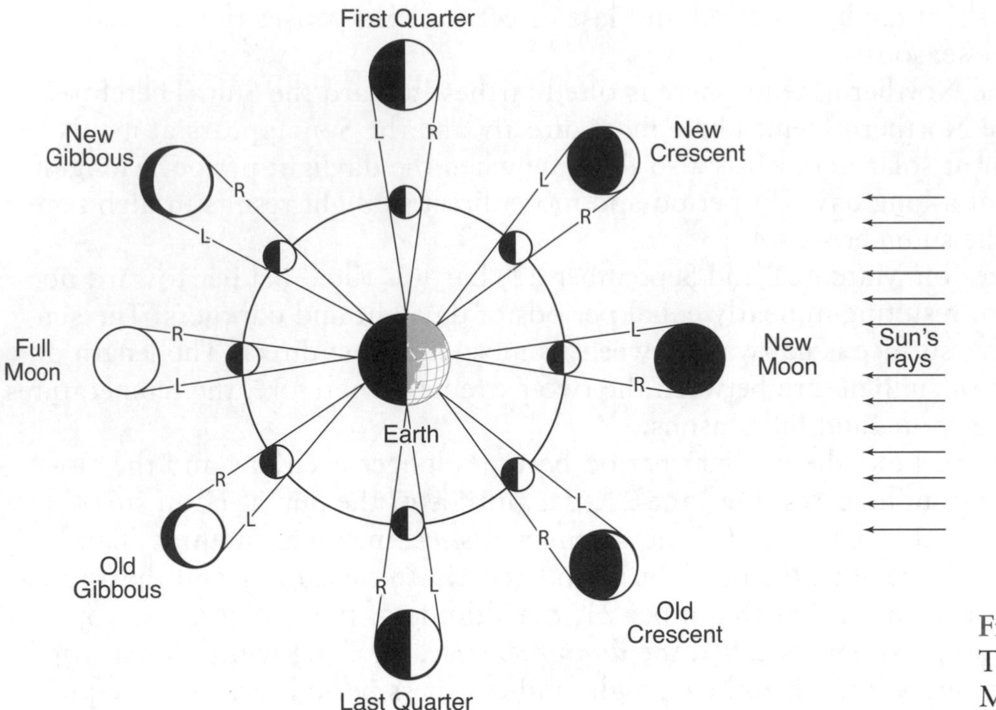

Figure 10.19
The phases of the Moon.

Eclipses of the Sun and Moon

A *solar eclipse* occurs when the Moon passes directly between Earth and the Sun, casting a shadow on Earth and blocking our view of the Sun. Both the Moon and Earth are illuminated by the Sun and cast shadows in space. Earth and its Moon are very tiny compared with the Sun, so the shadows they cast are extremely long and narrow. The Moon's shadow barely reaches Earth. The totally dark part of the shadow it casts on Earth's surface is never more than 269 km in diameter. The 5° tilt of the Moon's orbit together with the small size of its shadow makes it easy for the shadow to miss Earth at the full and new moon phases. Thus, total eclipses of the Sun are rare. See Figure 10.20.

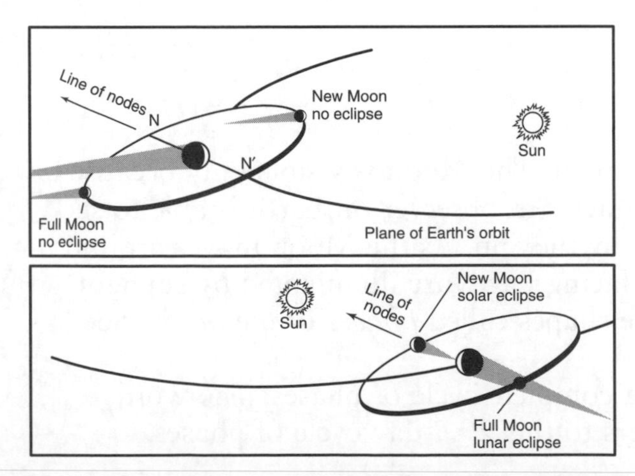

Figure 10.20
Eclipses of the Sun and Moon. Solar and lunar eclipses can occur when the Moon intersects the ecliptic in new-moon or full-moon position. If the Moon is above or below the ecliptic during these phases, no eclipse occurs.

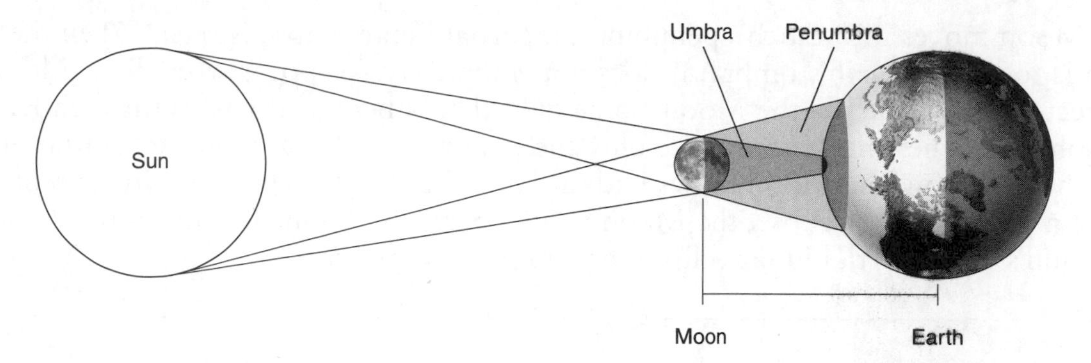

Figure 10.21
Solar eclipse showing umbra and penumbra. Sites that pass through the umbra experience a total solar eclipse; sites that pass through the penumbra experience a partial solar eclipse.

Whether an observer sees a total eclipse or a partial eclipse depends on what part of the Moon's shadow passes over the observer. See Figure 10.21. The Moon's shadow has two parts. The ***umbra*** is the part of the shadow in which all of the light has been blocked. In the ***penumbra***, only part of the light is blocked so the light is dimmed but not totally absent. An observer in the Moon's umbra would see a total eclipse. An observer in the penumbra would see a partial eclipse. An observer outside of the Moon's shadow would see no eclipse.

A ***lunar eclipse*** occurs when the Moon moves through Earth's shadow at full moon. See Figure 10.22. If the Moon moves into Earth's umbra, a total lunar eclipse is seen.

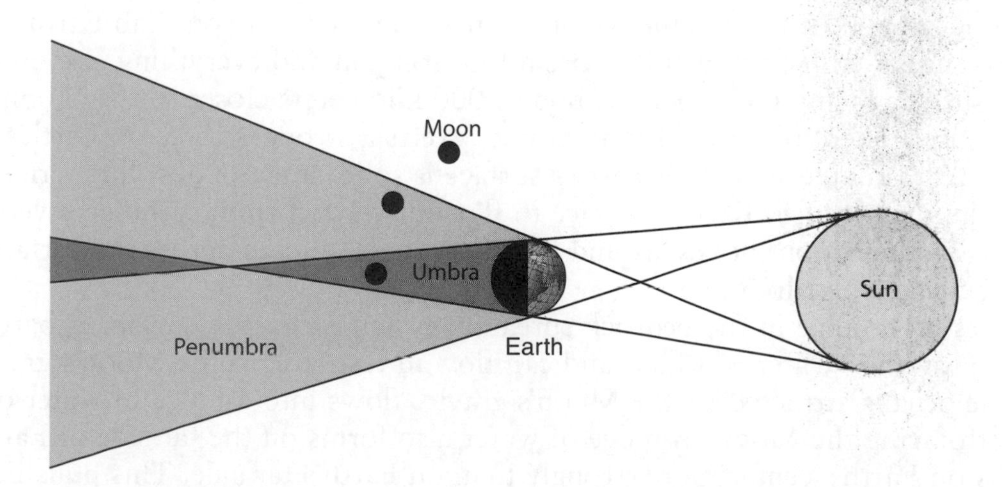

Figure 10.22
A Lunar Eclipse. If the moon passes only through Earth's penumbra, a partial eclipse of the Moon would be observed. If the Moon passes through Earth's umbra, a total eclipse of the Moon is observed.

If the Moon moves into Earth's penumbra, a partial lunar eclipse is seen. When the Moon is totally in Earth's umbra, it does not completely disappear from view. Although no direct sunlight reaches the Moon, some light that is bent as it passes through Earth's atmosphere reaches the Moon. Since only the long waves of red light are bent far enough to reach the Moon, the Moon glows with a dull red color during a total lunar eclipse. During a partial lunar eclipse, the Moon is only partially dimmed as Earth blocks some of the Sun's light. Partial lunar eclipses are not very impressive.

TRY THIS

A total solar eclipse was visible to observers in the southeastern United States on February 26, 1998. The diagram below shows the Sun and Earth as they were viewed from space on that date. On the diagram, draw the Moon (○) showing its position at the time of the solar eclipse.

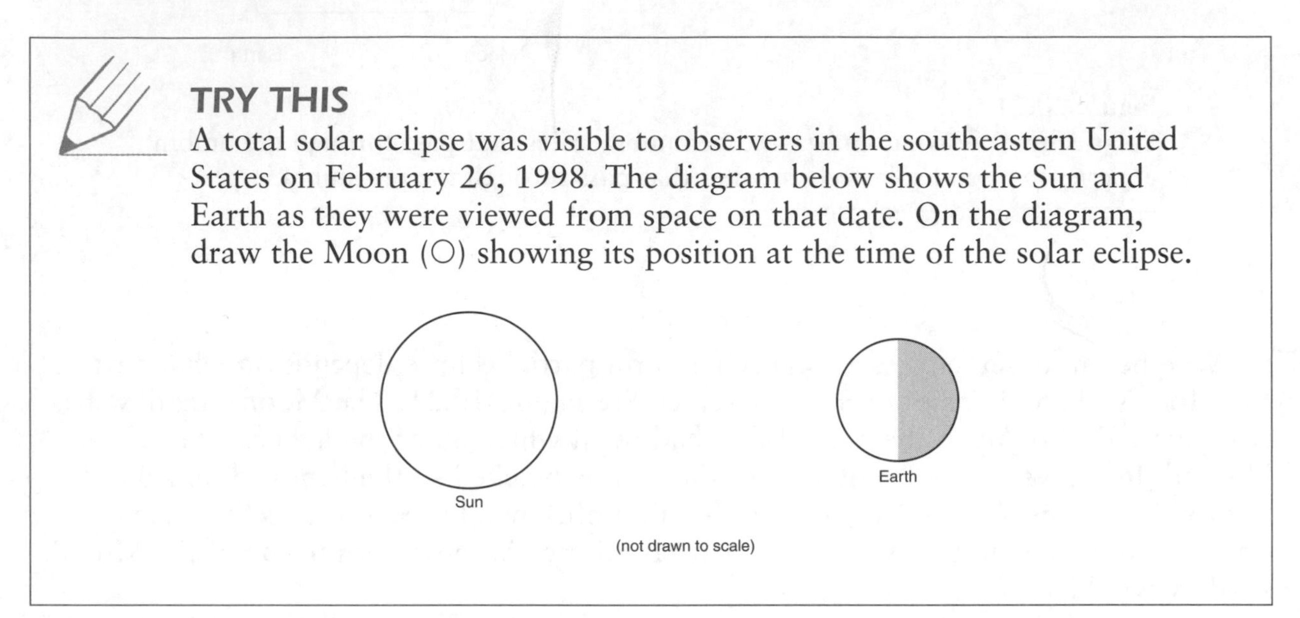

Tides

Every day, you feel the mutual attraction of gravity between your body and Earth pulling you downward with a force called your weight. However, Earth's gravity is not the only gravity acting on you. Although the Moon is farther away from you than Earth's center and has less mass than Earth, it still exerts a force on you and everything else on Earth.

Earth's surface facing the Moon is about 6,000 kilometers closer to the Moon than to Earth's center. Therefore, the Moon's gravity exerts a stronger force on Earth's surface than does Earth's center. Although Earth's surface is solid, it is not absolutely rigid. The Moon's gravity causes Earth's surface to flex outward, forming a bulge several inches high. As the Moon moves around Earth, the bulge moves across the surface as it remains beneath the Moon.

An inches-high bulge in the bedrock spread over half of Earth's surface is barely noticeable. However, water is a fluid and can flow in response to the Moon's gravity. Water in the oceans attracted by the Moon's gravity flows into a bulge of water on the side of Earth facing the Moon. A bulge of water also forms on the far side of Earth. The Moon pulls on Earth's center more strongly than on Earth's far side. This pulls Earth away from the oceans on the far side. The water flows into this space, creating a bulge. The water flows into these bulges from the area in between them, creating a deep region and a shallow region in the ocean waters.

As Earth rotates on its axis, the position of the tidal bulges remains lined up with the Moon. As the rotating Earth carries a location into a tidal bulge, the water deepens and the tide rises on the beach. As the location rotates out of the tidal bulge, the water becomes shallower and the tide falls. Since two bulges are on opposite sides of Earth, the tide rises and falls twice a day.

The Sun also produces tidal bulges in Earth's surface and oceans. At new moon and full moon, the Sun's tidal bulges and the Moon's tidal bulges align with one another and add together. The result is very high and very low tides. These are called *spring tides* because they spring so high, not because they happen during the spring season. Spring tides occur at every new and full moon no matter the season. When the Sun's tidal bulges and Moon's tidal bulges are at right angles to each other during the first-quarter and third-quarter moons, they nearly cancel each other out. *Neap tides* then occur in which there is very little difference between high and low tides. See Figure 10.23.

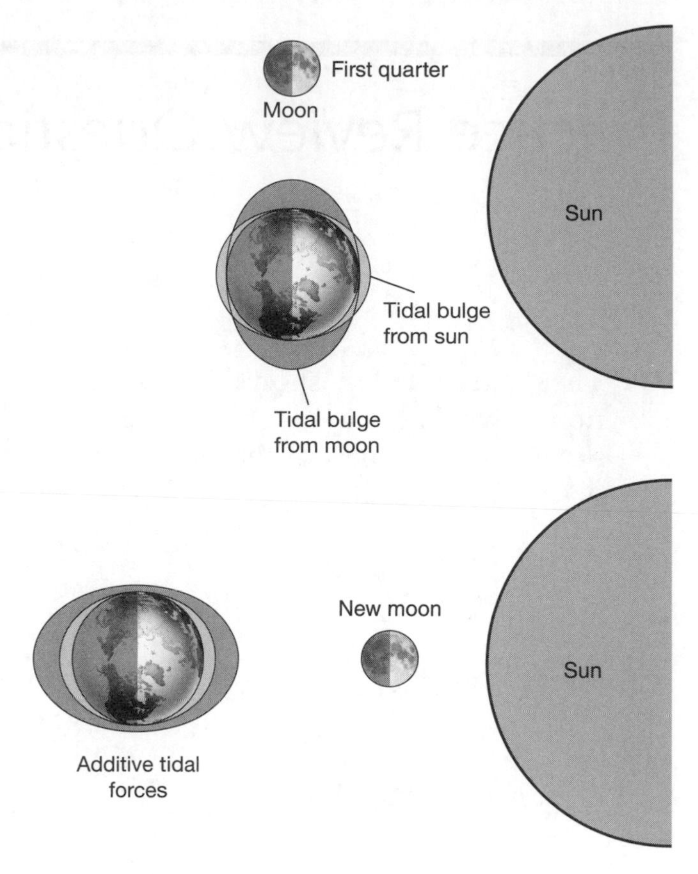

Figure 10.23
Spring tides and neap tides.

TRY THIS

The diagram to the right shows the Sun, Earth, and Moon's orbit as seen from space.

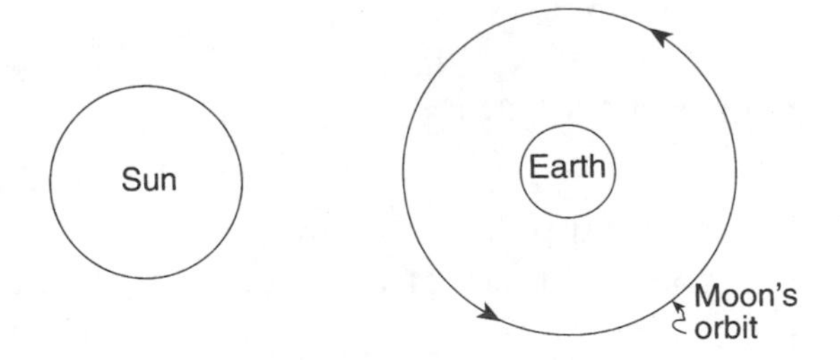

1. On the diagram, draw a circle about this size ○ to show the Moon's position in its orbit when a lunar eclipse is seen from Earth.
2. On the diagram draw two dots on Earth's surface to show where the highest ocean tides most likely occur at the time of the eclipse.
3. Approximately how many complete revolutions around Earth does the Moon make each month?

Practice Review Questions

1. The diagram below shows the position at which the Sun appears in the sky at various times of the day to an observer on Earth.

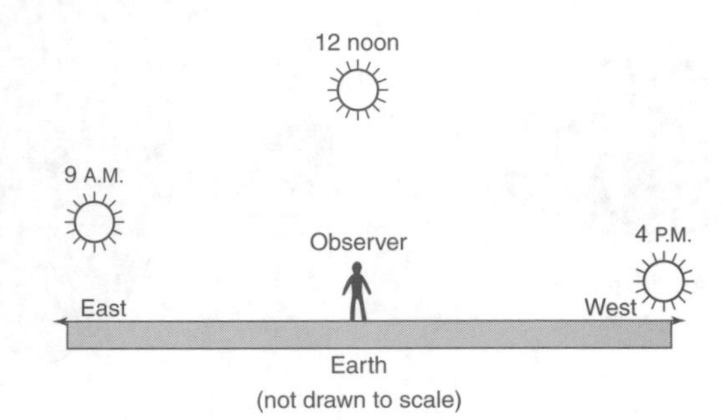

The best explanation for the apparent change in the position of the Sun is that Earth is

1. revolving around the Sun
2. tilted on its axis
3. attracted to the Sun
4. rotating on its axis

2. Why do stars appear to move through the night sky at the rate of 15° per hour?

1. Stars actually revolve around Earth at 15° per hour.
2. Earth actually revolves around the stars at 15° per hour.
3. Earth actually moves around the Sun at 15° per hour.
4. Earth actually rotates on its axis at 15° per hour.

3. How would a time exposure photograph of stars in the Northern Hemisphere appear if the camera had been aimed at Polaris?

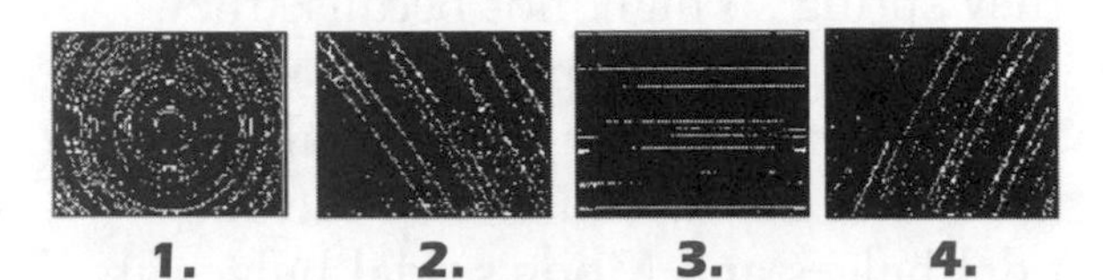

1. **2.** **3.** **4.**

4. Which model correctly shows the movement of the Earth (*E*) and the Moon (*M*) in relation to the Sun?

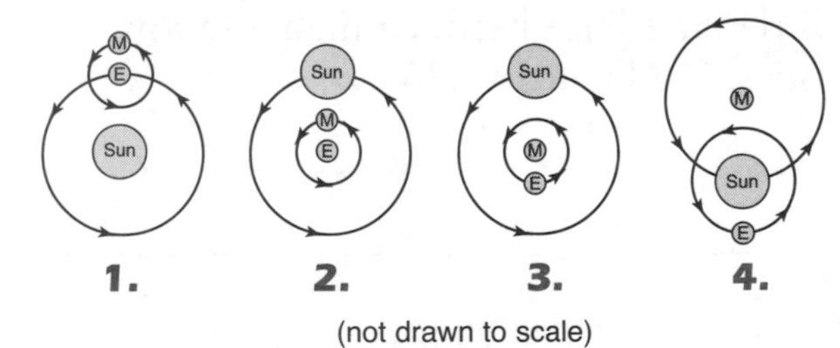

1. **2.** **3.** **4.**

(not drawn to scale)

5.* The Sun is an example of which of the following?

1. comet
2. planet
3. galaxy
4. star

* Reproduced from IEA TIMSS 1999 Science Released Items for Grade 8, TIMSS International Study Center, Boston College, MA

GO ON 

Refer to Table 10.1, Solar System Data on page 141 to answer questions 6–9.

6. Three planets that are relatively large, gaseous, and of low density are

1. Mars, Jupiter, and Uranus
2. Jupiter, Saturn, and Uranus
3. Venus, Jupiter, and Neptune
4. Mercury, Venus, and Earth

7. Which sequence correctly shows the relative size of the eight planets of our solar system?

1.

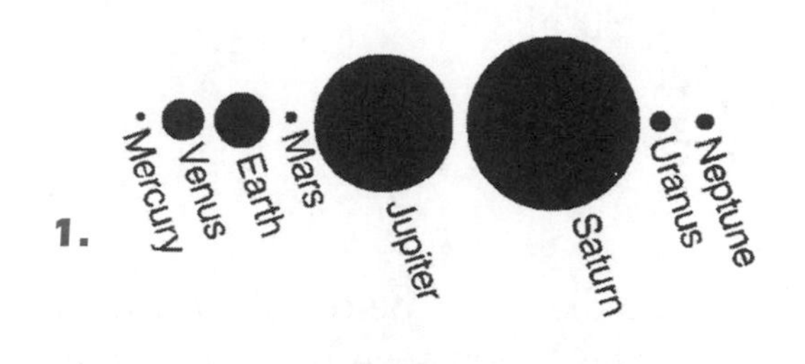

2.

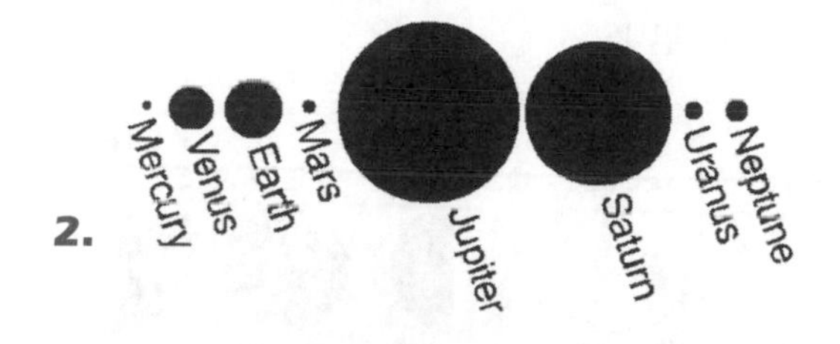

3.

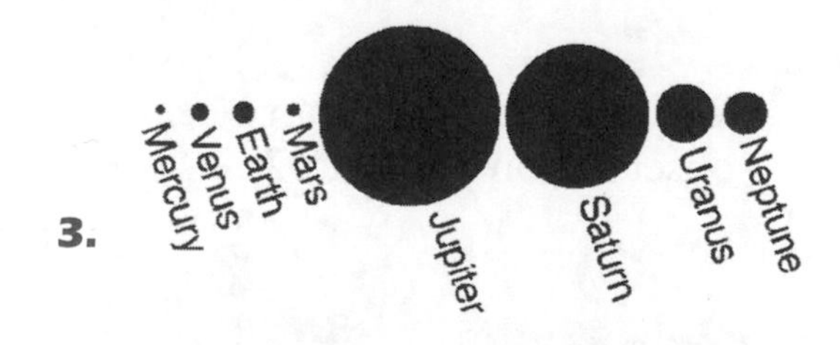

4.

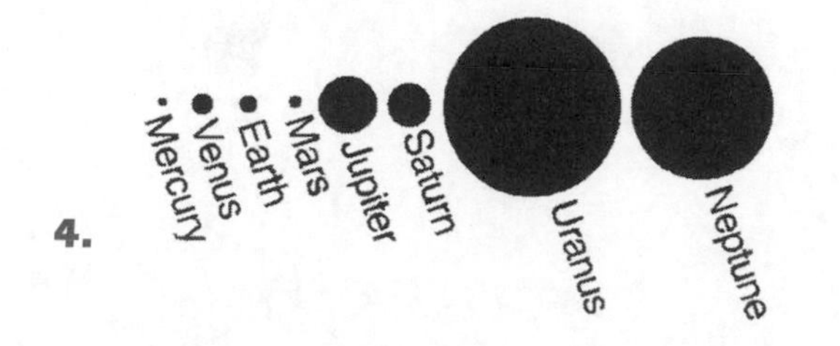

8. The period of time a planet takes to make one revolution around the Sun is most dependent on the planet's average

1. amount of solar radiation
2. mass
3. rotation rate
4. distance from the Sun

9. Which planet takes longer for one spin on its axis than for one orbit around the Sun?

1. Earth
2. Mars
3. Venus
4. Mercury

Base your answers to questions 10 and 11 on the map below, which shows a latitude-longitude grid of the world.

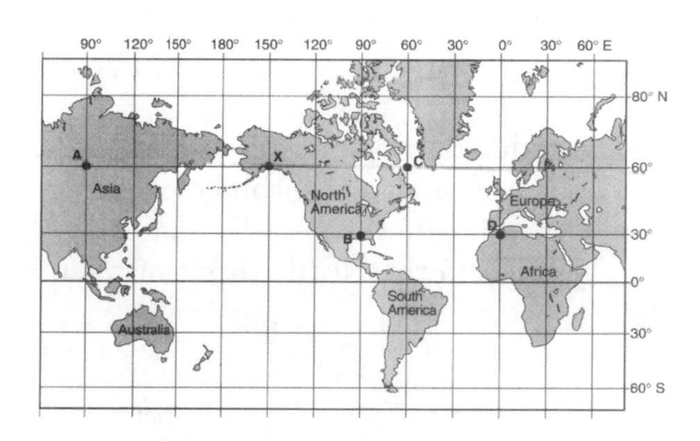

10. What is the approximate latitude and longitude of point *X*?

1. 60°S, 150°E
2. 60°N, 150°W
3. 150°S, 60°E
4. 150°N, 60°W

GO ON ➡

11. If it is solar noon at location *X*, at which location is the solar time 4 P.M.?

1. *A*
2. *B*
3. *C*
4. *D*

12. The length of an Earth day is based on the time needed to complete about one

1. Sun revolution
2. Earth revolution
3. Sun rotation
4. Earth rotation

13. The time required for one Earth revolution is about

1. 1 year
2. 1 month
3. 1 day
4. 1 hour

14. The diagram below shows Earth in four positions as it orbits the Sun. Which position shows Earth during summer in the Northern Hemisphere?

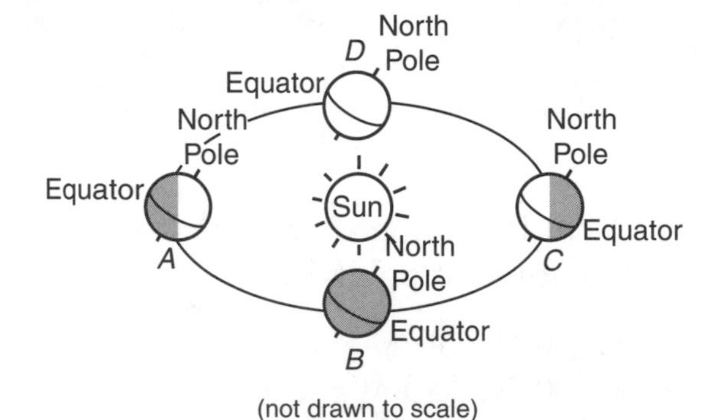

(not drawn to scale)

1. *A*
2. *B*
3. *C*
4. *D*

15. The diagram below shows Earth at a specific position in its orbit. Arrows show radiation from the Sun.

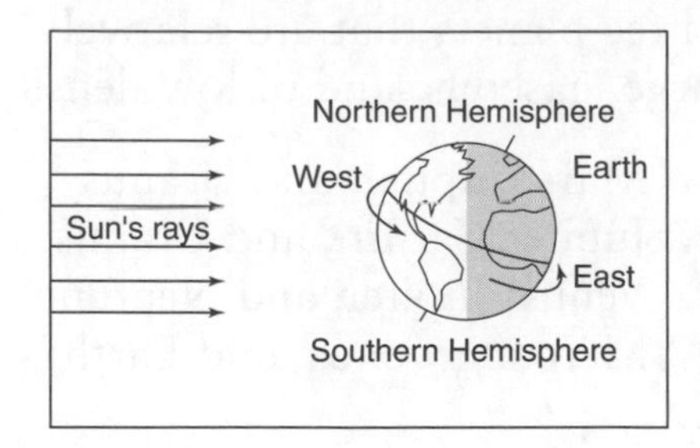

Which location would have the greatest number of daylight hours when Earth is in this position?

1. 90°N
2. 30°N
3. 90°S
4. 30°S

16. The diagrams below show the phases of the Moon seen by an observer in August.

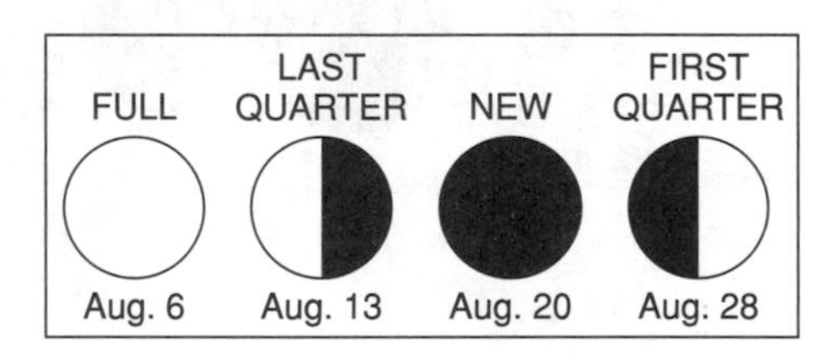

Which phase could have been observed on August 17?

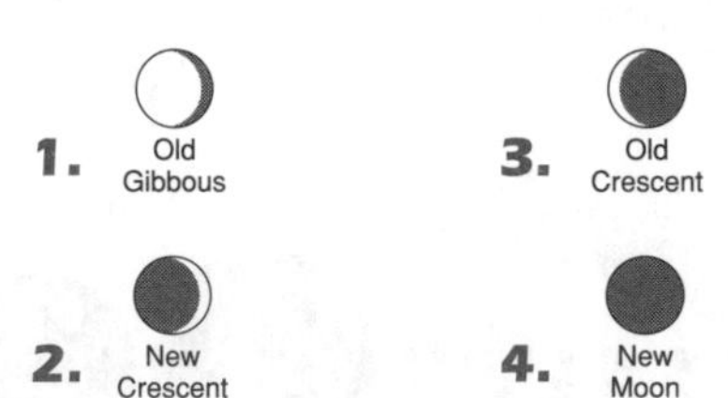

GO ON →

17. A cycle of Moon phases can be seen from Earth because the

1. Moon revolves around Earth
2. Moon spins on its axis
3. Earth spins on its axis
4. Moon's distance from Earth changes

18. Which model of the Sun, Earth (*E*), and Moon (*M*) best shows a position that would cause the highest ocean tides?

1. 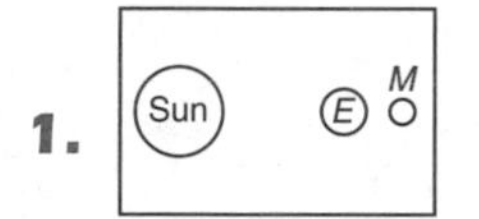

3.
M
Sun
E

2. 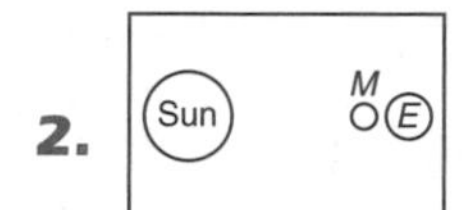

4.
Sun
E
M

Base your answers to questions 19 and 20 on the graph below showing the water levels caused by tides for 2 days at a location on the ocean.

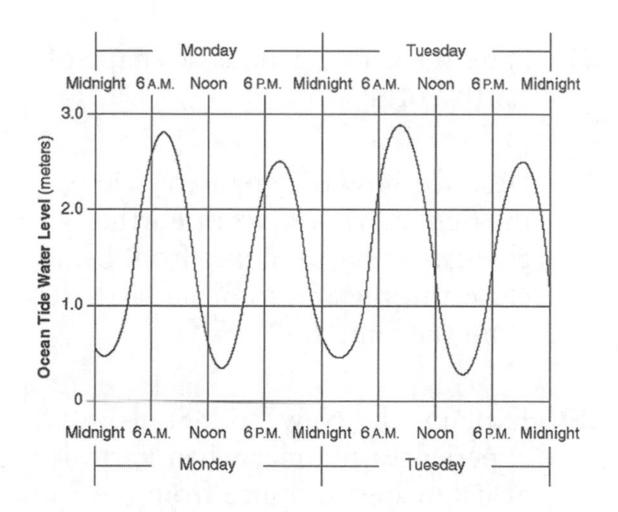

19. What was the approximate ocean water level at 6 A.M. on Monday?

1. 1.0 meters
2. 1.5 meters
3. 2.5 meters
4. 3.0 meters

20. According to the pattern given on the graph, the next high tide will occur on Wednesday at about

1. 3 A.M.
2. 10 A.M.
3. 12 noon
4. 6 P.M.

21. What is represented by the diagram below?

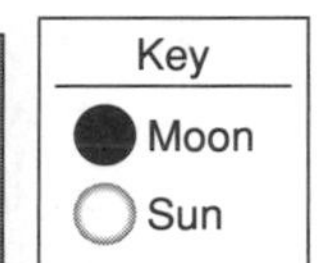

1. stages in an eclipse of the Moon
2. changing phases of the Moon
3. stages in an eclipse of the Sun
4. changing phases of the Sun

GO ON

22. The diagram below shows the positions of the Sun, Moon, and Earth when an eclipse was observed from Earth. Points *A* and *B* are locations on Earth's surface.

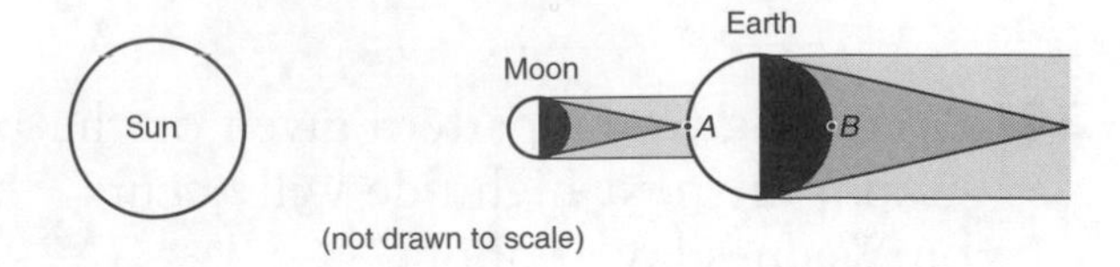

Which statement correctly describes the type of eclipse that was occurring and where the eclipse was observed?

1. A solar eclipse was observed from position *B*.
2. A lunar eclipse was observed from position *B*.
3. A solar eclipse was observed from position *A*.
4. A lunar eclipse was observed from position *A*.

CONSTRUCTED-RESPONSE QUESTIONS

Base your answers to questions 23–25 on the diagram below.

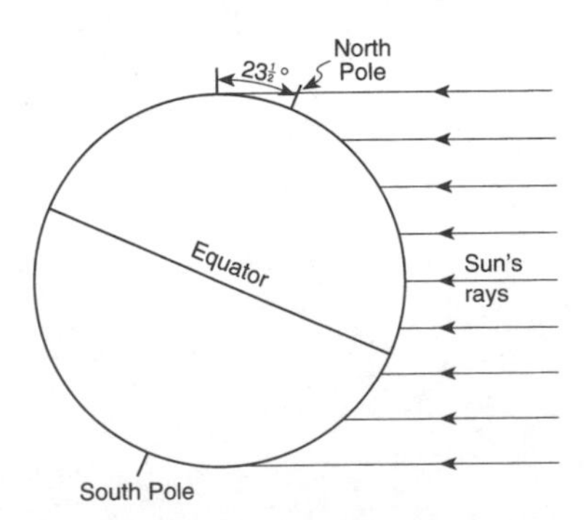

23. On the diagram, neatly shade the area of Earth that is in darkness.

24. Draw the line of latitude that is receiving the most direct rays of the Sun on this date.

25. What month of the year is represented by this diagram?

Base your answers to questions 26–29 on the diagram below showing the Sun, Earth, and Moon orbit as seen from space.

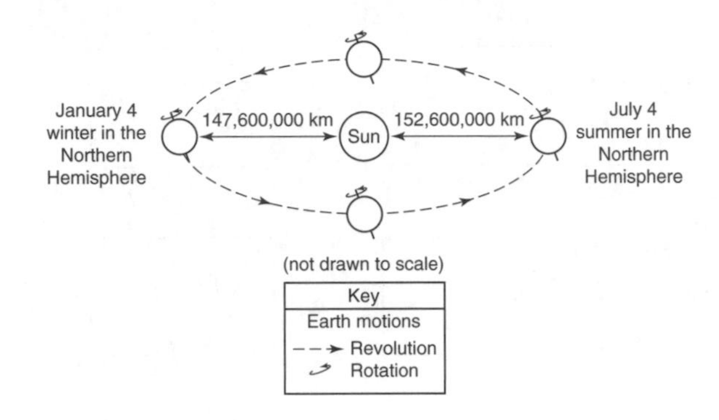

26. On the diagram, place an **X** on Earth's orbit to indicate Earth's position on May 21.

27. State the actual geometric shape of Earth's orbit.

28. Describe how the apparent size of the Sun, as viewed from Earth, changes as Earth moves from being closest to the Sun to being farthest from the Sun.

29. Explain why New York State experiences summer when Earth is at its greatest distance from the Sun.

Chapter 11

Geology

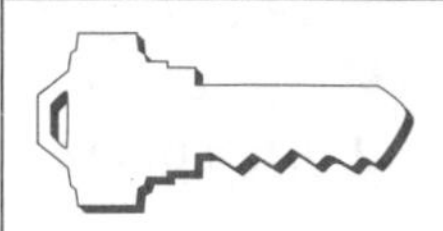

Key Idea: Many of the phenomena that we observe on Earth involve interactions among components of air, water, and land.

EARTH'S STRUCTURE

When looking back at Earth from near the Moon, the *Apollo 8* astronauts took photographs that show a hauntingly beautiful sphere, with deep blue seas and green lands veiled in swirling white clouds suspended against the vast blackness of space. See Figure 11.1. These images show that Earth is not made of a single substance but has distinct layers: a solid surface, a thin layer of water, and a layer of gases. Since Earth is a sphere, all three layers are spherical and are called, respectively, the *lithosphere*, *hydrosphere*, and *atmosphere*. See Figure 11.2. Earth is a dynamic planet. At Earth's surface, the lithosphere, hydrosphere, and atmosphere come into contact with one another and interact. Therefore, Earth's surface is constantly undergoing change.

THE LITHOSPHERE

Earth formed as part of a giant cloud of gas and dust particles that contracted due to gravity, creating our solar system. As the young Earth contracted and the particles collided and rubbed against each other, they got hot enough to melt. Like a mixture of oil and water, substances in the molten Earth then separated into layers due to density differences. Denser substances sank toward the center, and less dense substances floated to the top. Since Earth is a sphere, the layers that formed were spheres. The end result was a planet with a structure that is layered like an onion, with the densest substances at the center and the least dense ones at the surface. Over time, Earth has cooled. Earth's outer surface has solidified to a depth of about 100 kilometers,

Figure 11.1
Earth as seen from space.

forming a rigid shell of rock called the *lithosphere* (meaning "rock sphere").

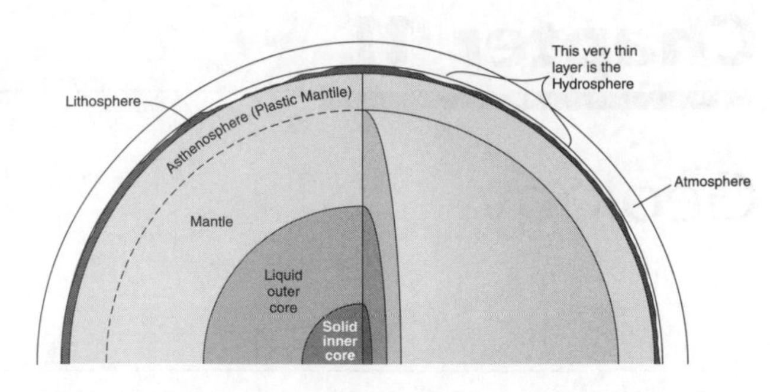

Figure 11.2
The atmosphere, hydrosphere, and lithosphere.

THE HYDROSPHERE

As Earth cooled, water vapor released from its interior condensed into liquid water that came to rest on the lithosphere, forming a thin layer of water called the ***hydrosphere***. The hydrosphere includes all Earth's oceans, lakes, streams, underground water, and ice. A tiny fraction exists in the atmosphere as water vapor. The hydrosphere covers about 71% of the lithosphere's surface to an average depth of about 3.8 kilometers. This is very thin compared with the diameter of Earth. If you dipped a basketball into water, the water wetting its surface would be deeper in many places than the hydrosphere is on Earth. Since the hydrosphere is water, it consists of the elements hydrogen and oxygen.

THE ATMOSPHERE

In addition to water vapor, many other gases have escaped from Earth's fiery interior or have been brought in by comets that hit Earth. Over time, these gases have built up to form a thin layer of gases bound to Earth by gravity. This layer is the ***atmosphere***.

The atmosphere is mostly gases, but it also contains water droplets, ice, dust, and other particles—a mixture called *air*. On average, dry air is about 78% nitrogen and 21% oxygen. The remaining 1% consists mostly of argon, with traces of carbon dioxide and a number of other gases. Water vapor is also present but varies from 0% over deserts and polar ice caps to as much as 4% in tropical jungles.

The atmosphere itself is stratified into layers that differ in composition and temperature. The densest, lowest layer contains 90% of the air and all of the water vapor in the atmosphere. Nearly all weather occurs in this lowest layer of the atmosphere.

EARTH'S COMPOSITION

The solid part of Earth is made of rocks. Rocks, in turn, are made of minerals, and most minerals are made of chemical elements. The properties of a mineral depend on how it formed and the elements of which it is made. Most properties of minerals can be explained in terms of the arrangement and properties of their atoms.

MINERALS

Minerals are naturally occurring, inorganic, crystalline substances with a fixed chemical composition. *Naturally occurring* means that minerals form as a result of natural processes in or on Earth. *Inorganic* means that minerals do not come from living things. *Fixed chemical composition* means that a chemical symbol or formula can be written for a mineral. Thus, minerals are either elements or compounds. *Crystalline* means that minerals are solids in which atoms or molecules are joined in fixed positions and form a definite pattern.

MINERAL IDENTIFICATION

Since minerals are elements or compounds, they have definite chemical and physical properties. Several easily observed properties commonly used to identify minerals are color, luster, streak, hardness, and cleavage.

Color is often the first property you notice about a mineral. However, color alone is an unreliable property by which to identify a mineral. Many minerals have almost the same color, while some minerals come in a variety of colors. Weathering and impurities can also change a mineral's color.

The ***luster*** of a mineral is the way light reflects from its surface. Luster can be metallic or nonmetallic. Nonmetallic lusters can be described as glassy, brilliant, greasy or oily, waxy, silky, pearly, or earthy.

The ***streak*** of a mineral is the color of its powder. To find its streak, a mineral is rubbed onto a piece of unglazed porcelain. A mineral's streak varies less than the color of a solid sample of the mineral, making the streak useful in identification.

Hardness is a mineral's resistance to being scratched. Harder minerals scratch softer minerals, and two minerals of the same hardness can sctrach each other. The hardness of a mineral is the same as the softest mineral that can scratch it. The hardness of minerals is usually stated in terms of the Mohs' scale. On this scale, ten common minerals are arranged in order from the softest to the hardest. See Table 11.1. For example, quartz is the softest mineral that will scratch tourmaline, so tourmaline's hardness is 7. Hardness is often tested by trying to scratch an unknown mineral with an object of known hardness, such as a fingernail, copper penny, or piece of glass.

Table 11.1 Mohs' Hardness Scale of Minerals

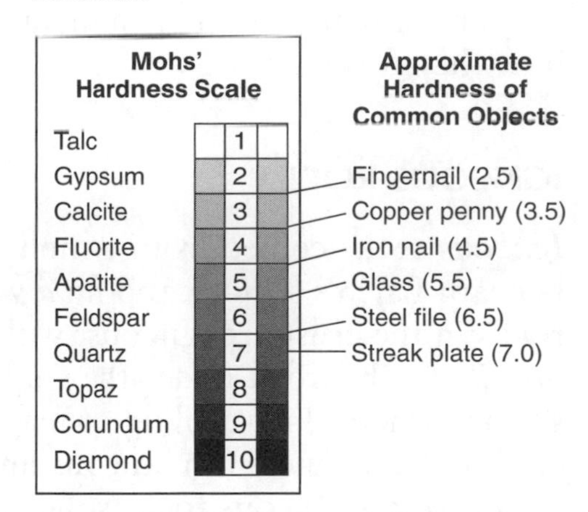

Mohs' Hardness Scale		Approximate Hardness of Common Objects
Talc	1	
Gypsum	2	Fingernail (2.5)
Calcite	3	Copper penny (3.5)
Fluorite	4	Iron nail (4.5)
Apatite	5	Glass (5.5)
Feldspar	6	Steel file (6.5)
Quartz	7	Streak plate (7.0)
Topaz	8	
Corundum	9	
Diamond	10	

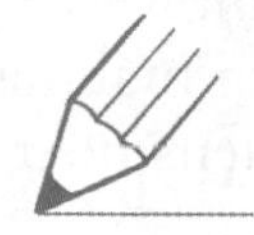

TRY THIS

1. According to Mohs' Hardness Scale, which of the following minerals can scratch the other three? (calcite, topaz, talc, quartz)
2. State one reason why color is often less useful than cleavage or hardness in mineral identification.

Cleavage is the tendency of a mineral to split along smooth surfaces that follow planes of weakness in the crystalline structure. Cleavage surfaces often occur at very specific angles to one another and can be helpful in identifying a mineral. See Figure 11.3.

Since no single property can be used to identify all minerals, mineral identification is usually a process of elimination. As each property is observed, one discovers what a mineral is not, rather than what it is. Step-by-step, the possibilities are narrowed down until the identity of the mineral is determined.

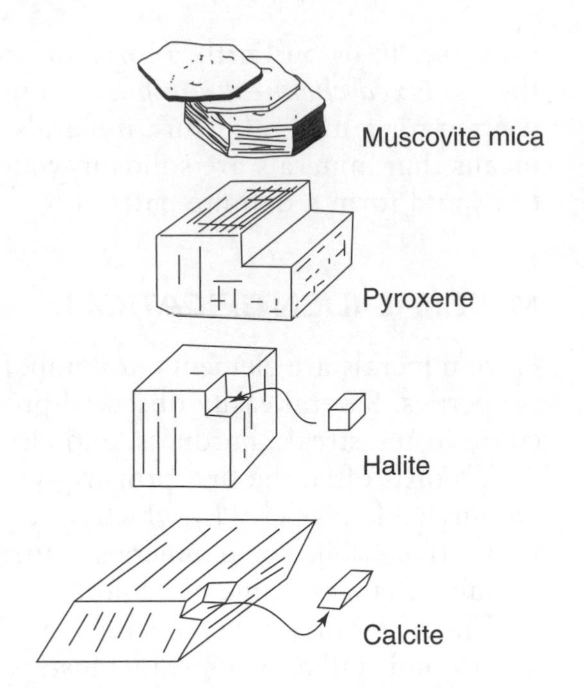

Figure 11.3
Some common minerals that display cleavage.

ROCKS

A ***rock*** is a naturally formed, nonliving Earth material that holds together in a firm, solid mass. Most rocks are aggregates, that is, they are made of lots of grains stuck together. The grains may be mineral crystals, tiny bits of broken rock, or even solid parts of once-living things. The grains may be small or large, tightly or loosely bound together, made of one mineral or a mixture of many different minerals. Rocks are grouped into three families based upon how they formed.

IGNEOUS ROCKS

Igneous rocks cooled from a molten state. Molten rock is called ***magma*** when it is beneath Earth's surface and ***lava*** when it pours out onto the surface. Two-thirds of the rocks in the crust are igneous, with *basalt* making up the bedrock under the oceans and *granite* the bedrock under the continents. ***Intrusive*** igneous rocks solidified beneath the surface where slow cooling of magma resulted in large crystals. Pools of magma that cool underground form igneous intrusions of many shapes and sizes.

Extrusive igneous rocks solidify after pouring out onto the surface from a volcano or other vent, and rapid cooling of the lava results in small crystals. The most common extrusive rock is *basalt*. A ***volcano*** is both the opening in the crust through which magma erupts and the mountain built by the erupted material. Since lava cools and hardens near the opening, a cone-shaped mound builds up around the opening.

Two properties useful in identifying rocks are texture and mineral composition. ***Texture*** is the size, shape, and arrangement of the mineral crystals or grains in a rock. ***Mineral composition*** is simply the minerals that comprise the rock. Some common igneous rocks are classified by their composition and texture in Figure 11.4.

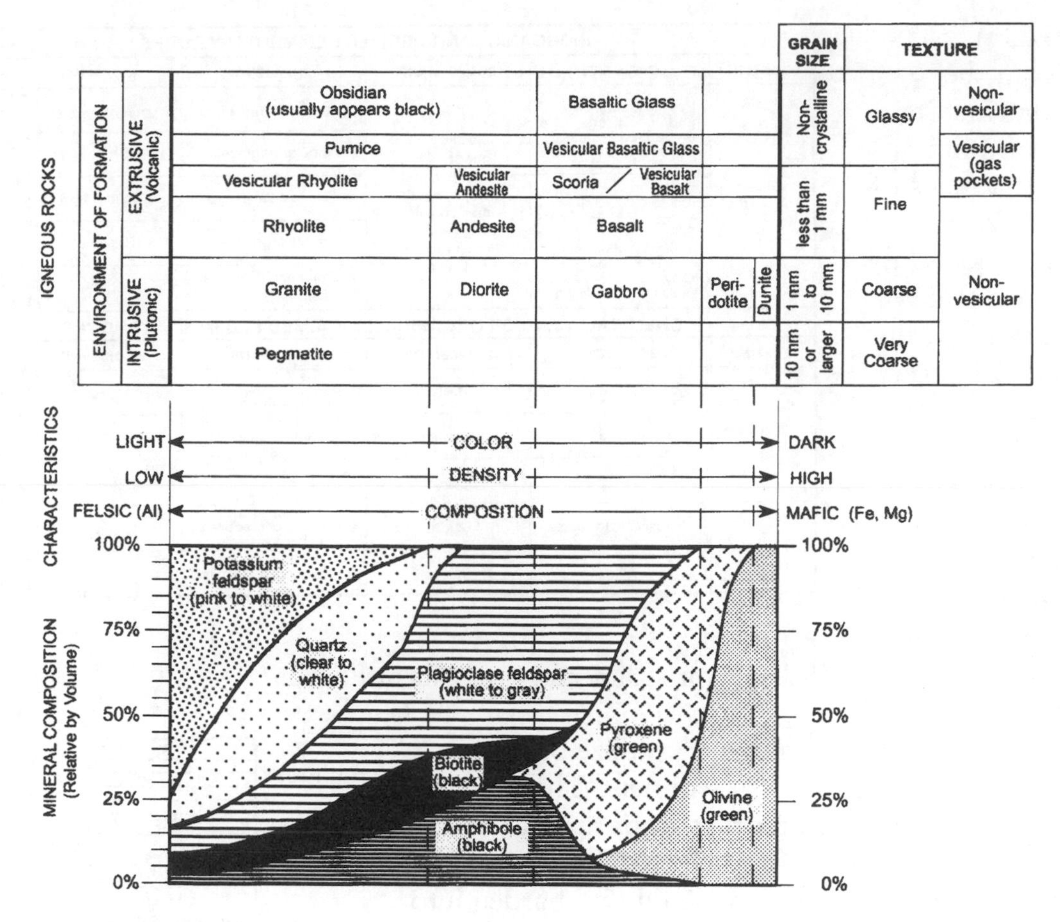

Figure 11.4
Scheme for igneous rock identification.

TRY THIS

1. In one or more complete sentences, explain the difference between magma and lava.
2. In one or more complete sentences, explain why the rock that forms from igneous intrusions usually has large mineral crystals.

SEDIMENTARY ROCKS

Most ***sedimentary*** rocks are made of fragments of other rocks that have been eroded by running water, wind, or glaciers. Shale, sandstone, and conglomerate are examples of rocks made of cemented-together fragments, or ***clastic*** rocks. Other sedimentary rocks are made of materials that precipitated from a solution, were left behind when water evaporated, or formed from the shells of marine life. Rock salt and limestone are examples of such *chemical* or *bioclastic* rocks. Only 8% of the crust is sedimentary rock, but these make up about three-quarters of all surface rocks.

Lithification is the hardening of layers of loose sediment into rock. Coarse sediments harden mainly by ***cementation***, the binding together of sediment particles by substances that crystallize or fill in spaces between loose particles of sediment. The cement crystallizes out of groundwater that seeps through the sediments. The crystals that form between the grains of sediment hold them together. Fine sediments harden into rock mainly by ***compaction***, the squeezing together of sediments due to the sheer weight of layers on top of them. Slow chemical reactions in compacted sediment can form new minerals, which also help harden them into rock. Some of the more common sedimentary rocks are classified by their origin in Figure 11.5.

INORGANIC LAND-DERIVED SEDIMENTARY ROCKS					
TEXTURE	GRAIN SIZE	COMPOSITION	COMMENTS	ROCK NAME	MAP SYMBOL
Clastic (fragmental)	Pebbles, cobbles, and/or boulders embedded in sand, silt, and/or clay	Mostly quartz, feldspar, and clay minerals; may contain fragments of other rocks and minerals	Rounded fragments	Conglomerate	
			Angular fragments	Breccia	
	Sand (0.2 to 0.006 cm)		Fine to coarse	Sandstone	
	Silt (0.006 to 0.0004 cm)		Very fine grain	Siltstone	
	Clay (less than 0.0004 cm)		Compact; may split easily	Shale	
CHEMICALLY AND/OR ORGANICALLY FORMED SEDIMENTARY ROCKS					
TEXTURE	GRAIN SIZE	COMPOSITION	COMMENTS	ROCK NAME	MAP SYMBOL
Crystalline	Varied	Halite	Crystals from chemical precipitates and evaporites	Rock Salt	
	Varied	Gypsum		Rock Gypsum	
	Varied	Dolomite		Dolostone	
Bioclastic	Microscopic to coarse	Calcite	Cemented shell fragments or precipitates of biologic origin	Limestone	
	Varied	Carbon	From plant remains	Coal	

Figure 11.5
Scheme for sedimentary rock identification.

Since 70% of Earth's surface is covered with water, most sediment is deposited into water. When sediments settle out of water, they form horizontal layers. These layers are preserved when the sediments change into solid rock. Each layer forms atop the one that is already there. Therefore, each layer is younger than the one under it and older than the one on top of it, a principle called ***superposition***. The relative age of any two layers can be determined based upon which layer is on top and which is on the bottom. See Figure 11.6.

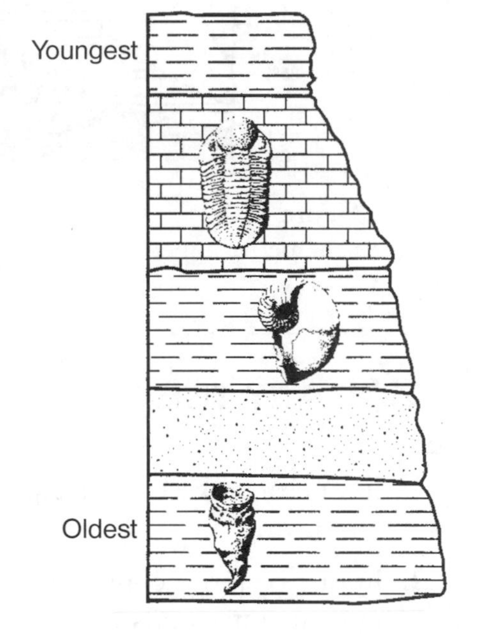

Figure 11.6
The Principle of Superposition. The bottom layer of a series of sedimentary layers is the oldest unless it has been overturned or had older rock thrust over it. Fossils are the same age as the rock layers in which they are found. Therefore, fossils found in the bottom layer are the oldest and those at the top are the youngest.

Sedimentary rocks also often contain traces of their watery origin, such as *ripple marks* created by waves in shallow water or *mud cracks* that formed when fine sediments dried out in the Sun. Sedimentary rocks may also contain ***fossils***, or remains of once living things. Plants and animals often live in or near water in which sediments are being deposited. When the remains or traces of living things get buried in layers of sediment, they may be preserved as fossils. A fossil may be an entire organism or parts of an organism, such as bones, teeth, and shells. Traces of living things, such as leaf *imprints*, the *footprints* of walking animals, or the *tracks and trails* of crawling animals such as snails, are also fossils.

Fossils are useful in piecing together Earth's history. The record of life preserved as fossils mirrors the evolution of life in response to changes in the environment. Fossils can be arranged from youngest to oldest based on the age of the rock layers in which they are found. The type of fossil in each layer tells us something about the local environment at the time the organism was alive. For example, it can tell us if the region was dry land or covered by an ocean. It can also tell us if the weather was warm or cold. Fossil coral, for example, indicate a region covered by water in a warm climate; fossil mammoth teeth indicate a region of dry land with a cold climate.

METAMORPHIC ROCKS

Metamorphic rocks formed from igneous, sedimentary, or other metamorphic rocks under the influence of *heat*, *pressure*, or *chemical action*. Metamorphic rocks get their name from the Greek words *meta*, meaning "change," and *morph*, meaning "form."

Sometimes the changes occur in the minerals that make up the rock. Heat and pressure change the environment of the rock. In response, the minerals rearrange their structure and form new minerals that are more stable in the new environment. When rocks recrystallize under great pressure, minerals can grow at right angles to the stress. This forms parallel layers of crystals, or ***foliation***. This is what happens when shale becomes slate, or schist. The minerals may also separate by density into layers of dark and light minerals, or become ***banded***. See Figure 11.7. This is what happens when granite becomes gneiss. Minerals can also be changed when hot, watery fluids rich in dissolved chemicals move through the rock. As these fluids move through openings in a rock, their chemicals react with minerals in the rock to form new minerals.

Sometimes the changes are to the characteristics of the grains in a rock. When softened by heat and squeezed under pressure, the grains may change in size or shape. They may become more compacted or even fuse together, forming a denser, harder rock. This is what happens when limestone is metamorphosed into marble or sandstone into quartzite.

You might ask, "When has a rock changed enough to be called metamorphic?" Generally, if a change can be recognized in a rock, it is considered metamorphic. Figure 11.8 lists some common rocks and the rocks they can become when metamorphosed.

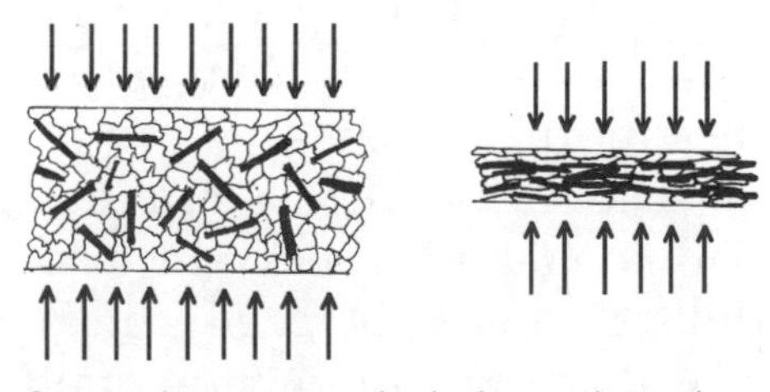

Compression causes randomly dispersed crystals to become oriented in a direction perpendicular to the pressure. The oriented crystals form thin sheets or layers called foliation.

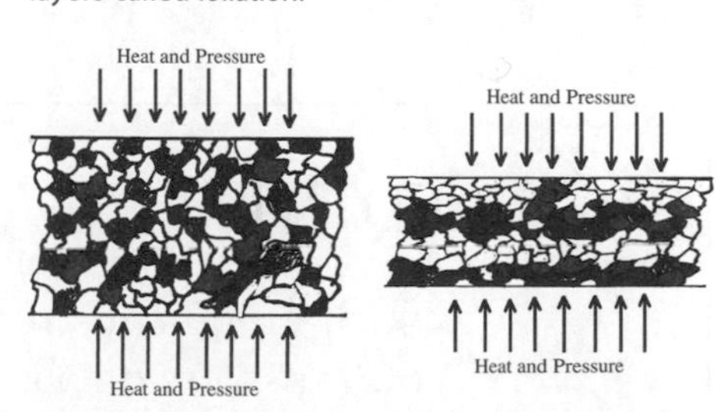

Migration of mineral crystals due to density differences produces banding—regions of light- and dark-colored minerals.

Figure 11.7
Foliation and banding.

THE ROCK CYCLE

TEXTURE		GRAIN SIZE	COMPOSITION	TYPE OF METAMORPHISM	COMMENTS	ROCK NAME	MAP SYMBOL
FOLIATED	MINERAL ALIGNMENT	Fine	MICA, QUARTZ, FELDSPAR, AMPHIBOLE, GARNET, PYROXENE	Regional	Low-grade metamorphism of shale	Slate	
		Fine to medium		(Heat and pressure increase with depth)	Foliation surfaces shiny from microscopic mica crystals	Phyllite	
					Platy mica crystals visible from metamorphism of clay or feldspars	Schist	
	BAND-ING	Medium to Coarse			High-grade metamorphism; some mica changed to feldspar; segregated by mineral type into bands	Gneiss	
NONFOLIATED		Fine	Variable	Contact (Heat)	Various rocks changed by heat from nearby magma/lava	Hornfels	
		Fine to coarse	Quartz	Regional or Contact	Metamorphism of quartz sandstone	Quartzite	
			Calcite and/or dolomite		Metamorphism of limestone or dolomite	Marble	
		Coarse	Various minerals in particles and matrix		Pebbles may be distorted or stretched	Metaconglomerate	

Figure 11.8
Scheme for identifying metamorphic rocks.

Where do rocks come from? You might say igneous rock comes from magma, sedimentary rock comes from sediments, and metamorphic rock comes from other rocks that have changed. Consider, though, that magma is melted rock, sediments are broken rock, and there had to be an already existing rock to be metamorphosed. So, basically, rocks come from other rocks.

Rocks are constantly changing from one type to another in a never-ending cycle called the ***rock cycle***. Figure 11.9 shows the rock cycle. The outside circle shows the different forms in which rock matter can exist: magma, igneous rock, sediment, and so on. Arrows leading from one to another are labeled to show the change taking place. Arrows inside the circles show some alternate paths in the rock cycle. For example, an igneous rock does not always break down into sediments. If buried, it may be metamorphosed. So an arrow inside a circle leads from igneous rock to metamorphic rock.

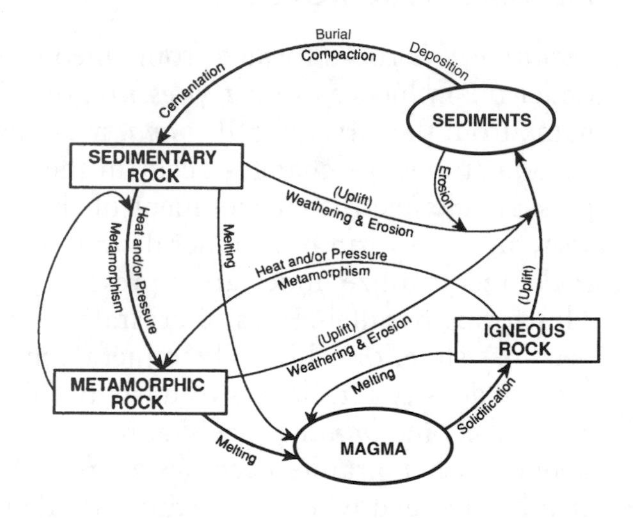

Figure 11.9
The rock cycle in Earth's crust.

The rock cycle diagram has no beginning and no end. Of course, the rock cycle really did start somewhere. Much evidence indicates that Earth was originally totally molten. No solid rock existed, only magma. Thus, it is thought that the original rocks formed as this magma cooled, and the rock cycle began with igneous rocks.

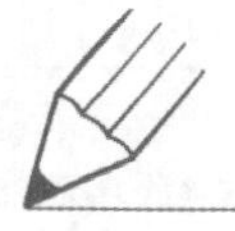

TRY THIS

According to the rock cycle diagram in Figure 11.9, what are two ways in which an igneous rock could be formed from a sedimentary rock? Model your answers on the sample below, which shows one way an igneous rock could become a metamorphic rock.

Sample: Igneous rock → Heat and/or Pressure → Metamorphic Rock

THE WATER CYCLE

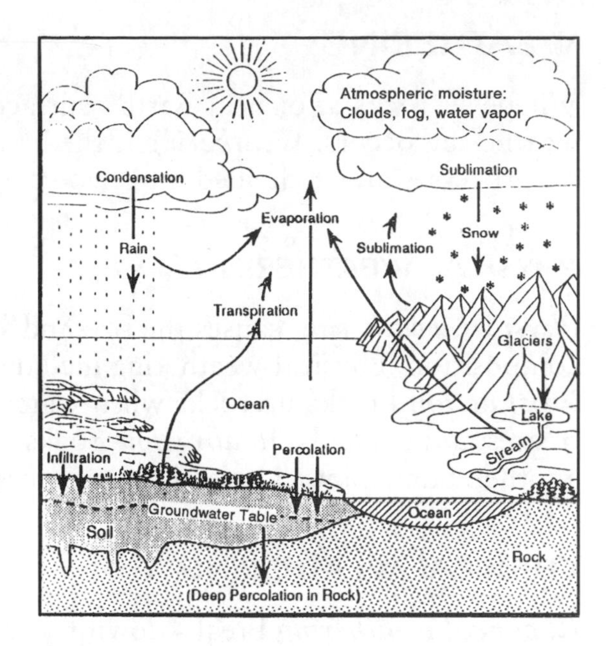

Figure 11.10
The water cycle.

Water is one of the most abundant substances on Earth, covering nearly three-quarters of its surface. Water moves between hydrosphere, atmosphere, and lithosphere in a process called the ***water cycle***, as shown in Figure 11.10.

Each day, sunlight striking the oceans provides energy to water molecules at the surface. The molecules heat up, and trillions of tons of water change from a liquid to a gas, or ***evaporate***. The resulting water vapor enters the atmosphere. In the atmosphere, the water vapor is circulated by winds and carried upward by rising warm air.

When air is cooled to its dew point, the water vapor changes from a gas back to a liquid, or ***condenses***, forming tiny water droplets. These tiny water droplets suspended in air form clouds. When the water droplets become too large to remain suspended, they fall back to the surface as ***precipitation***. Precipitation may fall onto the oceans or the land. Water that falls onto the oceans has completed its cycle and may evaporate back into the atmosphere, beginning the entire process anew.

A number of things may happen to precipitation that falls onto land. It may become ***runoff***, water that flows downhill into rivers and streams that eventually carry it back into the ocean or another body of water. It may seep into the soil to become ***groundwater*** and move slowly back to the oceans through the ground. However, most of it seeps into rivers and streams that carry it back to the oceans.

Plants play a role in the water cycle as well. Water taken in by plant roots is transported to the leaves and released back into the atmosphere by a process called ***transpiration***. In a forest, transpiration returns more than one-third of precipitation back to the atmosphere as water vapor. Some of the water may evaporate again to be carried further over land or be blown back over the oceans.

Once the water is back in the oceans, the cycle begins all over again. Many variations occur in the water cycle. Sometimes the water that evaporates over the oceans condenses and falls directly back into them as rain. Water falling onto land may evaporate almost immediately. In some cases, precipitation evaporates before it ever reaches the ground. Other variations are possible, but all are part of the endless water cycle that receives its energy from the Sun.

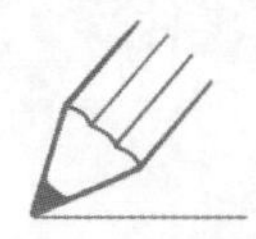

TRY THIS

You are a molecule of water in the ocean. In a few brief paragraphs, describe your adventures as part of the water cycle before eventually returning to the ocean.

WEATHERING

Whenever rocks at or near Earth's surface interact with the air, water, and living things, weathering occurs. ***Weathering*** is the breakdown of rocks into smaller particles by natural processes and is divided into two general types: physical and chemical.

PHYSICAL WEATHERING

Physical weathering causes the size and shape of rocks to change but not their chemical composition. Physical weathering includes processes such as frost action and abrasion. ***Frost action*** breaks up rocks when water that seeps into cracks and crevices in rocks freezes and expands. In ***abrasion***, rocks break down as a result of rubbing against one another. Some physical weathering processes are shown in Figure 11.11.

CHEMICAL WEATHERING

Chemical weathering breaks down rocks by changing their chemical composition. This changes the properties of the minerals in the rocks and almost always weakens their structure. As a result, the rocks either fall apart or are more easily broken down by physical weathering. Chemical weathering includes the combining of minerals with oxygen and water and the dissolving of minerals by acids.

Surface water is slightly acidic due to both carbon dioxide dissolving in water to form carbonic acid and the presence of acids from decaying plants and animals. The weakly acid water attacks many common rock-forming minerals and completely dissolves the calcite in limestone. Groundwater seeping through limestone bedrock can

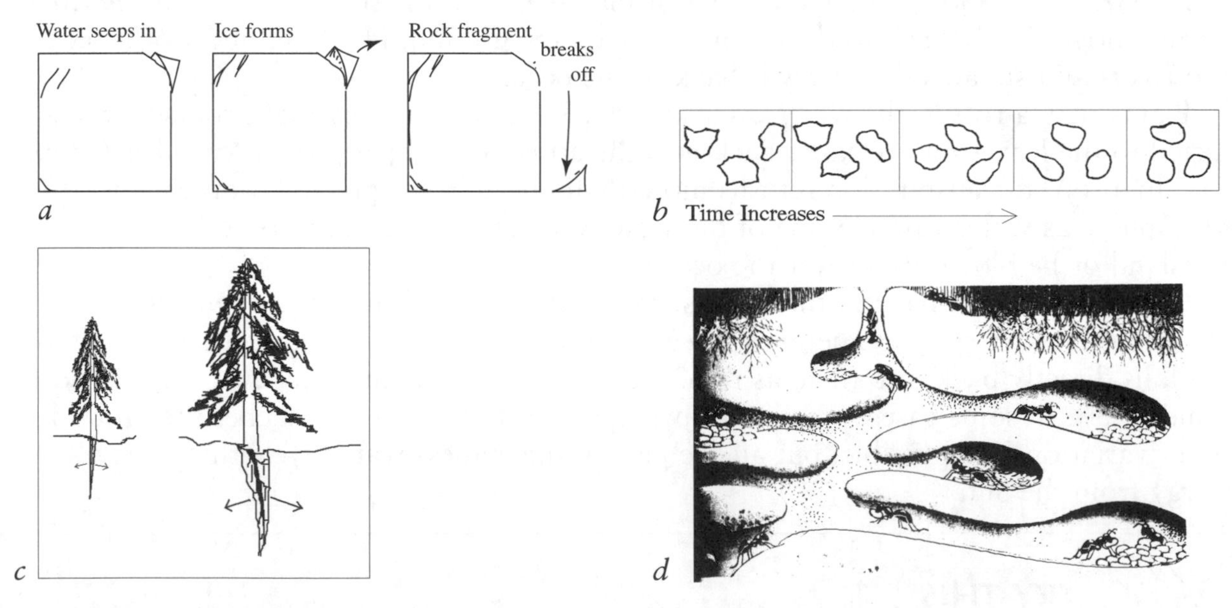

Figure 11.11
Physical weathering processes. (*a*) Frost action—water expands as it freezes in cracks. (*b*) Abrasion—particles rub against one another and become rounded. (*c*) Plant action—plant roots grow in rock crevices. (*d*) Burrowing animals—animals expose rocks and allow air and water to penetrate surfaces.

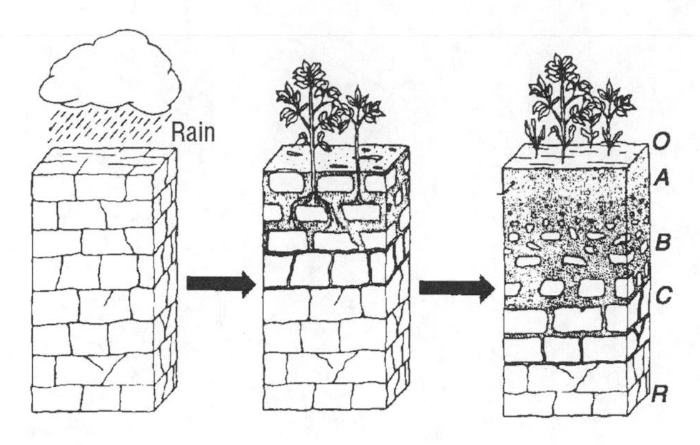

Horizon O—Organic. Leaf litter, twigs, roots, and other organic material lying on the surface of the soil.

Horizon A—Topsoil. Loose and crumbly with varying amounts of organic matter; generally the most productive layer of soil.

Horizon B— Subsoil. Usually light colored, dense, and low in organic matter.

Horizon C—Parent Material. The partly weathered bedrock or transported sediments from which soil forms.

Horizon R— Bedrock. The solid rock that underlies the soil.

Figure 11.12
Development of a mature soil.

dissolve huge holes in the bedrock, forming spectacular caverns. New York State's Howe Caverns were formed in this way.

SOIL: THE END PRODUCT OF WEATHERING

The end result of physical and chemical weathering is the same—solid rock is broken into fragments. The fragments or particles of rock produced by weathering are called ***sediments.*** Over time, weathering and plant growth change exposed rock and sediments into ***soil***—a mixture of weathered rock, water, air, bacteria, and decayed plant and animal material (humus).

When soil forms directly from the weathering of the underlying bedrock, it is called a *residual soil.* In contrast, a *transported soil* is deposited by agents of erosion, such as ice and water, and does not come from the underlying bedrock. Erosion during the last ice age affected much of New York State. So the state has mostly transported soil.

As soil forms, the processes of weathering and plant growth cause recognizable layers, or *soil horizons*, to form in the soil. A soil that has been forming long enough to have developed distinct horizons is called a mature soil. The last cross section in Figure 11.12 shows the horizons in a mature soil from surface to bedrock with a description of each horizon.

EROSION

Any natural process that moves sediments from place to place is called ***erosion.*** In order to move anything, a force is needed. The driving force of erosion is gravity. Gravity can move sediments by acting on them *directly*, as when a rock breaks loose and falls from a cliff. Gravity can also move sediments by acting on them indirectly, with ***agents of erosion.*** For example, water runs downhill under the direct influence of gravity. The running water, in turn, can exert a force onto sediments in its path, causing them to move. The running water is an agent of erosion. Some other agents of erosion are winds, glaciers, waves, and currents. Each of these agents of erosion produces changes in the material that it transports and creates surface features and landscapes.

GRAVITY

Mass wasting is the downhill movement of sediments under the direct influence of gravity. The rate at which it occurs depends on the steepness of the slope. On steep slopes, rock may fall, bounce, or slide downhill. On gentle slopes, sediments may ooze or creep slowly downhill. Figure 11.13 shows some examples of mass-wasting processes.

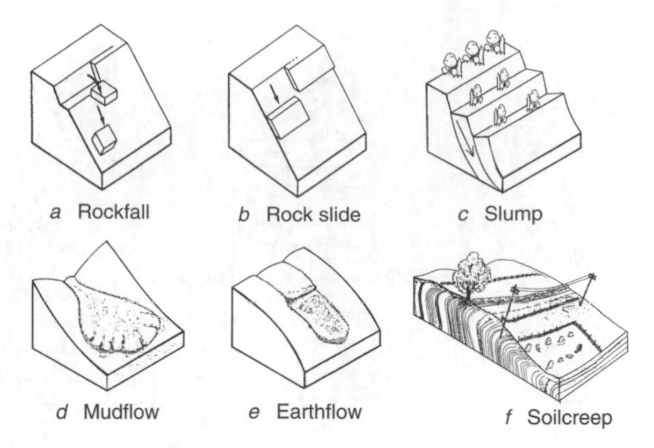

Figure 11.13
Mass-wasting processes.

MOVING WATER

Moving water is responsible for more erosion than all other agents of erosion combined. Water that falls to Earth as precipitation and does not seep into the ground or evaporate becomes runoff. ***Runoff*** is water that flows downhill under the influence of gravity. As runoff flows downhill, it drags sediments along with it. Most of these sediments end up in streams. ***Streams*** are bodies of water that flow downhill in a relatively narrow but clearly defined channel. See Figure 11.14*a*.

Water flowing downhill in streams can be very powerful. Have you ever seen white-water rafters shooting a rapid? The water flowing downhill exerts quite a force! Any loose sediment in the path of such streams is likely to be carried away. Streams transport different sediments in different ways. See Figure 11.14*b*. The amount and type of sediments that a stream can carry depend on its speed and volume. Water can carry more and larger sediments as its speed increases. At any speed, the more water there is in a stream, the more sediments it can carry.

Water carrying sediments is like a cutting tool. Sediments carried in a stream collide with and knock chips off each other and the streambed. They crush and grind up smaller particles between them. The process of crushing, grinding, and wearing away rock due to the impact of sediments is called ***abrasion***. Abrasion wears away a stream's channel and causes the sediments carried by a stream to become smaller and rounded. Stream abrasion can cut through solid bedrock. Prolonged erosion by streams can carve deep canyons and valleys into Earth's surface. The Grand Canyon is an amazing example of what sediment-laden water can do.

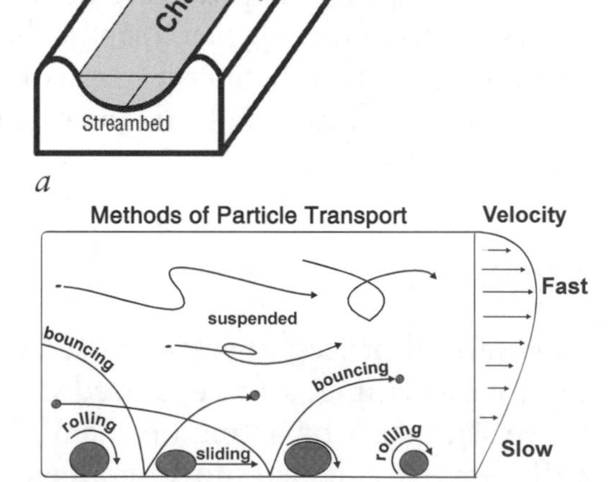

Figure 11.14
(*a*) A stream channel. (*b*) Transport in a stream.

More often, though, erosion by runoff and streams lowers the surface of the surrounding land. Steep slopes are worn back and become more rounded and less steep. Eventually, the surrounding landscape is worn down to a flat plain.

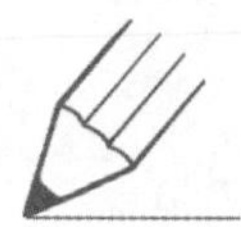

TRY THIS

1. State three ways in which streams transport sediments.
2. What happens to the size and shape of sediment particles as they are carried in a stream.

WAVES AND CURRENTS

Ocean water is constantly in motion. Waves and currents are examples of the movement of ocean water. Moving ocean water can transport sediments and erode the land. ***Waves*** can be very powerful. A single wave sends tons of water crashing against a shore. The force of waves can break up rock. Loose fragments are stirred up and carried along in the turbulence of breaking waves.

Currents are movements of water. When waves strike a shore at an angle, they are reflected and interfere with incoming waves, forming a ***longshore current*** parallel to the shore. Longshore currents can move sediments along just like a stream. Along some coasts, tides flowing into and out of narrow openings form swift currents. These, too, can move large amounts of sediments.

WIND

Wind that blows fast enough can pick up and carry loose particles of rock. Wind is an important agent of erosion mainly in arid regions where there is little plant cover and the soil, loose sediments, and weathered bedrock are dry and exposed. Wind erodes the land in two ways: deflation and abrasion. In *deflation*, the wind picks up and carries away loose particles much as a stream carries sediments. In *abrasion*, exposed rock is worn away by wind-driven particles. These worn away rock particles are then carried away by the wind.

GLACIERS

Glaciers are large masses of ice that form where snowfall exceeds melting for extended periods of time. As snow piles up, its sheer weight causes the snow to recrystallize into ice. Ice under pressure behaves like a fluid. Under the influence of gravity, the ice in a glacier will flow slowly downhill and outward.

As a glacier moves outward or downhill, it eventually reaches warmer regions. The ice begins to melt. If the ice flows outward faster than it melts back, the glacier's leading edge advances. If it melts faster than it flows outward, the leading edge retreats. As climates have changed, glaciers have advanced and retreated over much of Earth's surface. Glaciers shaped many of the landscapes in New York State. They carved out the Finger Lakes, the U-shaped Hudson River Valley, and the beautiful lakes and valleys of the Adirondacks.

TRY THIS

In one or more complete sentences, describe what is happening when a glacier advances and when a glacier retreats.

DEPOSITION

Deposition is the process by which sediment is dropped into a new place when an agent of erosion slows down and can no longer carry it. Imagine you are flying a kite and the wind slows down. The slower the wind blows, the less carrying power it has. The kite will begin to settle to the ground. Deposition occurs in much the same way. As the agent of erosion slows, sediment settles to the ground, or is deposited. Each agent of erosion forms distinctive landforms when it deposits sediments.

MASS WASTING

Sediment is deposited by mass-wasting processes when it stops moving at the base of a slope. Most deposits resulting from mass wasting are an unsorted jumble of rock fragments and do not show any distinct layering. Refer back to Figure 11.13.

STREAMS

Streams deposit sediment when they slow down or decrease in volume. Sediments settle through water at different rates. Large, dense, rounded sediments settle faster, while small, light, flat particles settle slower. Therefore, deposition by streams results in layers that are *sorted* according to particle size, shape, and density. See Figure 11.15.

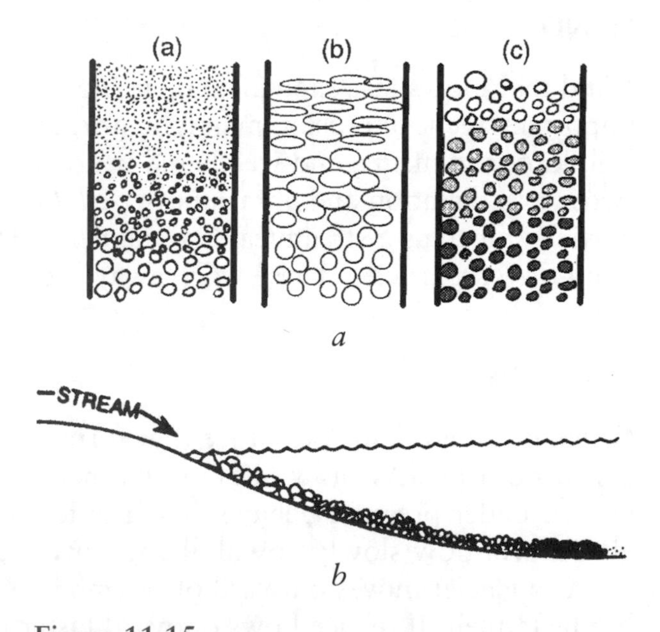

Figure 11.15
Vertical and horizontal sorting. (*a*) Larger, rounder, and denser particles settle faster. (*b*) Large particles are dropped first as a stream slows down.

When a stream flows into a quiet body of water such as an inland sea, ocean, or lake, it stops moving. Most of the stream's sediment is deposited. The result is a ***delta,*** a large, flat, fan-shaped pile of sediment at the mouth of a stream.

Whenever waves and currents slow down, they deposit their load. ***Beaches*** form as waves wash up against the land, slow down, and deposit sediment along the shoreline. ***Sandbars*** are long, narrow piles of sand deposited in open water. Sandbars may form where waves wash beach sediments into deeper, quieter waters. They may also form where the shoreline curves away from a longshore current. As the current curves away from the shoreline, it flows into deeper, quieter water and deposits its sediments.

WIND

When a wind slows or stops moving, the particles it is carrying settle to the ground and are deposited. ***Dunes*** are mounds of sand deposited by wind. They often form when windblown sand meets an obstacle such as a rock or a bush. Sand hits the obstacle, falls to the ground, and builds up in a pile. In time, the sand forms a mound that slopes gently up toward the tip of the obstacle. Wind then pushes sand grains up this slope and over the top of the dune. Once over the top, the sand is deposited on the other side. The dune grows in size. As wind continues to blow sand up one side of the dune and deposit it on the other, the entire dune moves.

GLACIERS

When a glacier melts, sediments frozen in or on the ice are released. Rock fragments of all shapes and sizes drop to the ground in a confused jumble, forming an ***unsorted*** deposit called ***till***. As the ice melts back, the jumbled piles of sediment that were pushed up along its edges are also left behind. Some glacial sediment is deposited into streams of meltwater. These sediments are then sorted by the running water and deposited in layers. Fine sediments carried into lakes by meltwater may settle out in sorted layers on the lake bed. The water and piles of sediment left behind by melting glaciers form a number of distinctive landforms such as moraines, drumlins, outwash plains, and glacial lakes. See Figure 11.16.

TRY THIS
In one or more complete sentences, explain two ways in which a delta and an alluvial fan are similar.

EARTH'S INTERIOR

With current technology, we cannot drill more than a few kilometers into the solid Earth. From this, we know that temperatures rise inside the Earth. How, though, can we know anything of Earth's internal structure? Earthquakes provide the key. Earthquake waves, or seismic waves, literally travel around and through Earth. By analyzing records of many earthquakes and earthquake wave behavior, scientists have been able to piece together a picture of Earth's internal structure.

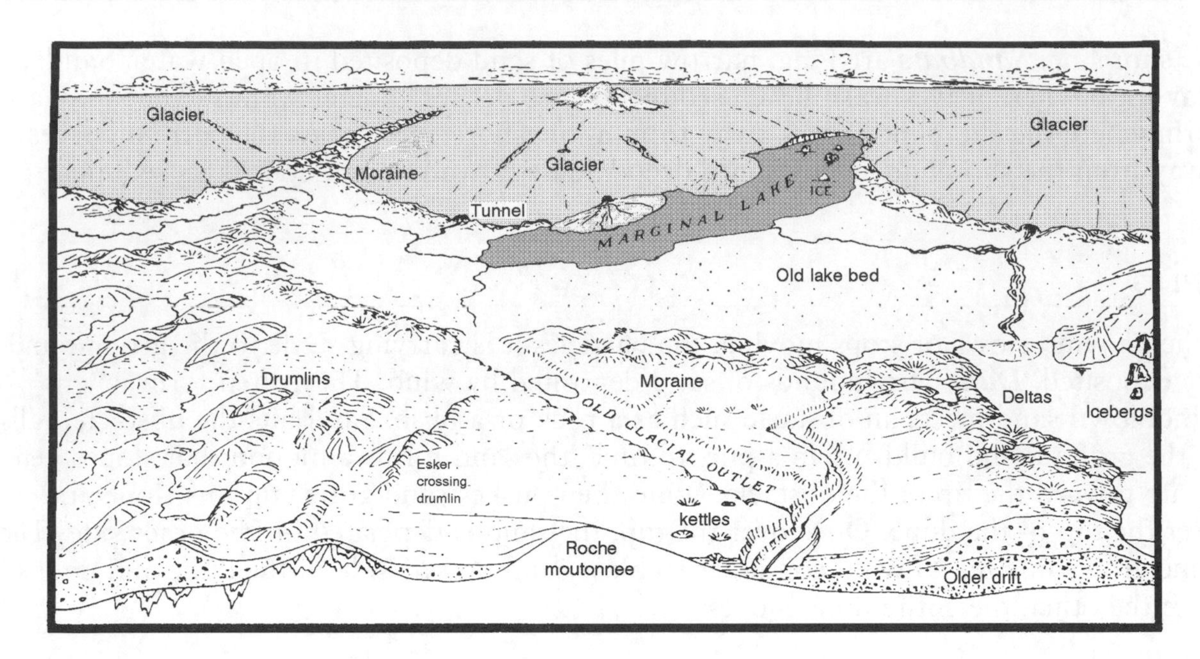

Figure 11.16
Landforms resulting from glacial deposits.

INTERNAL HEAT

Beneath its cool, solid lithosphere, Earth is glowing hot. At the surface, we find much evidence of Earth's internal heat: molten rock pours from volcanoes, and boiling hot water and steam rise from hot springs and jet out of geysers. Further evidence comes from mines and deep wells. As we drill deeper into the crust, the temperature of the rock steadily increases—about 30°C for every 1 kilometer of depth.

Considering that Earth's center is more than 6,000 kilometers down, it is not unlikely that temperatures there exceed 6,000°C. Some of Earth's internal heat is left over from when it formed. Some is produced by the decay of radioactive minerals. Although Earth's interior is very hot, it is not entirely molten. Pressure influences melting temperature by holding crystals together. Although temperatures within Earth are great enough to melt rock, the great pressures there prevent the rock from turning into a liquid.

EARTHQUAKES AND SEISMIC WAVES

The majority of earthquakes are caused by sudden movements of blocks of crust along a fault, which relieve stresses that have built up over a period of time. When stressed, rock can bend, much as a wooden ruler can bend. Eventually, though, a point is reached where the rock can bend no further without breaking. The rock snaps, and great masses of rock suddenly scrape past each other. The shock of this wrenching action jars the crust and sets an earthquake in motion. The point where the rock breaks is called the *focus* of an earthquake. Vibrations called earthquake waves, or ***seismic waves,*** spread out from the focus, traveling through the surrounding rock and to Earth's surface. The

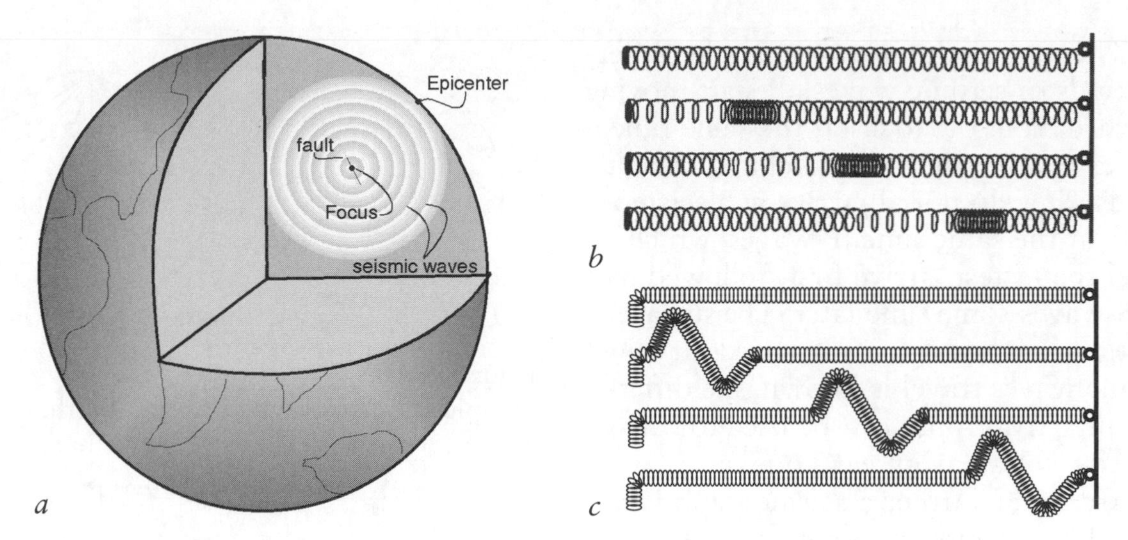

Figure 11.17
(*a*) The focus and epicenter of an earthquake. (*b*) P-waves. (*c*) S-waves.

earthquake is first felt at the ***epicenter***, a point on Earth's surface directly above the focus. See Figure 11.17*a*.

Earthquakes give rise to several types of seismic waves. ***Primary*** (or P) waves vibrate back and forth in the same direction as the wave travels. P-waves are like the waves that travel down a coiled spring when one end is pulled out. See Figure 11.17*b*. P-waves can travel through solids, liquids, and gases because all three can be compressed and expanded. ***Secondary*** (or S) waves move back and forth at right angles to the direction the wave travels. S-waves are like those that form when you shake the end of a string. See Figure 11.17*c*. S-waves can travel only through solids. This is because liquids and gases do not return to their original shape after being moved by an S-wave. ***Surface*** (or L) waves are like water waves, in which matter moves in circles. They occur only at Earth's surface.

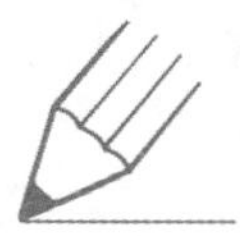

TRY THIS

1. In one or more complete sentences, explain the difference between the focus of an earthquake and the epicenter of an earthquake.
2. Compare and contrast the ways in which P-waves and S-waves vibrate and the speeds at which they move through Earth.

When an earthquake occurs, the different kinds of seismic waves all start moving outward from the focus at the same time. However, since they all travel at different speeds, they do not all arrive at a seismograph at the same time. P-waves, which travel the fastest, arrive first, followed by the S-waves some time later. The surface waves arrive last. Since the speeds at which seismic waves travel is known, the difference in arrival times can be used to calculate how far away an earthquake occurred. The farther a seismograph is from the epicenter, the greater the difference between the arrival times of the P-waves and the S-waves.

The speed of P- and S-waves changes at the boundary between regions with different properties. So these waves bend, or *refract*. They also travel at different speeds in substances that have different densities. See Figure 11.18.

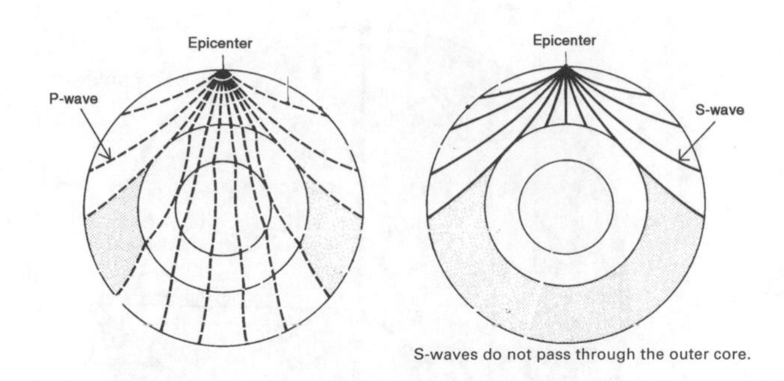

Figure 11.18
Seismic waves traveling through Earth.

CRUST, MANTLE, CORE

Scientists have analyzed seismic waves received at observing stations around the world over many years. Their studies show that Earth's interior is made of three main layers: the crust, the mantle, and the core. See Figure 11.19. The densest layer is the *core*, which is made mostly of iron and nickel. It is about 3,400 kilometers thick and has two parts: a solid inner core and a liquid outer core. The outer core is thought to be liquid because these S-waves disappear when they reach it. Above the core is the *mantle*, a 2,900-kilometer-thick layer of rock made mostly of dark, dense minerals. Mantle rock is still so hot that it is not completely solid. Like softened butter, rock in the mantle can flow under pressure. Since hot rock is less dense than

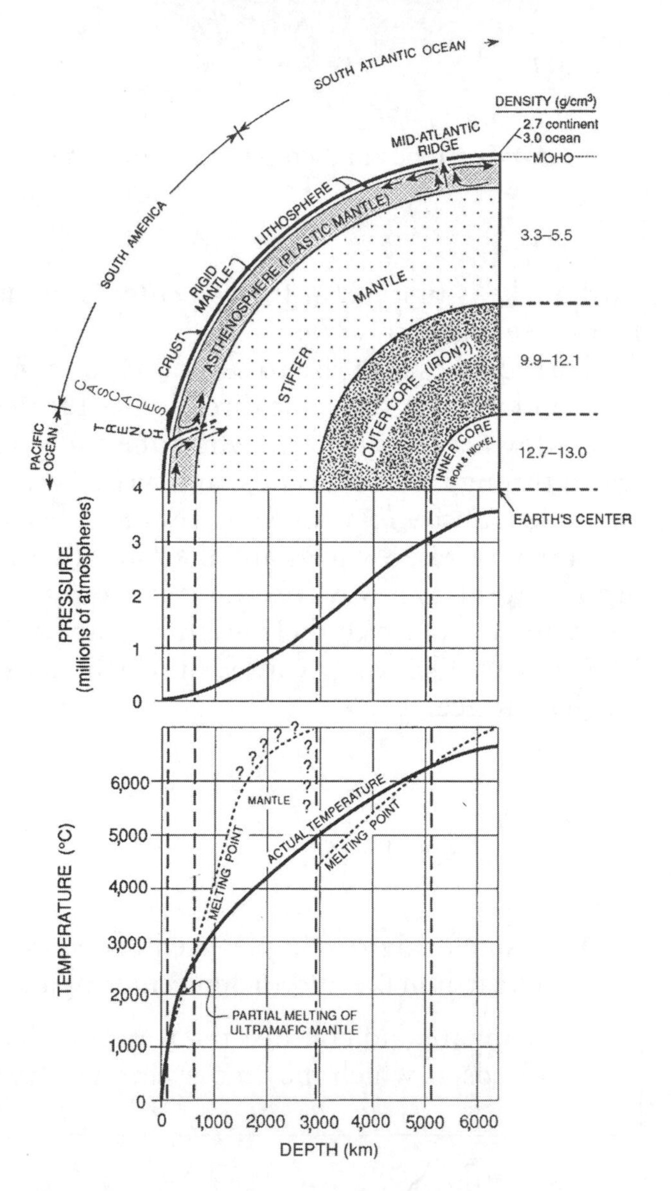

Figure 11.19
Inferred properties of Earth's interior.

cool rock, the hot rock in the mantle rises, and the cool rock sinks, forming huge density currents. Over the mantle is the least dense layer—a thin, solid ***crust***. The crust's thickness ranges from an average of 5 kilometers under the oceans to 20–50 kilometers under the continents. The crust is made mostly of compounds of about eight to ten abundant elements, such as silicon, oxygen, calcium, potassium, aluminum, iron, magnesium, and sodium.

LITHOSPHERE AND ASTHENOSPHERE

If, instead of density, the *rigidity* of rock is considered, another pattern is seen. The crust and the outermost part of the mantle make up a rigid shell of rock about 50 to 100 kilometers thick, which is called the ***lithosphere***. A 100-kilometer-thick layer of the mantle just below the lithosphere, called the ***asthenosphere***, is able to flow like modeling clay squeezed in your hand. Unlike the rigid lithosphere and the rest of the mantle, when pressure (such as the weight of a continent) is exerted on the asthenosphere for a long period of time, the asthenosphere slowly flows. The lithosphere actually floats on the denser asthenosphere like ice floats in water.

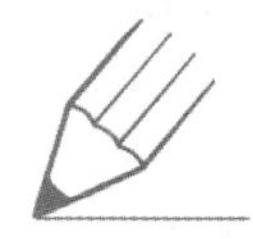

TRY THIS

1. According to the information in Figure 11.19, what is the density of the upper portion of the lithosphere that forms the ocean floors? The continents?
2. How does the density of Earth's interior change with depth?

PLATE TECTONICS

Geologists have used the lithosphere/asthenosphere pattern to devise a new model of Earth's structure—***plate tectonics*** (tectonics is the study of crustal movements). In the plate tectonic model, the lithosphere is seen not as solid, unbroken rock but as broken into several large slabs called plates—hence the name *plate tectonics*. The plates are believed to be floating on a layer of liquid rock—the asthenosphere—and set into motion by the convection currents in the mantle. Figure 11.20 shows Earth's major plates and the directions in which they are moving.

As the plates move sideways around Earth, their edges interact in one of three ways. They spread apart, they collide, or they slide past each other. See Figure 11.21.

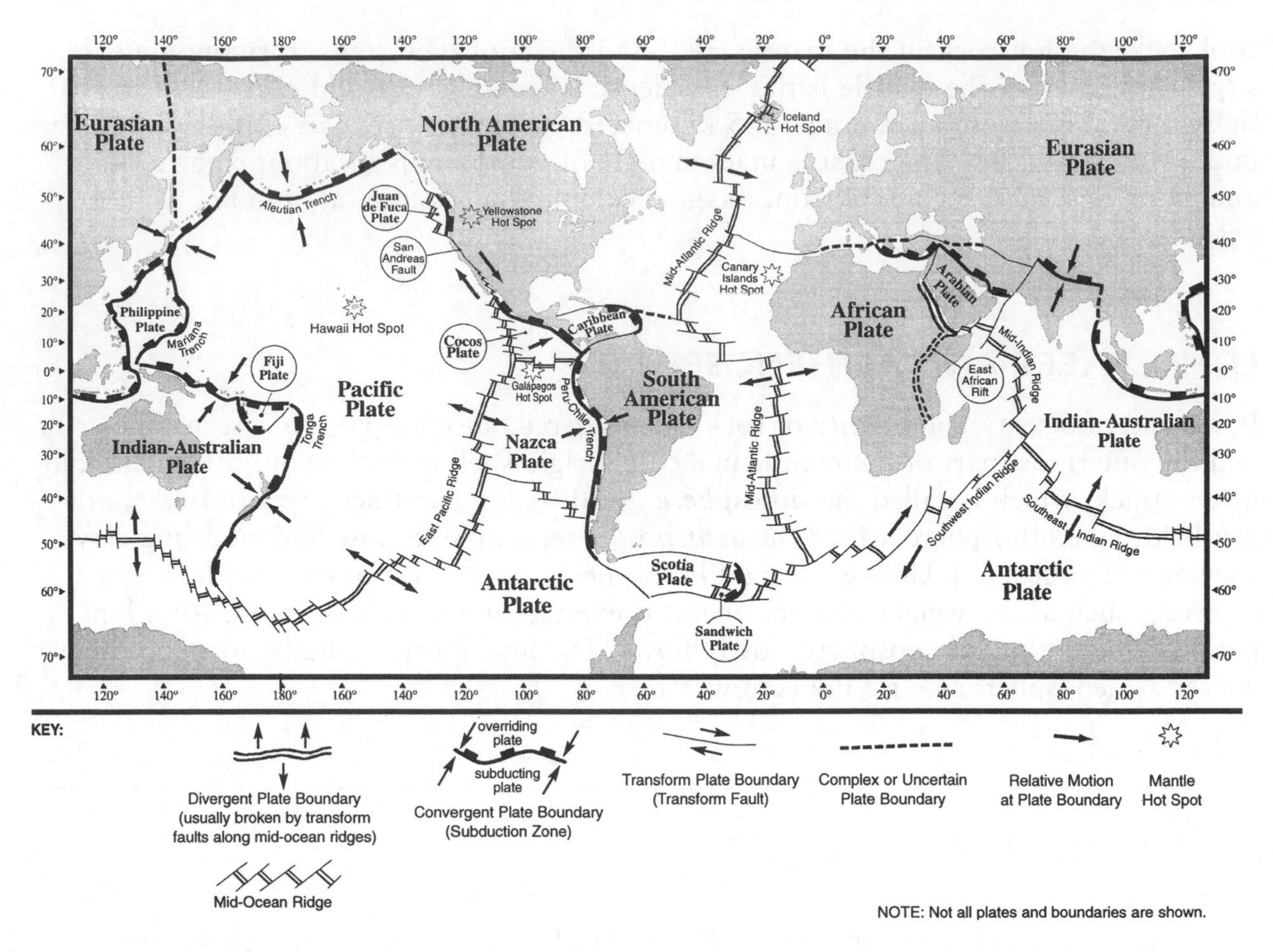

Figure 11.20
Plate tectonic world map.

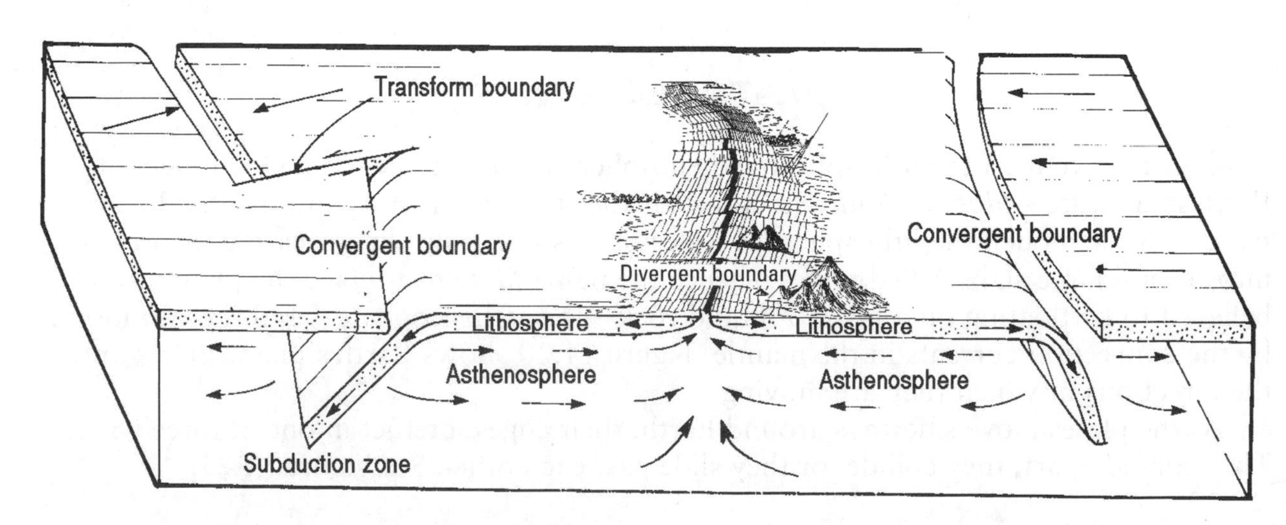

Figure 11.21
Convergent, divergent, and transform boundaries.

Divergent boundaries are places where plates are moving apart. This causes earthquakes and opens cracks through which magma can rise to the surface and solidify to form new lithosphere. Most of Earth's oceans have a divergent boundary running down their center. These boundaries form a continuous series of cracks called the ***midocean rift***. Magma welling up through the midocean rift has built a line of volcanic mountains called the ***midocean ridges***. It also adds new rock to the edges of the plates. As new rock rises and hardens in the rift, older rocks are pushed aside. Thus, the age of ocean floor rock increases as you move away from the rift. As the new rock hardens, it is magnetized in the same direction as Earth's magnetic field. Since Earth's magnetic field reverses several times per million years, alternating strips of rock with normal and reversed magnetism are seen on both sides of the rift. See Figure 11.22.

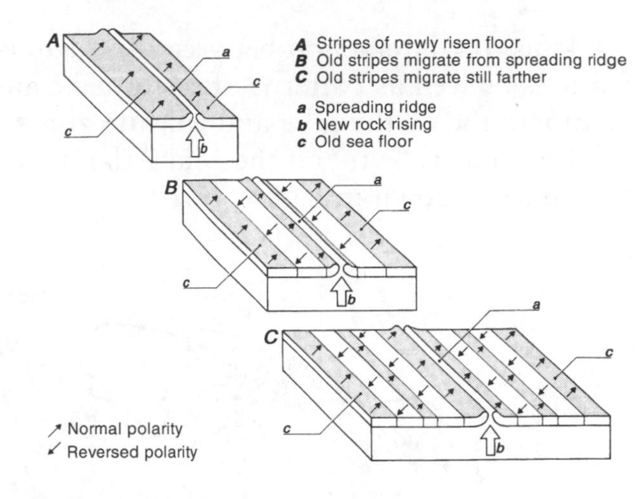

Figure 11.22
Paleomagnetism and the ages of rocks on both sides of a rift zone.

Convergent boundaries are places where plates move toward each other and collide. When plate edges collide, squeezing and twisting forces cause rocks to fold, fracture, and move along faults. Ocean plates and continental plates interact differently along divergent boundaries. Ocean crust is denser than continental crust. When these plates collide, the denser ocean plate tends to get pushed under the continental plate and is forced into the mantle where it melts—a process known as ***subduction***. This is why Earth stays the same size even though new crust is constantly forming at divergent boundaries. Friction between the two plates produces frequent earthquakes and melts rock, which rises to the surface, forming volcanoes. The plunging plate also pulls the plate edges downward, forming a deep ***trench***. Trenches are the deepest places in the oceans, reaching depths of more than 11 kilometers.

If two plates carrying continental crust collide, they hit head-on. The edges of both crumple and fold as well as fracture and move along faults. Folding and faulting thrusts the crust upward, forming mountain ranges along such boundaries. The Andes and Himalaya Mountains formed along convergent boundaries.

Transform boundaries are places where plates slide sideways past one another. This shearing motion causes rock to fracture, forming numerous faults. The plates on either side of the faults smash and rub against each other like the sides of two ships that came too close together and grinded past each other. The rock along these faults is intensely shattered, and the sliding motion of the plates causes many earthquakes. If the plates separate a little, some magma may leak through the boundary, causing small-scale volcanism. The San Andreas fault is part of a huge transform boundary along which the Pacific Plate is moving past the North American Plate.

Thus, the interaction between the edges of moving plates explains many of Earth's features as well as patterns of earthquake and volcanic activity and crustal movements. Compare the earthquake and volcano zones, such as the Pacific Ring of Fire, the Mediterranean Belt, and the Mid-Atlantic Ridge shown in Figure 11.23 with the plate boundaries shown in Figure 11.20.

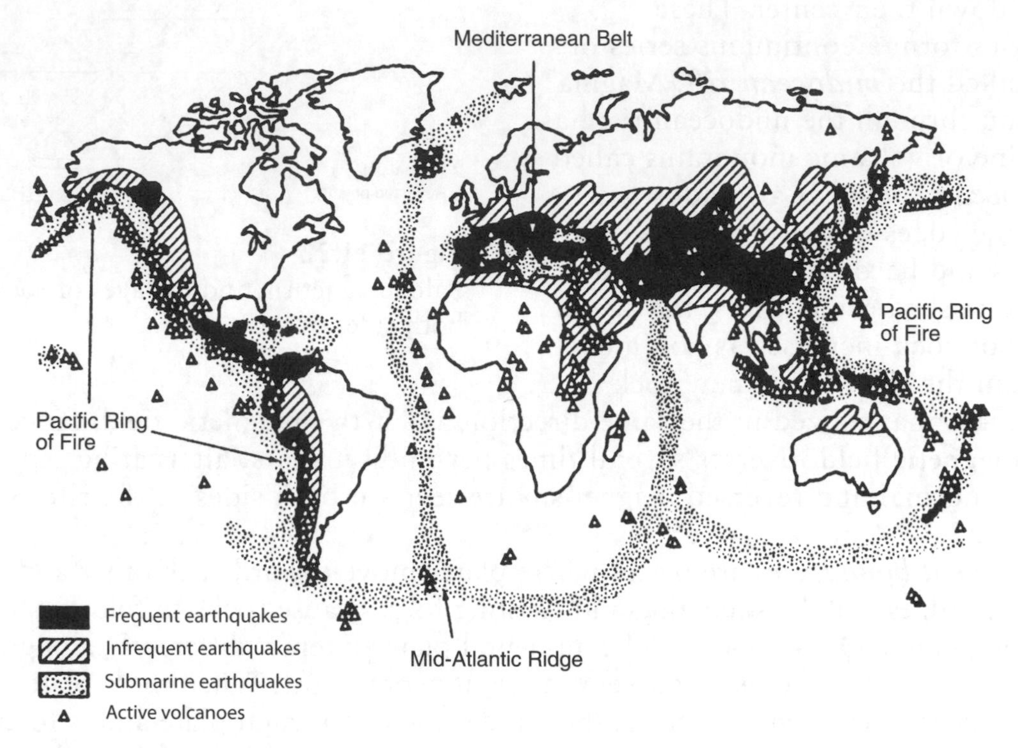

Figure 11.23
Worldwide earthquake and volcano belts.

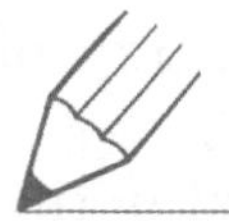

TRY THIS

1. On which tectonic plate are each of the following located:
   1. the continental United States
   2. India
   3. Hawaii
   4. Greenland
2. The Nazca Plate is roughly rectangular in shape. Along which two edges is new plate material forming? Along which edge is plate material being destroyed?

Practice Review Questions

1. Earth's oceans, lakes, streams, underground water, and ice are part of the

1. atmosphere
2. biosphere
3. hydrosphere
4. lithosphere

2. Which line best identifies the boundary between the lithosphere and the troposphere?

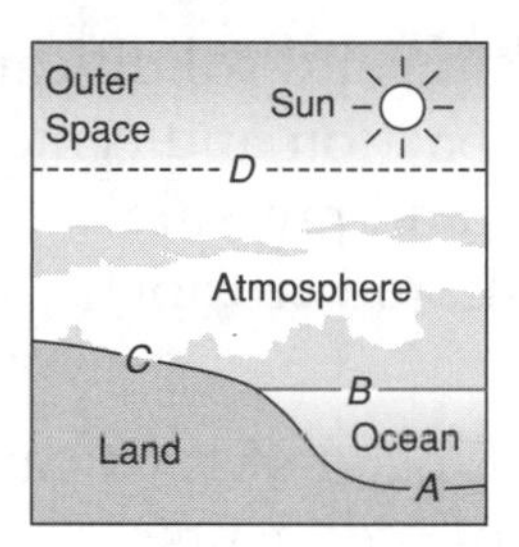

1. line *A*
2. line *B*
3. line *C*
4. line *D*

3. The physical properties of a mineral are largely due to its

1. melting point
2. internal arrangement of atoms
3. volume at room temperature
4. age when discovered

Questions 4 and 5 refer to the following.

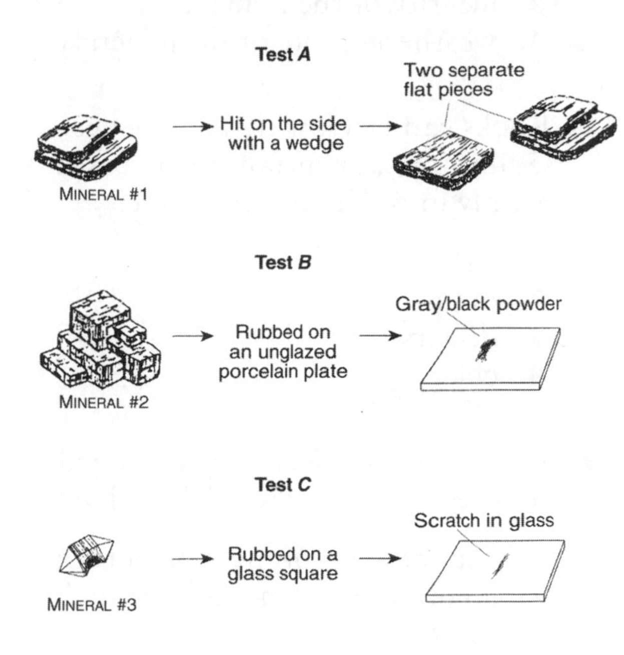

4. Which sequence correctly matches each test, *A*, *B*, and *C*, with the mineral property being tested?

1. *A*—streak; *B*—hardness; *C*—cleavage
2. *A*—cleavage; *B*—hardness; *C*—streak
3. *A*—cleavage; *B*—streak; *C*—hardness
4. *A*—streak; *B*—cleavage; *C*—hardness

GO ON

5. The results of *all* three physical tests would be *most* useful for determining the

1. environment in which the minerals formed
2. geologic age of the minerals
3. identity of the minerals
4. weathering rate of the minerals

6. Rocks can be classified as igneous, sedimentary, or metamorphic based mainly upon differences in their

1. age
2. origin
3. density
4. color

7. The size of the mineral crystals found in an igneous rock is directly related to the

1. cooling time of the molten rock
2. amount of sediments cemented together
3. density of the minerals
4. color of the minerals

8. A rock that contains fossil seashells most likely formed as a result of

1. volcanic activity
2. sedimentation in shallow water
3. deposition from a melting glacier
4. mass wasting on a mountain slope

9. Most metamorphic rocks are formed when

1. lava flows cool rapidly
2. rocks are subjected to heat and pressure
3. sediments are compacted and cemented
4. magma cools slowly, deep underground

Questions 10 and 11 refer to the following:

Rock Cycle in Earth's Crust

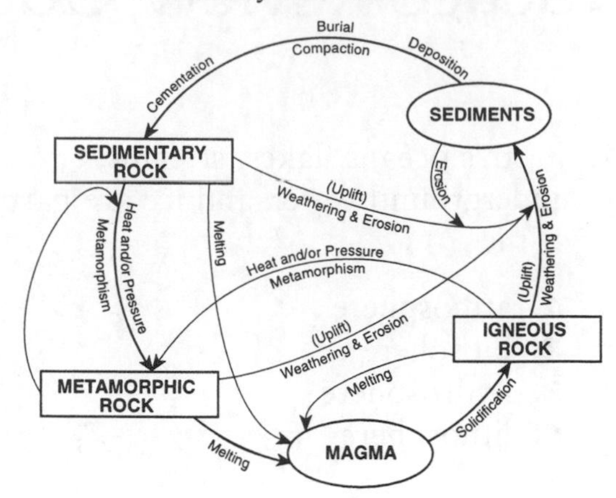

10. Which two processes result in the formation of sedimentary rocks?

1. melting and solidification
2. compaction and cementation
3. heat and pressure
4. uplift and erosion

11. Which type(s) of rock can be the source of sediments?

1. igneous and metamorphic rocks only
2. metamorphic and sedimentary rocks only
3. igneous, metamorphic, or sedimentary rocks
4. sedimentary rocks only

GO ON ➡

12. The flowchart below shows part of Earth's water cycle. Which process should be shown in place of the question marks to best complete the flowchart?

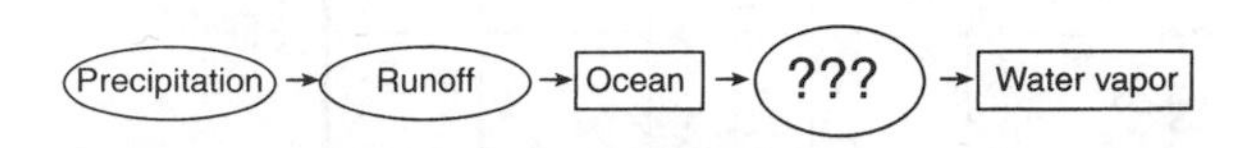

1. condensation
2. precipitation
3. evaporation
4. infiltration

13. The diagram below shows the stump of a tree whose root grew into a small crack in bedrock and split the rock apart.

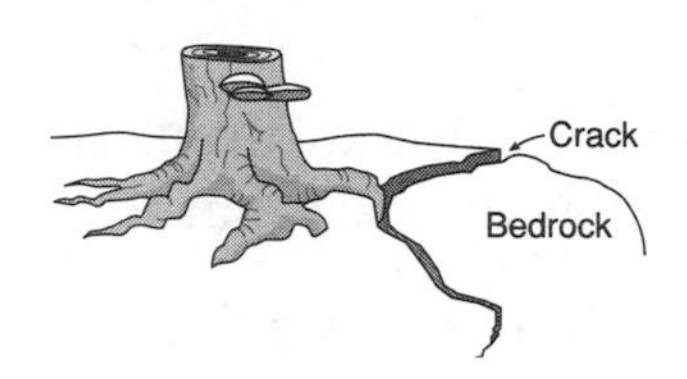

The action of the root splitting the bedrock is an example of

1. deposition
2. erosion
3. physical weathering
4. chemical weathering

14. Rocks chemically weather fastest in a climate that is

1. hot and dry
2. cold and dry
3. hot and moist
4. cold and moist

15. Which erosional force acts alone to produce landslides and rockfalls?

1. running water
2. sea waves
3. winds
4. gravity

16. What change is usually seen in a pebble as it is carried a great distance by streams?

1. It will become rounded, and its mass will increase.
2. It will become rounded, and its size will decrease.
3. It will become jagged, and its size will increase.
4. It will become jagged, and its mass will decrease.

GO ON ➡

Questions 17 and 18 refer to the map below. The dots on the map represent major earthquakes. The letter *A* marks a location on the surface.

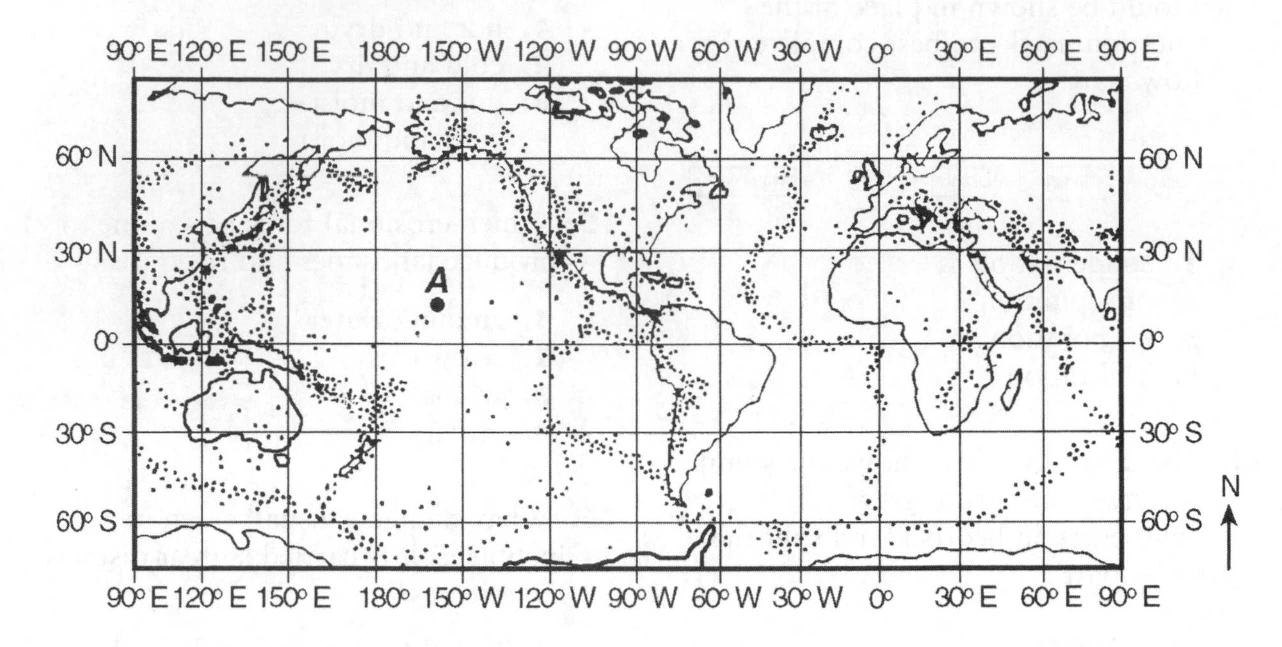

17. This map would be most useful in finding the location of

1. boundaries between crustal plates
2. the boundary between the mantle and the core
3. major ocean currents
4. prevailing wind patterns

18. Location *A* in the diagram is best described as an area that is

1. near a midocean ridge
2. near the center of a crustal plate
3. at the boundary of two converging plates
4. at a divergent plate boundary

Questions 19 and 20 refer to the diagram below, which shows a cross section of part of Earth's crust and mantle.

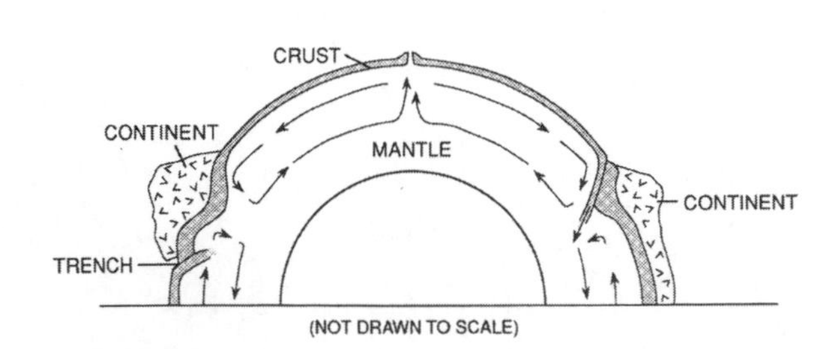

19. The arrows shown in the mantle represent the slow circulation of heated mantle rock by a process called

1. radiation
2. conduction
3. evaporation
4. convection

GO ON ➡

20. The movement of the mantle shown by the arrows moves crustal plates and results in

1. erosion by glaciers
2. some earthquakes and volcanoes
3. river deposits on the ocean floor
4. sand deposits on ocean beaches

21. The rock layers in this cross section contain trilobite fossils and have not been displaced by crustal movement.

Which fossil trilobite is considered the oldest in the cross section shown?

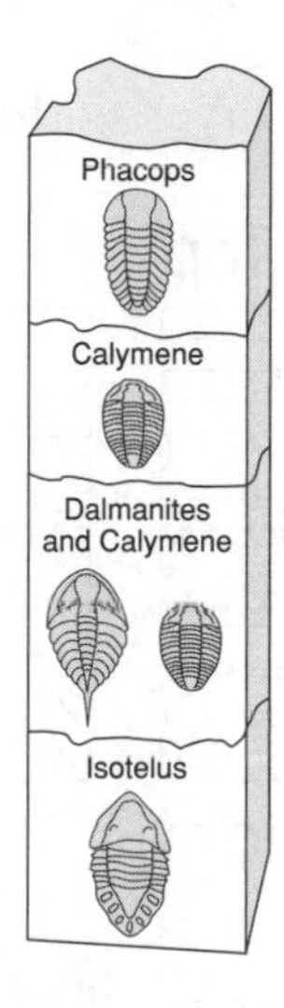

1. *Calymene*
2. *Dalmanites*
3. *Isotelus*
4. *Phacops*

22. In the diagram below, the arrows show the direction of forces that are compressing rock layers in Earth's crust.

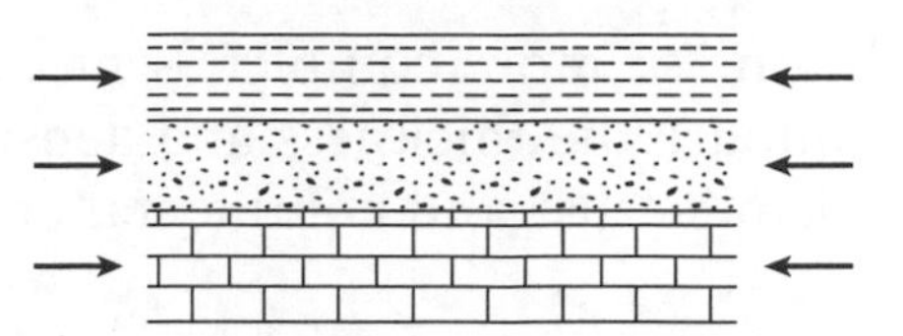

Which diagram shows the most likely result of these forces?

1. 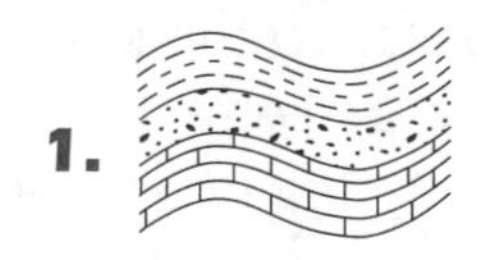

3. 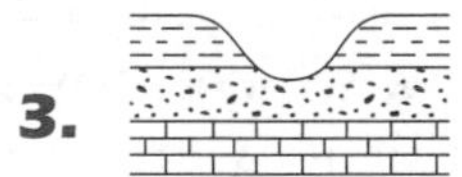

2. 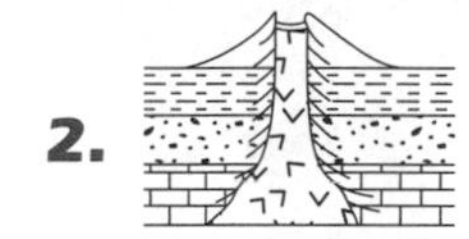

4. 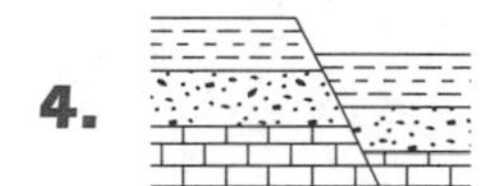

GO ON

CONSTRUCTED-RESPONSE QUESTIONS

23. In one or more complete sentences, compare the thickness and density of oceanic crust and continental crust.

24. State three characteristics of a particle that will affect its settling rate.

Questions 25 and 26 refer to the map below, which shows where some continents were probably located at one time in the past and where fossils of four organisms are found today.

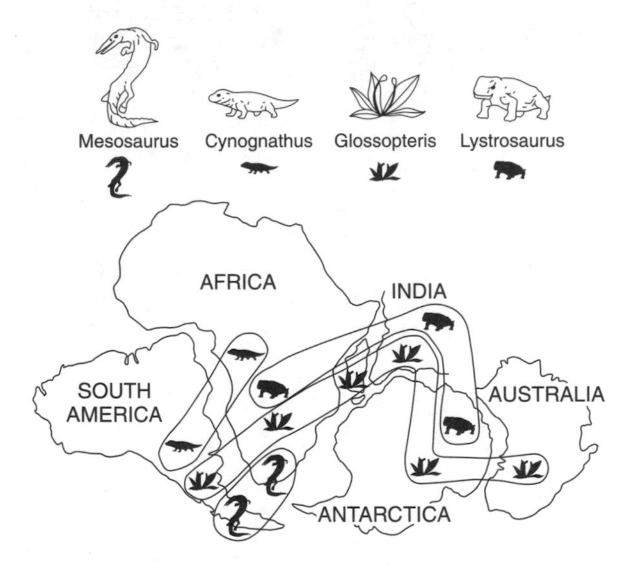

25. Which fossil shown in the diagram would be most useful for matching rocks among all of the continents shown?

26. Name two pieces of evidence shown in this diagram that suggest that these continents were once joined.

27. The diagram below shows layers of sedimentary rock found in four different locations. Four layers are identified as *A*, *B*, *C*, and *D*. No layers have been disturbed by crustal movement.

Which rock layer is the *youngest?*

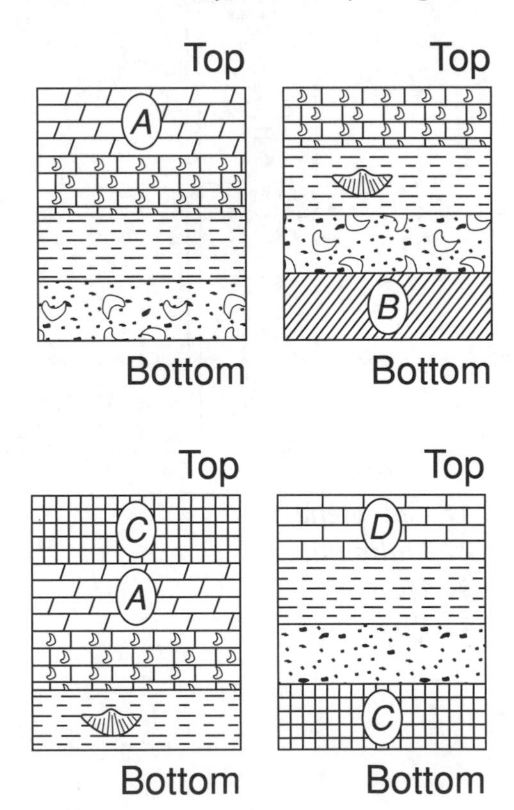

Chapter 12

Meteorology

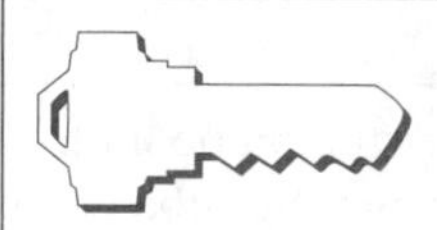

Key Idea: Many of the phenomena that we observe on Earth involve interactions among components of air, water, and land.

WEATHER

Weather describes the conditions in the atmosphere at a given place for a short period of time. Everything we call weather takes place in the lowest layer of the air that surrounds Earth. The major cause of weather is unequal heating of the atmosphere, hydrosphere, and lithosphere by the Sun's rays. As heat moves between the three, the air becomes warmer in some places and cooler in others. Warm air is less dense than cold air. Gravity causes the dense, cold air to sink below the warmer air, producing circulation of the air in the atmosphere. This circulation causes the air at any one place to be moved away constantly and replaced by air from a different place—causing the weather to change.

WEATHER VARIABLES

The constantly changing weather is described in terms of the changing characteristics of the air, or *weather variables*. The weather variables used to describe the atmosphere include air temperature, air pressure, humidity, wind speed and direction, clouds, and precipitation. Weather instruments are used to measure weather variables. See Table 12.1.

AIR TEMPERATURE

Air temperature is related to the amount of heat energy present in the atmosphere at a place. A little of the heat energy in the atmosphere is absorbed directly from sunlight. However, practically all of the heat that affects the air close to Earth comes from Earth itself. This does not mean that Earth makes the heat. The heat comes from solar radiation that is absorbed by the hydrosphere and lithosphere. This heat then goes into the air as it passes over their surfaces. The heating of Earth's surface depends on what the surface is made of and the angle at which sunlight strikes it.

Table 12.1 **Weather Variables, Weather Instruments, and How the Instruments Work**

Weather Factor	Instrument	How It Works
Air temperature	Thermometer	Liquid expands when heated and rises in tube.
Air pressure	Barometer	Air exerts pressure on pool of mercury, forcing it up tube.
Humidity	Sling psychrometer	Evaporation from wet-bulb thermometer causes cooling. The drier the air, the more evaporation, the more cooling, and the greater the difference between wet-bulb and dry-bulb temperatures.
Wind speed	Anemometer	Wind exerts force on cups, causing anemometer to spin. The higher the wind speed, the faster it spins.
Wind direction	Wind vane	Wind exerts more force on tail than on tip of arrow, causing vane to swing and point in direction from which wind is blowing.
Precipitation	Rain gauge	Measures volume of liquid precipitation collected.

Some kinds of surfaces get hotter than others when they absorb sunlight. For example, dark-colored, rough surfaces absorb more heat, and get hotter, than light-colored, smooth surfaces. More heat energy is also needed to change the temperature of water than to change the temperature of land. Therefore, water does not get as hot or as cold as nearby land. Since plants are mostly water, they also do not get as hot. That is why the pavement and sand at the beach get much hotter than the water or grass.

As you learned earlier, because Earth is a sphere, sunlight does not strike all places on its surface at the same angle. Near the equator, the Sun's rays strike the surface most directly. Near the poles, they strike at more of a slant. As you saw in Figure 10.17, direct rays are concentrated in a smaller area while slanted rays are spread out over a wider area. Concentrated rays heat the surface more than spread-out rays. That is why the equator is hotter than the poles. The angle of the Sun's rays also changes during the course of each day. The Sun's rays strike most directly around noon and are most slanted at dawn and sunset. That is why the warmest part of each day usually occurs in the middle of the day—the more direct noontime rays heat Earth's surface more than the slanting early-morning or evening rays.

TRY THIS

Earth is closest to the Sun in January, yet January is a winter month in New York State. In one or more complete sentences, explain why this is so.

AIR PRESSURE

Air is a mixture of gases. Gases are made of tiny molecules that whiz to and fro. When these moving molecules hit a surface, they exert a force on that surface. ***Air pressure*** is the amount of force exerted by the air's gas molecules hitting a given surface area. Air pressure is measured in units called ***millibars***. At sea level, normal air pressure is 1,013.2 millibars. Since air molecules move randomly in all directions, they exert pressure in all directions. See Figure 12.1.

When air heats up, its molecules move faster. You might think that this would increase air pressure because fast-moving molecules hit a surface harder than slow-moving molecules, but in an open system like the atmosphere, it does not. When air heats up, its molecules also spread farther apart. This means that fewer molecules will hit the surface. The number of molecules hitting the surface has more of an effect on the force they exert on a surface than their speed, so heating air causes air pressure to decrease. See Figure 12.2.

Air pressure also changes as water vapor enters and leaves the air. When water vapor enters air, it displaces molecules of other gases, such as nitrogen and oxygen. Water vapor

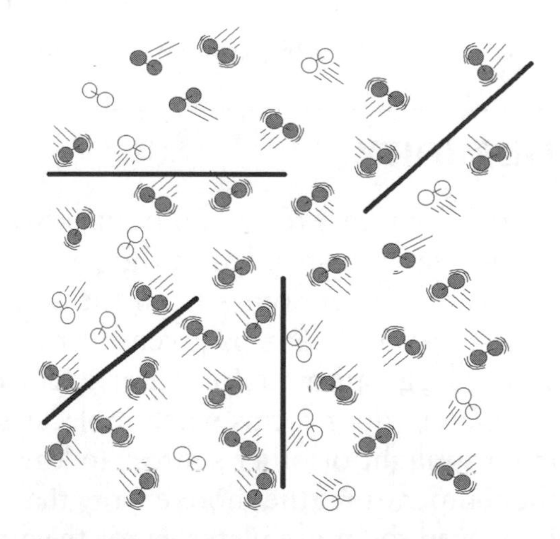

Figure 12.1
Air exerts equal pressure in all directions. Wherever the surface is placed, air molecules strike it with the same amount of force.

molecules have less mass than molecules of nitrogen or oxygen. The less mass a molecule has, the less force it exerts when it hits a surface. Therefore, increasing the amount of water vapor in air decreases air pressure. Displacing nitrogen and oxygen molecules with water vapor molecules that have less mass also decreases the mass of a given volume of air, that is, decreases its density. Thus, moist air is less dense than dry air.
See Figure 12.3.

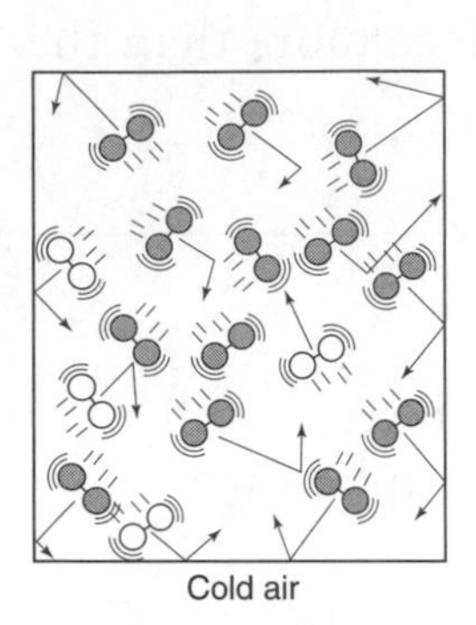

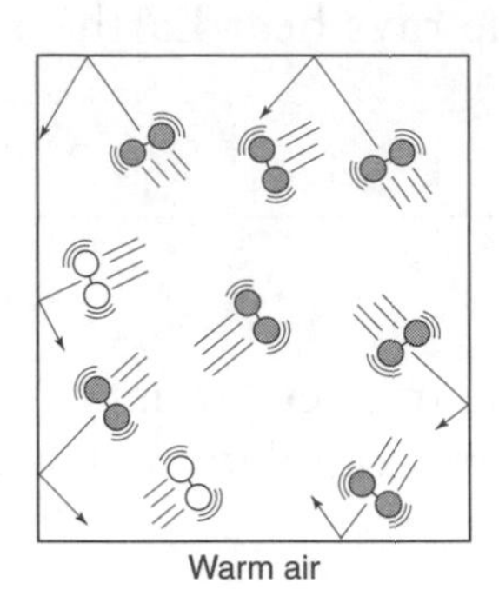

Figure 12.2
Increasing air temperature decreases air pressure.

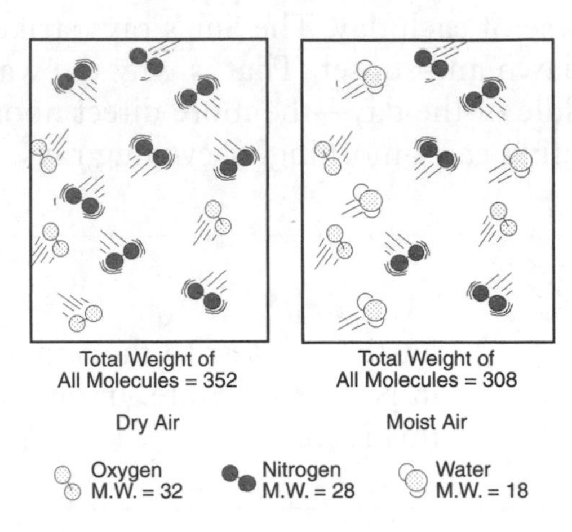

Figure 12.3
Moist air is less dense than dry air.

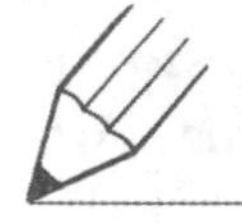

TRY THIS

In the summer, the southwest monsoon winds carry warm, moist air over India. In the winter, the wind direction reverses and carries cold, dry air from Siberia over India. In one or more complete sentences, explain how the air pressure in India would change from summer to winter.

HUMIDITY

Humidity refers to the amount of water vapor in the air. Humidity is not droplets of liquid water suspended in the air. It is water vapor—a colorless, odorless gas. Most of the water vapor comes from the oceans, lakes, and streams. A lot of it comes from rain and snow, and some is given off by plants and animals. Where there are large areas of water, large amounts of water vapor enter the air as water evaporates.

Water vapor forms when molecules of liquid water ***evaporate***, or gain enough energy from sunlight or other sources to leave the liquid state and enter the air as a gas. The higher the temperature, the more energy that molecules have, so the more molecules that evaporate. However, the more water vapor the air contains, the more likely that water vapor molecules in the air that arrive at the surface of the liquid will be trapped there and return to the liquid state, or ***condense***. At any given temperature, there is a maximum amount of water vapor that can exist in the air before water vapor is condensing as fast as it evaporates.

The actual amount of water vapor in the air is often expressed as a percentage of this maximum amount, or ***relative humidity***. For example, a relative humidity of

75% means the air contains 75% of the maximum amount of water vapor that can exist in the air at the current temperature. The temperature at which the amount of water vapor in the air equals the maximum that can exist in the air is called the ***dew point.*** When the air temperature is at its dew point, relative humidity is 100%. If the air temperature cools to the dew point, the excess water vapor condenses out of the air, forming clouds, fog, or precipitation. See Figure 12.4.

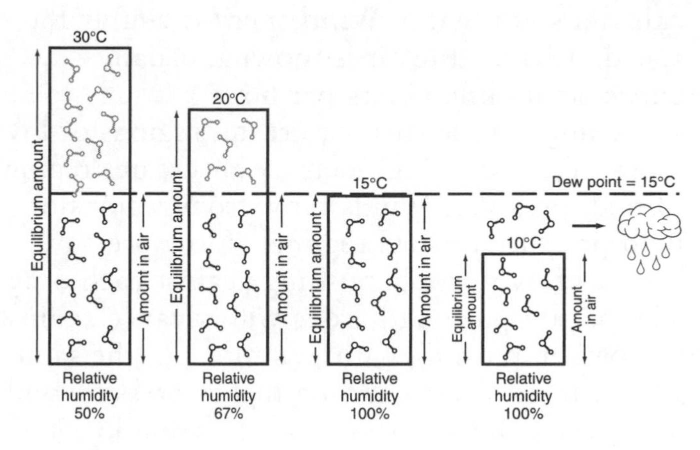

Figure 12.4
Condensation at the dew point temperature.

TRY THIS

In one or more complete sentences, explain what will happen to the probability of rain as the air temperature approaches the dew point temperature.

WIND

When the air in the atmosphere is heated, its molecules move faster and spread farther apart. This causes the heated air to become less dense than the surrounding air, and the heated air rises. As the air rises, it expands and cools. This causes the air molecules to slow down and move closer together. This now-cooler air becomes denser than the surrounding air and sinks. As air rises and sinks, surrounding air moves in to replace it, setting up a circular pattern known as a ***convection cell.*** See Figure 12.5. The horizontal movement of air is called ***wind***; the up and down motions of air are called ***air currents.***

Winds are described by their speed and direction. ***Wind direction*** is the direction from which a wind blows. Just as a person who comes from the south is called a southerner, a wind that blows from the south is

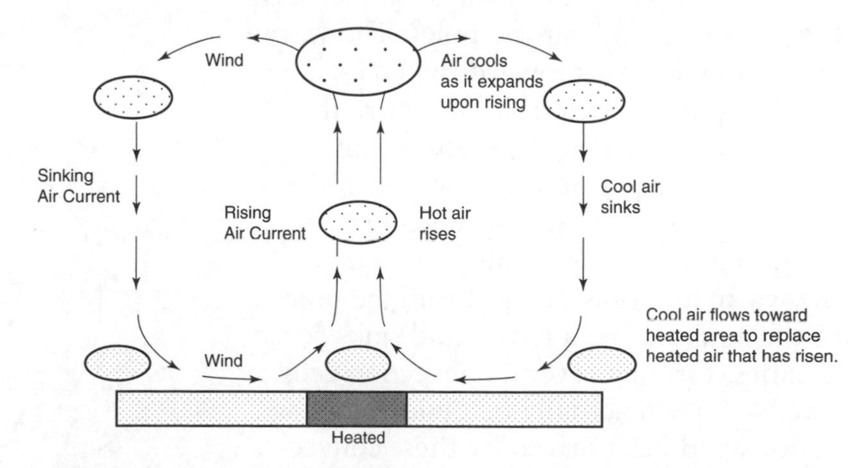

Figure 12.5
Heated air expands and rises, then cools and sinks, forming a circular convection current.

called a south wind. ***Wind speed*** is simply the speed at which the air is moving, usually expressed in kilometers per hour.

Rising low-density air creates regions of low pressure; sinking high-density air creates regions of high pressure. Winds blow from regions of high pressure toward regions of low pressure. Think of two people pushing against each other. The person pushing harder will advance against the person pushing with less force. In the same way, a mass of air exerting higher pressure will advance against a mass of air exerting lower pressure. The movement of the advancing air creates a wind. The greater the difference in pressure between the two masses of air, the faster the high-pressure air advances and the greater the wind speed.

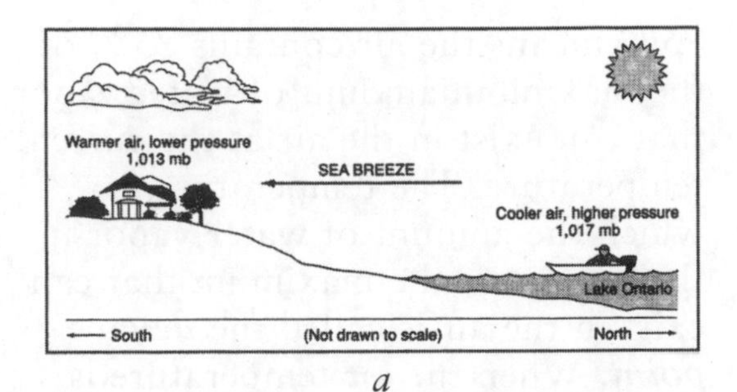

a

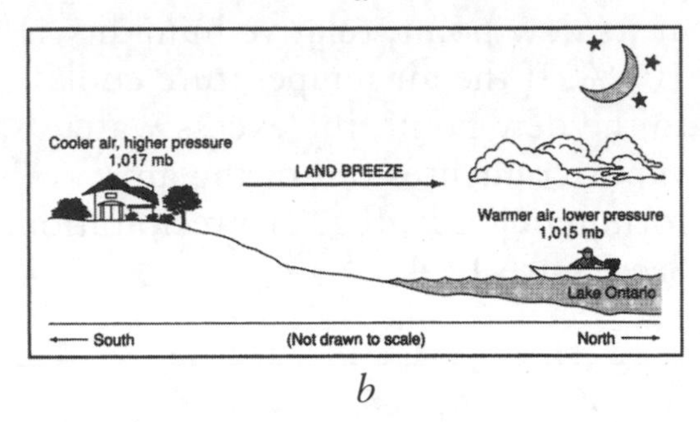

b

Figure 12.6
Land and sea breezes.

LAND AND SEA BREEZES

On a small scale, we see pressure effects in land and sea breezes. Land and water heat up and cool off at different rates. This causes differences in air pressure between the air over the land and the air over the water. During the day, sunlight heats up the land more than the water. The air over the land becomes warmer and exerts less pressure than the air over the water. This gives rise to a wind blowing from the sea toward the land—a sea breeze. At night, the situation is reversed, giving rise to a land breeze. See Figure 12.6.

GLOBAL WIND BELTS

On a global scale, it is always warmer near the equator than near the poles. The warm air rises and moves toward the poles. The cold air pours down to fill in where the warm air used to be. This flow of air between the equator and poles sets air in motion all over the world. Of course, along the way, the cool air warms up and the warm air cools down. From the poles to the equator, the air rises and sinks several times, forming several convection cells. The world can actually be divided into six major wind belts driven by these convection cells. See Figure 12.7. Notice that the three wind belts in the Southern Hemisphere are *exactly the reverse* of those in the Northern Hemisphere.

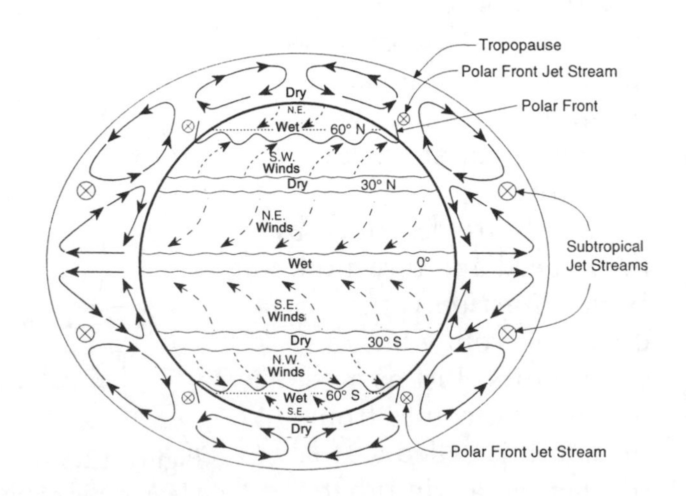

Figure 12.7
Global wind belts.

TRY THIS

In one or more sentences, explain how convection currents in the atmosphere cause the regions 30°N and 30°S of the equator to have hot, dry climates.

THE CORIOLIS EFFECT

Notice in Figure 12.7, that the global winds do not blow in a straight path between the global high- and low-pressure belts. Instead, the winds curve to the right in the Northern Hemisphere and to the left in the Southern Hemisphere. This does not only happen to winds. It also happens to ocean currents, airplanes, rockets, or any other matter moving over Earth's surface. Moving objects appear to curve as they move over Earth's surface because Earth is rotating while the matter is moving over it. The apparent change in direction of a moving object due to Earth's rotation is called the ***Coriolis effect*** (after the French mathematician Gaspard Coriolis, who first studied the curving of global winds).

To understand the Coriolis effect better, look at Figure 12.8. The turntable in Figure 12.8*a* is not moving. A ball covered with ink is rolled across the turntable, striking the target. When you look at the ink trail, you see that the ball's path is a straight line. In Figure 12.8*b*, the turntable is rotating. Again the ball is rolled across the turntable. It rolls away from you in a straight line. This time, though, when you look at the ink trail, you see that the ball's path is curved. The ball rolls off the turntable to the right of the target. The Coriolis effect due to the turntable's rotation caused the ball to make a curved path instead of a straight one.

The Coriolis effect causes global winds to appear to curve to the right in the Northern Hemisphere and to the left in the Southern Hemisphere. A good way to remember this is to recall that *south*paws are *left*-handed and the letter *r* is *right* in the middle of no*r*th.

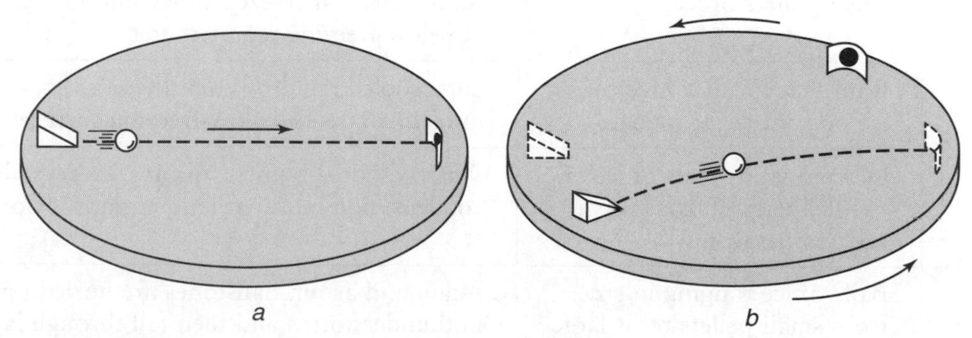

Figure 12.8
The Coriolis effect. (*a*) An ink-covered ball shoots across a turntable that is not rotating. The ball leaves a straight path and hits the target. (*b*) An ink-covered ball shoots across a rotating turntable. The ball leaves a curved path and misses the target.

TRY THIS

In one or more complete sentences, explain how Earth's global wind belts would change if Earth's direction of rotation were reversed.

CLOUDS AND PRECIPITATION

In the lowest part of the atmosphere, air temperature drops steadily with altitude. This causes warm air to cool as it rises. When it has been cooled to its dew point, the water vapor in the air condenses. This condensation creates a ***cloud*** made of a great many tiny water droplets or ice crystals. We see a cloud by the light that reflects from these droplets or crystals.

Clouds are important weather variables for three reasons. First, they are the source of precipitation, such as rain, snow, sleet, and hail. Second, they reflect a lot of sunlight, causing unequal heating of Earth's surface. Third, they are key indicators of overall weather conditions.

Not all clouds produce precipitation. The water droplets or ice crystals in clouds are very tiny (~1/50 millimeter) and often remain suspended in the air. Precipitation occurs only when cloud droplets or ice crystals join together and become heavy enough to fall. Table 12.2 summarizes the different types of precipitation and how they form.

Table 12.2 **Types of Precipitation**

Name	Description	Origin
Rain	Droplets of water up to 4 mm in diameter	Cloud droplets coalesce when they collide
Drizzle	Very fine droplets falling slowly and closely together	Cloud droplets coalesce when they collide
Sleet	Clear pellets of ice	Raindrops that freeze as they fall through layers of air at below-freezing temperatures
Glaze	Rain that forms a layer of ice on surfaces it touches	Supercooled raindrops that freeze as soon as they come into contact with below-freezing surfaces
Snow	Hexagonal crystals of ice or needlelike crystals at very low temperatures	Water vapor sublimes, forming ice crystals on condensation nuclei at temperatures below freezing
Hail	Balls of ice ranging in size from small pellets to as large as a softball with an internal structure of concentric layers of ice and snow	Again and again, hailstones are hurled up by updrafts in thunderstorms and then fall through layers of air that alternate above and below freezing—each cycle adds a layer to the hailstone—the more violent the updrafts, the larger and heavier the hailstone can become before falling

WEATHER PATTERNS

When weather variables are plotted on a map, large-scale patterns can be seen. Tracking weather patterns over time reveals how and where the air and its weather are moving. This information can then be used to predict, or *forecast*, weather changes. A map that summarizes weather variables measured at many places at the same time is called a ***synoptic weather map*** (a synopsis is a summary).

THE SYNOPTIC WEATHER MAP

A synoptic weather map is made by measuring weather variables at thousands of weather stations around the world four times a day. This data is then used to create maps that reveal large-scale weather patterns. By looking at a series of synoptic weather maps, the development and movement of weather systems can be tracked and predictions can be made.

THE STATION MODEL

Synoptic weather maps use a symbol called a ***station model*** to show a summary of the weather conditions at a weather station. Figure 12.9 shows a typical station model and explains what each of its elements represents.

Weather Map Symbols

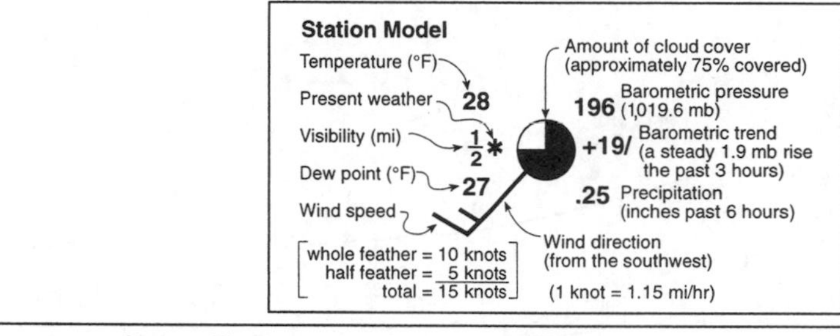

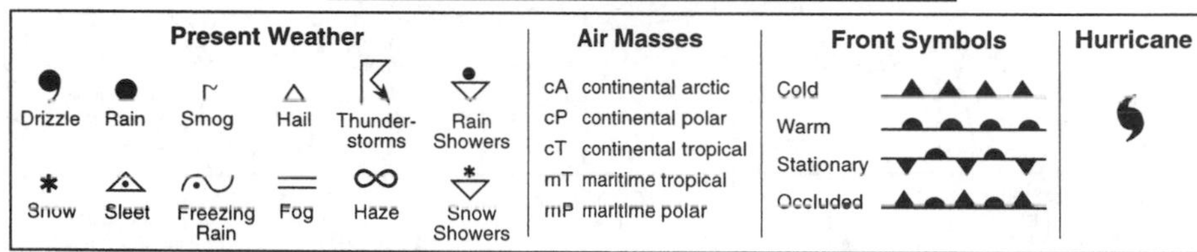

Figure 12.9
A station model.

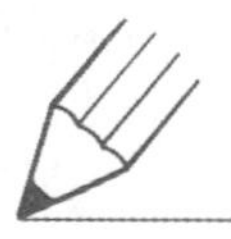

TRY THIS

The weather bureau in the city of Oswego, New York, reported the weather conditions shown in the box. The station model shown here is partially completed.

Air temperature: 65°F
Wind direction: from the southeast
Wind speed: 20 knots
Barometric pressure: 1,017.5 mb
Dew point: 53°F

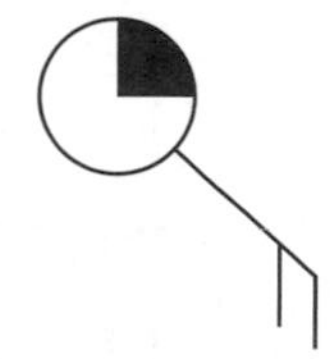

1. Using the weather conditions given, complete the station model by recording the air temperature and dew point in the proper locations.
2. State the sky conditions or amount of cloud cover over Oswego as shown by the station model.

FIELD MAPS AND ISOLINES

Station models plotted on a map summarize a wide range of weather data and can be used to create many different field maps. A ***field*** is a region of space that has a measurable quantity at every point. Field maps can be used to represent any quantity that varies in a region of space. One way to represent field quantities on a two-dimensional field map is to use isolines. ***Isolines*** connect points of equal field values. For example, a temperature field map would contain lines connecting points of equal temperature, or ***isotherms***. A pressure field map would contain lines connecting points of equal pressure, or ***isobars***. Field maps clearly show the weather patterns in the atmosphere. See Figure 12.10.

As you can see in Figure 12.10*c*, distinct regions in the atmosphere have similar conditions. For example, a region of high pressure, cooler temperatures, and clear skies is centered near Salt Lake City, Utah. A region of low pressure, warmer temperatures, and cloudy skies is centered near Cincinnati, Ohio.

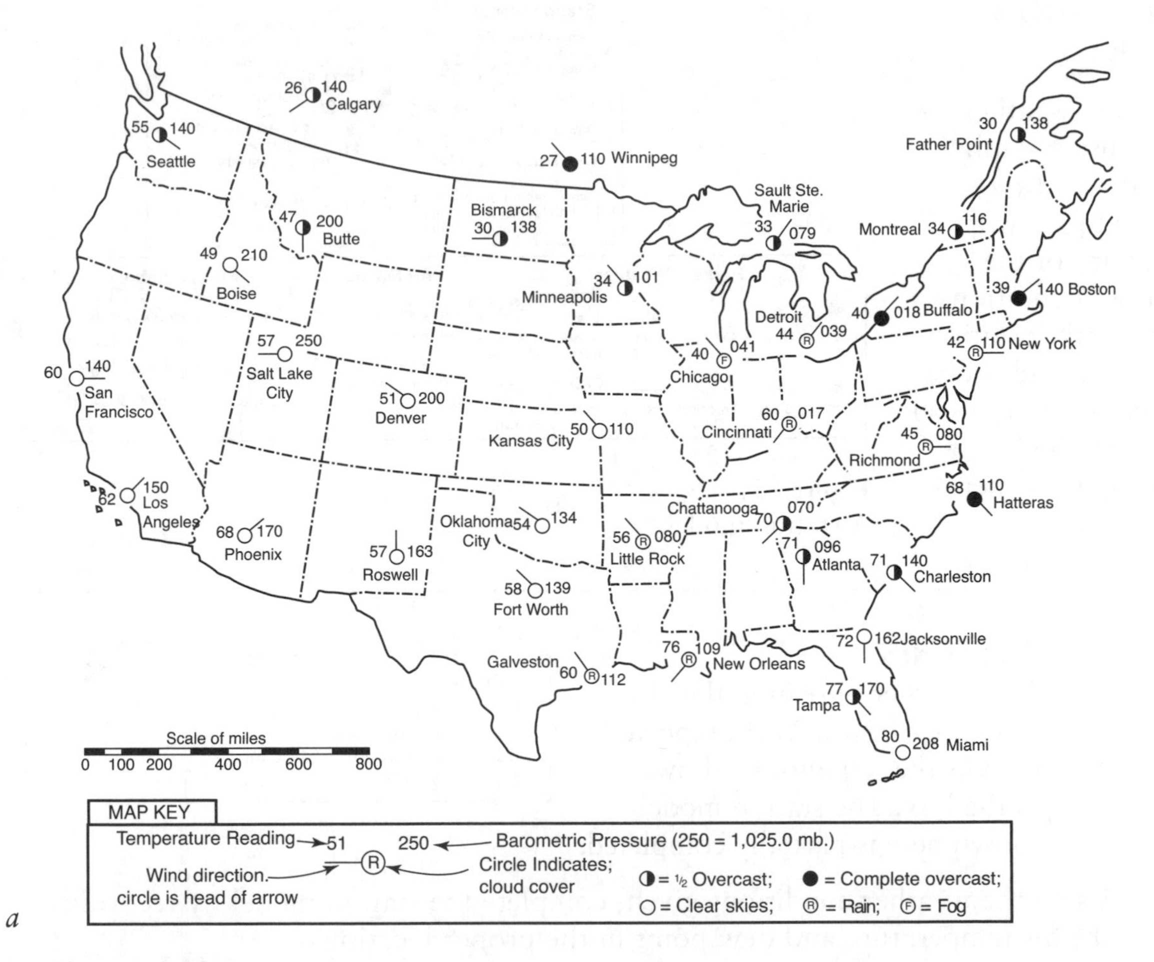

Figure 12.10
(*a*) Synoptic weather map. (*b*) Isotherms. (*c*) Isobars.

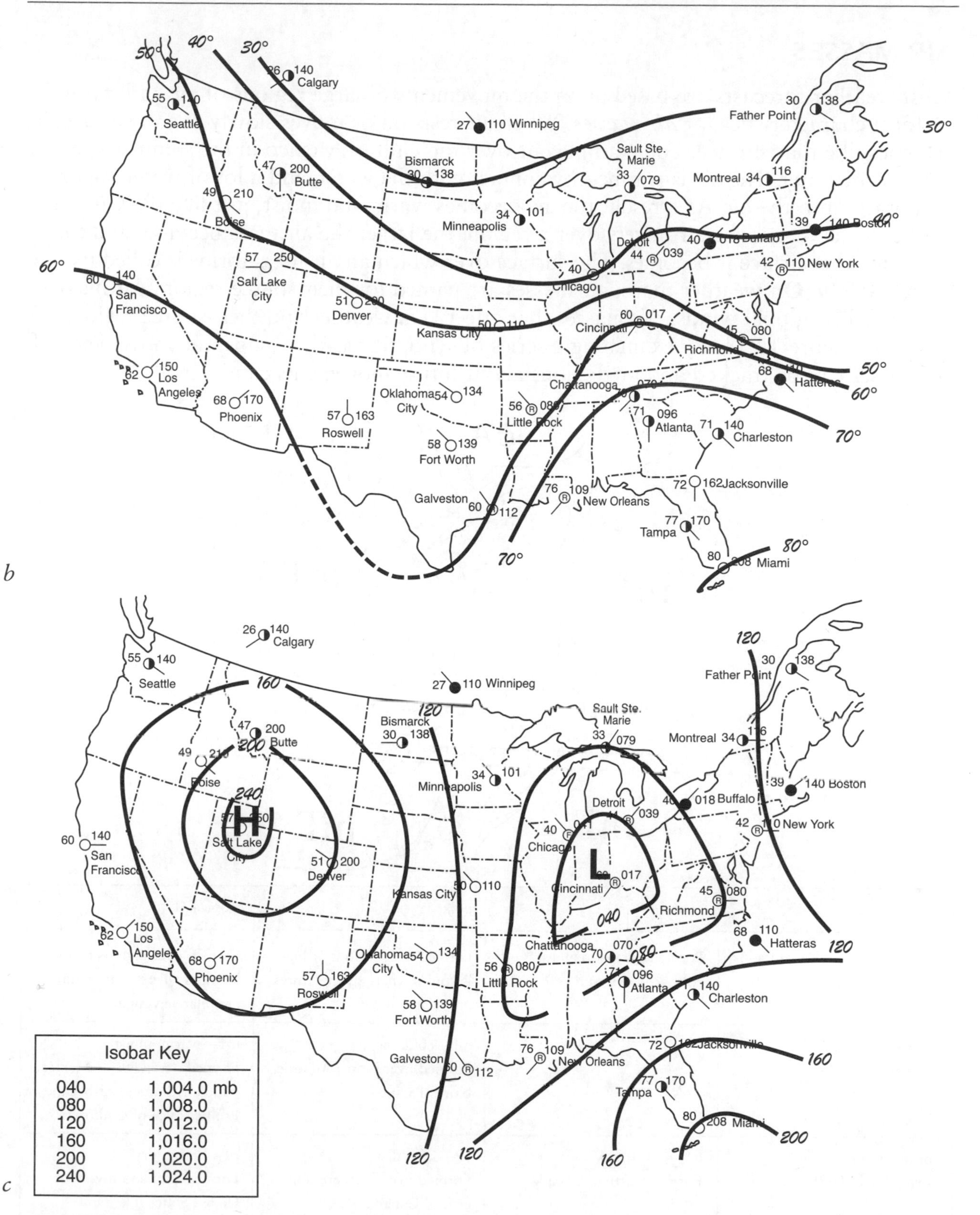

Figure 12.10
Continued

AIR MASSES

Most weather forecasts are based upon the movements of large regions of air with fairly uniform characteristics, or ***air masses***. When air rests on or moves slowly over a surface, it becomes like that surface. For example, air over the Gulf of Mexico in the summer rests on very warm water. The air is warmed by contact with the water, and a lot of water vapor evaporates into the air. As a result, the air becomes warm and moist, just like the Gulf of Mexico. The longer the air lingers over a region, the larger the air mass becomes and the more like the surface it becomes. The surface over which an air mass forms is called its ***source region***. On weather maps, air masses are named for their source region. Air masses are moved by global winds. Air masses that affect U.S. weather and their source regions are shown in Figure 12.11. By examining a series of synoptic weather maps, the movements of air masses can be tracked and predictions about future movements can be made.

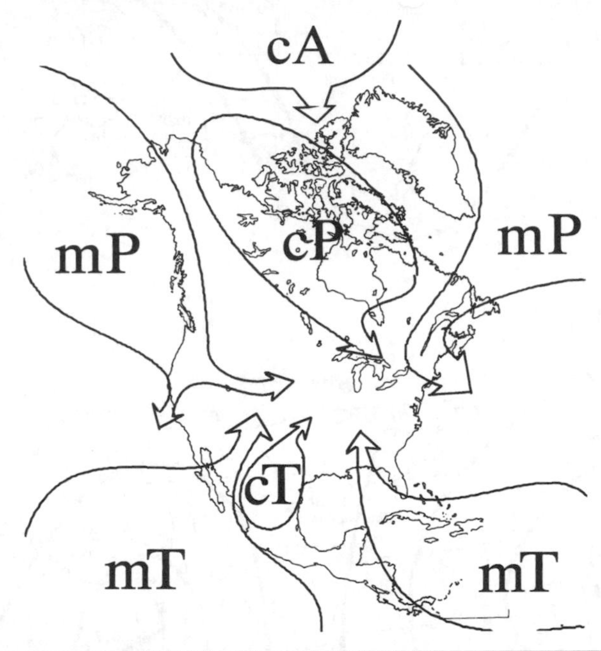

	arctic A	polar P	tropical T
	Formed over extremely cold, ice-covered regions	Formed over regions at high latitudes where temperatures are relatively low	Formed over regions at low latitudes where temperatures are relatively high
maritime m Formed over water, moist		mP—cold, moist Formed over North Atlantic, North Pacific	mT—warm, moist Formed over Gulf of Mexico, middle Atlantic, Caribbean, Pacific south of California
continental c Formed over land, dry	cA—dry, frigid Formed north of Canada	cP—cold, dry Formed over northern and central Canada	cT—warm, dry Formed over southwestern United States in summer

Figure 12.11
Air mass source region map and chart of air mass names.

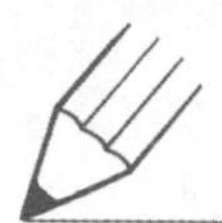

TRY THIS

1. Compare and contrast the air temperature, air pressure, and humidity in a maritime tropical air mass and a continental polar air mass.
2. Name a possible source region for a maritime polar air mass that has moved over New York State.

WEATHER FRONTS

At any given time, several air masses are usually moving across the United States. They generally move from west to east driven by global winds. When different air masses meet, very little mixing of the air takes place. The whole surface along which the air masses meet is called the *frontal surface*; the line on the ground marking the boundary between the air masses is called a ***front***. See Figure 12.12. It is called a front because it marks the leading edge, or front, of an air mass that is pushing against another air mass. Fronts are places where weather changes rapidly and often have unsettled and rainy weather.

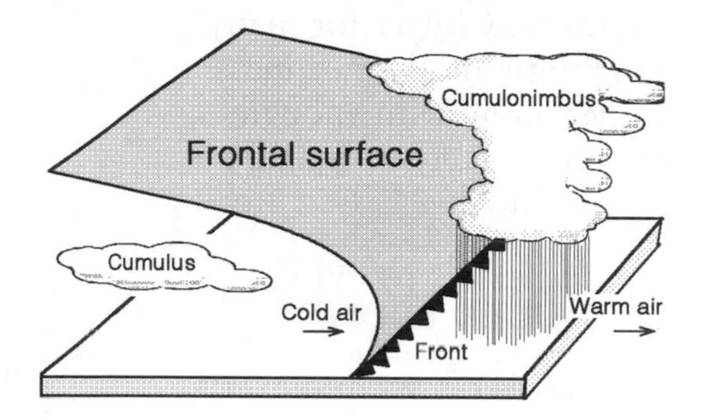

Figure 12.12
View of a frontal surface and a front line on the ground.

COLD FRONTS

A cold front forms along the leading edge of a cold air mass advancing against a warmer air mass. The cold air mass is denser than the warmer air ahead of it, so the cold air pushes against and under the warmer air like a wedge. This *forces* the warmer air upward rapidly and results in turbulence; rapid condensation forming heavy, vertically developed clouds; and heavy precipitation, or thunderstorms. The temperature drops sharply and the pressure rises rapidly as a cold front passes. See Figure 12.13*a*.

WARM FRONTS

A warm front forms where a warm air mass pushes up and over a cold air mass ahead of it. The warm air mass rides up and over the cold air because warm air is less dense than the cold air. As the warm air rides up, it rises, expands, and cools. This causes condensation to occur over the wide, gently sloping boundary. The result is thickening, lowering clouds, and widespread precipitation. See Figure 12.13*b*.

STATIONARY FRONTS

A stationary front forms between a warm air mass and a cold air mass when neither has enough force behind it to move the other. A stationary front slowly takes on the shape and characteristics of a warm front as the denser cold air slowly slides beneath the warmer, less dense air. A stationary front has the widespread rain and cloudiness of a warm front. However, the rain and clouds may linger for many days until another air mass comes along with enough impetus to get the stalled air masses moving.

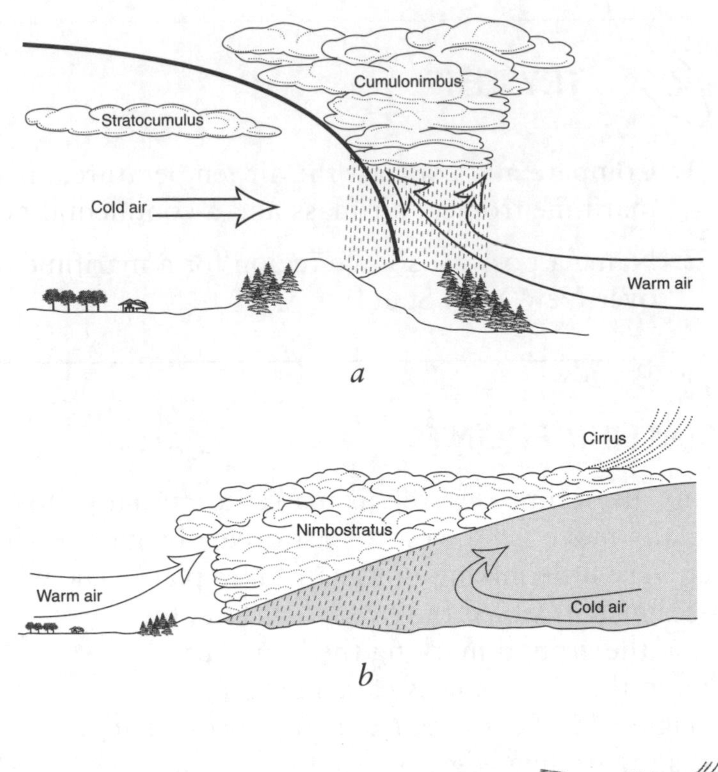

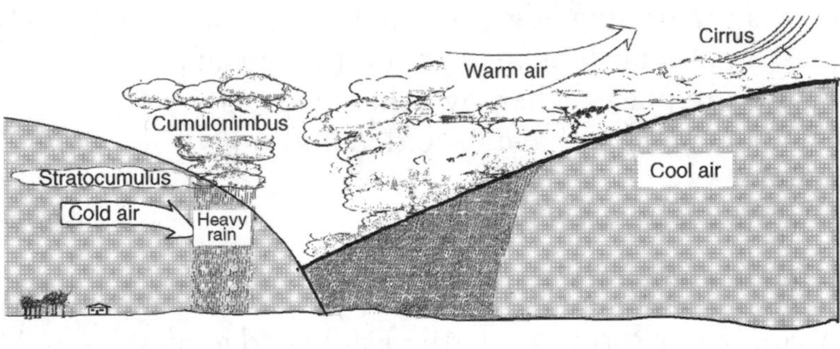

Figure 12.13
Types of fronts. (*a*) A cold front. (*b*) A warm front. (*c*) An occluded front.

OCCLUDED FRONTS

Cold air is denser and exerts more pressure than warm air. Therefore, cold air masses and cold fronts tend to move faster than warm air masses and warm fronts. Therefore, a cold front will sometimes overtake a warm front. Then the warm air mass is trapped between two cold air masses and is lifted completely off the ground, forming an ***occluded front***. See Figure 12.13*c*. The lifting of the warm air mass causes a lot of condensation and precipitation, resulting in widespread rain and thunderstorms. Depending on which cold air mass is denser, a cold front occlusion like the one shown in Figure 12.13*c* may form, or the advancing cold air mass may ride over the surface of the warm front forming a warm front occlusion (not shown).

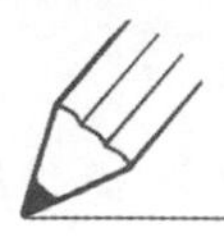

TRY THIS

Base your answers to the following questions on the diagram at right. The diagram shows a side view of a cold air mass meeting a warm air mass.

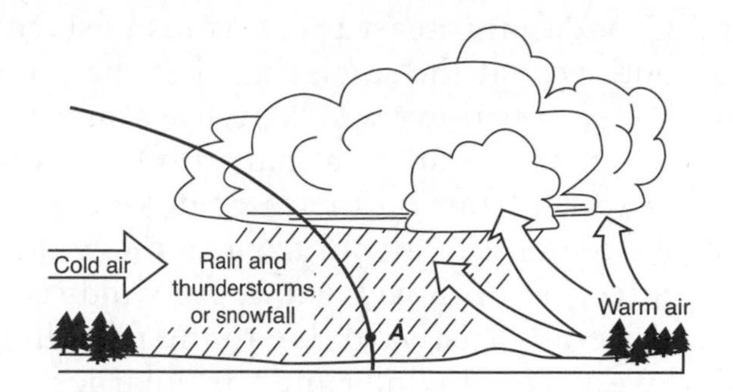

1. Why does the warm air rise upward over the cold air?
   1. The warm air is less dense than the cold air.
   2. The warm air is attracted to moisture in the clouds.
   3. The cold air is lighter than the warm air.
   4. The cold air contains more moisture than the warm air.

2. In the diagram, point *A* is located
   1. at the base of a cloud
   2. at a frontal boundary
   3. in the funnel of a tornado
   4. near the end of a rainbow

CYCLONES AND ANTICYCLONES

Winds in the middle latitudes form weather systems called cyclones and anticyclones. These systems may stretch for several hundred to a thousand or more miles across and move from west to east. At the center of a ***cyclone***, the air pressure is low. Air rushing into it curves to the right in the Northern Hemisphere and toward the left in the Southern Hemisphere. As a result, winds in a cyclone blow in a counterclockwise spiral in the Northern Hemisphere and a clockwise spiral in the Southern Hemisphere. See Figure 12.14*a*. An ***anticyclone*** is centered on a high-pressure region from which air moves outward. The Coriolis effect therefore causes winds in an anticyclone to blow outward in a clockwise spiral in the Northern Hemisphere and a counterclockwise spiral in the Southern Hemisphere. See Figure 12.14*b*. These spirals can be clearly seen in cloud formations photographed from satellites.

In general, cyclones bring unstable weather with clouds, rain, strong winds, and sudden temperature changes. On the other hand, weather associated with anticyclones is pleasant, with clear skies and little wind.

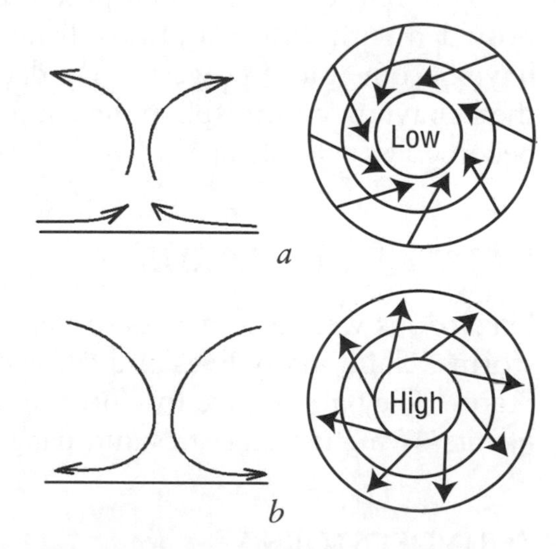

Figure 12.14
(*a*) A cyclone and (*b*) an anticyclone in the Northern Hemisphere.

Near the border between the United States and Canada, northeast polar winds push cold air southwest at the same time that the southwest prevailing winds push warm air northeast. The result is a midlatitude cyclone, with a warm front on the east side of the storm's center and a cold front to the west. The direction of movement of the winds is then generally eastward due to the prevailing southwest global winds and the polar jet stream. See Figure 12.15.

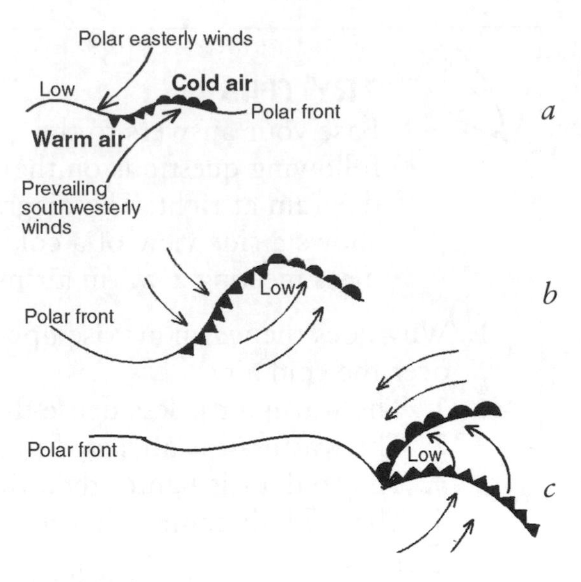

Figure 12.15
How cyclones form.

WEATHER FORECASTING

By looking at a series of synoptic weather maps, one can track the speed and direction in which fronts are moving. This information can be used to predict the path of a storm and make a basic weather forecast.

Weather forecasts based on synoptic weather maps are fairly accurate for large-scale, short-term (1–3 day) forecasts. However, local conditions can strongly affect weather. For example, the concrete and pavement of cities causes them to get hotter than the forests and fields of the surrounding countryside. This creates urban heat islands that can affect the path of a storm, the times when rain will begin and end, and the temperatures the area will experience.

Most modern forecasting is based on computer models that use complex equations to predict the behavior of air based on the laws of physics. Different computer models follow different computational procedures in generating their forecast. They may also be provided with different atmospheric, oceanographic, and geographic data. As computers have become more powerful, the detail with which these computer models can simulate the behavior of atmosphere has improved. However, whichever model is used, forecasts become less reliable as they try to predict further and further into the future.

WEATHER HAZARDS

Hazardous weather, such as tornadoes, severe thunderstorms, hurricanes, and winter storms, claim many lives and cause much property damage each year in the United States. They all involve low pressure, clouds, precipitation, and strong winds, but each has its own characteristics and dangers.

THUNDERSTORMS

Thunderstorms are likely to occur wherever and whenever there is strong heating of Earth's surface. This causes warm air to rise rapidly, forming a cloud that grows larger and larger as more and more warm, moist air is carried upward. The strong updrafts in the rising air support water droplets and ice crystals in the cloud so that they grow in size. When the updrafts cannot support the moisture anymore, rain or even hail occurs. The falling rain sets up downdrafts that cause internal friction with updrafts. The internal friction builds up static electric charges that may discharge as lightning. See Figure 12.16.

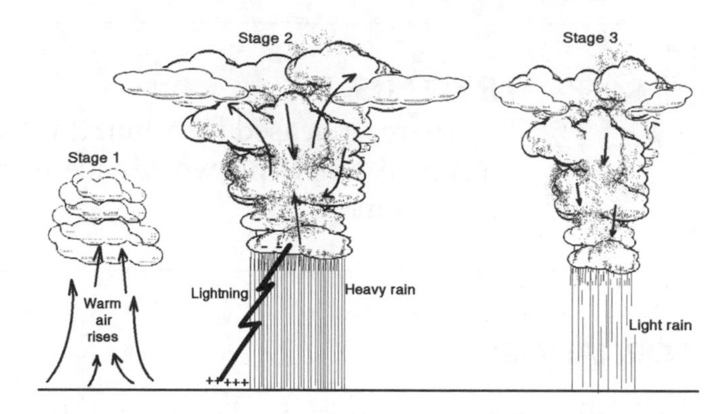

Figure 12.16
Formation of a thunderstorm.

HURRICANES

Hurricanes are huge cyclonic storms that form over oceans near the equator during summer months, when the ocean surface is the warmest. Both heat and water vapor enter the air from the warm ocean, decrease air's density, and lower air pressure. The combination of the two causes air pressure to get very low and many strong convection cells to form. With them, many thunderstorms occur. If the widespread thunderstorms merge, they can form a large updraft that is part of a huge convection cell. As strong winds carry heat and moisture toward the low-pressure center, the convection cell gets larger and stronger. When winds reach 119 kilometers per hour, the storm is called a hurricane. Fully formed hurricanes are huge cyclones, often exceeding 500 kilometers in diameter. In the center of the cyclone, the air is rising, not moving horizontally, so winds are calm. The rising air spreads out, carrying clouds outward, so skies are clear. This calm, clear area at the center of the hurricane is called the ***eye of the hurricane***. See Figure 12.17.

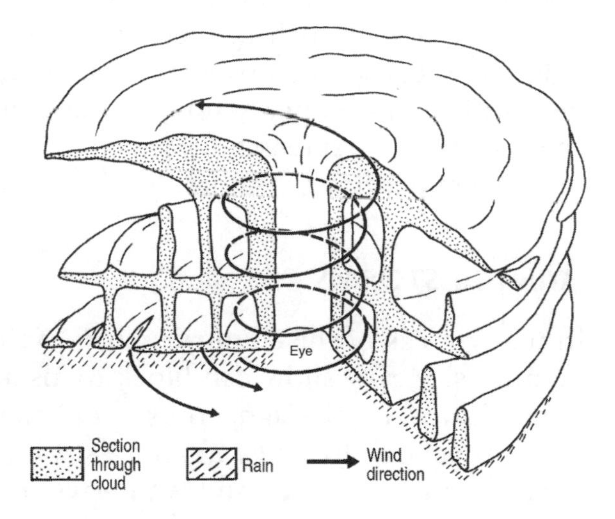

Figure 12.17
Cross section of a hurricane.

Anything that cuts off the supply of heat or moisture will weaken a hurricane, so hurricanes lose strength as they move over land or cool water. However, even after moving over land, hurricanes can last for many days and cause much damage due to high winds and flooding.

TRY THIS

The energy released by a hurricane in one day is greater than the energy consumed by people worldwide during an entire year. Where does all this energy come from?

TORNADOES

Tornadoes are small (most are less than 100 meters in diameter), brief (most last only a few minutes) disturbances that usually form over land from intense thunderstorms. Tornadoes usually form late in the day, when Earth's surface is the warmest. When heating is very intense, warm air rises in strong convection currents. The upward movement of the air causes a sharp decrease in pressure. Air rushes into the low-pressure region from the sides and is given a spin by the Coriolis effect. Wind speed gets very high near the center of the updraft due to the big difference in air pressure. The rapidly moving air decreases the pressure even more, which further feeds the updraft. The whole process spirals upward in intensity, and a funnel forms that eventually touches the ground. Wind speeds near the center of a tornado may reach 500 kilometers per hour or more. See Figure 12.18.

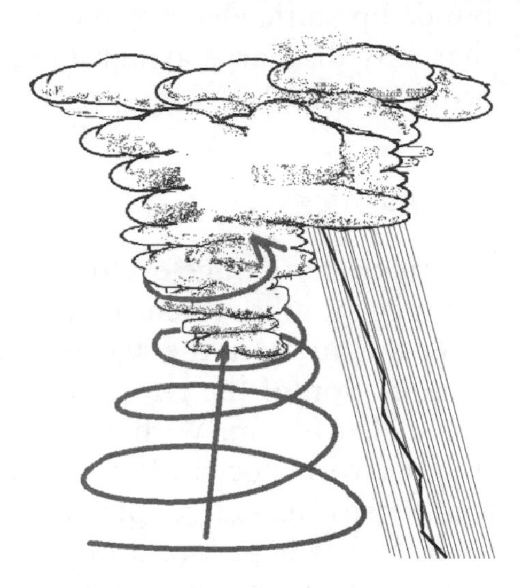

Figure 12.18
Formation of a tornado.

WINTER STORMS

In many areas of the country, winter cyclones bring heavy snowfall and very cold temperatures. Heavy snow can block roads and cause power lines to fall down. The cold temperatures can be dangerous if a person is not properly dressed. Winter storms include ice storms and blizzards. In an *ice storm*, rain freezes when it hits the ground, creating a coating of ice on roads and walkways. Rain that turns to ice pellets before reaching the ground is called sleet. Sleet also causes roads to freeze and become slippery. In a *blizzard*, heavy snow and strong winds produce a blinding snow, near-zero visibility, deep drifts, and life-threatening wind chills. A major winter storm can be lethal. So if possible, stay indoors and do not travel.

EMERGENCY PREPAREDNESS

We can do certain things to reduce the risk of damage and loss of life from hazardous weather. Advance planning and quick response are the keys. Toward this end, the National Weather Service issues two levels of alerts about life-threatening weather such as tornadoes, severe thunderstorms, winter storms, and hurricanes: watches and warnings. A ***watch*** means that hazardous weather conditions are likely to develop in your area. If a watch is issued for your area, you should stay alert to the weather by listening to the radio or television and be prepared to take shelter. A ***warning*** means that hazardous weather has been sighted or shown on radar in your area. If a warning is issued, the danger is very serious and everyone should go to a safe place. Once there, listen to a battery-operated radio or television for further instructions and an official all clear before resuming normal activities. The Federal Emergency Management Agency has developed severe thunderstorm, tornado, and hurricane fact sheets that describe specific steps to be taken when watches and warnings are issued for these weather hazards. These fact sheets are available online at *www.fema.gov*

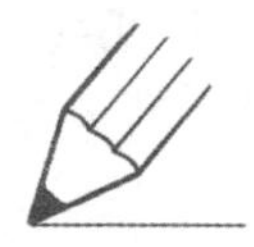

TRY THIS

1. In one or more complete sentences, explain the difference between a tornado watch and a tornado warning.
2. What should you do when the National Weather Service issues a hazardous weather warning?

CLIMATE

Climate is the term used to describe the general weather conditions in a given area over a long period of time. One cannot determine the climate of a region by looking at the weather over a single year. In 1988, the Midwest suffered a nearly rainless summer. Fields of grain in the nation's heartland turned brown and shriveled. Water levels dropped in the Mississippi River. Wildfires blazed through millions of acres of forested land. Yet, in the summer of 1993, many of these same places were drenched by torrential rains. Farmers watched floodwaters wipe out their fields, and the Mississippi River flooded communities all along its course. Many people wondered, are we seeing Earth's global climate change? One or two extreme summers cannot answer that question. Our picture of climate develops slowly as we watch dozens of seasons pass. Some winters are warmer than others, some summers dryer, some falls colder. We can get a sense of the shifting patterns of climate only by comparing measurements taken over many years and decades.

THE MAIN ELEMENTS OF CLIMATE

Climate is the result of the interplay of a number of factors. One of the most important is energy. Earth's main source of energy is sunlight, which warms the land. In turn, this heats the atmosphere. A key way of tracking energy flow in a region is by monitoring its temperature. Therefore, scientists who study climate keep track of air temperatures over land and sea, air temperatures at various altitudes, and ocean water temperatures around the globe.

The other major climate factor is water. Water is important to all living things. To a large extent, it controls the type of plant and animal life that can live in a region. It is also important in weathering and is the main agent of erosion and deposition on Earth. Some of the measurements used to track water and climate changes are precipitation, the amount of water vapor in the air, the amount of snow and ice cover on land, and the extent of sea ice.

SOME FACTORS AFFECTING CLIMATE

The main elements of a region's climate are its temperature and precipitation patterns. These are controlled by various factors, such as latitude, nearness to a large body of water, elevation, mountain ranges, and vegetation.

LATITUDE

Latitude is the main factor affecting a region's temperature patterns. Latitude determines both the angle at which sunlight strikes Earth's surface and the length of the daylight period during which sunlight can warm the region. Differences in these two things result in three main temperature zones on Earth. The always cold ***frigid zone*** is near the poles. The always hot ***torrid zone*** is near the equator. The seasonally changing ***temperate zone*** is in between.

NEARNESS TO LARGE BODIES OF WATER

Land surfaces heat up and cool off faster than water surfaces. Therefore, air temperatures are usually colder in the winter and warmer in the summer when over landmasses than when over oceans at the same latitude. Large bodies of water moderate the temperatures of nearby land by warming it in the winter and cooling it in the summer. Therefore, cities near a large body of water tend to have warmer winters and cooler summers than cities far inland.

ELEVATION

In the troposphere, a gradual but steady decrease in temperature occurs with elevation. Temperature decreases about 1°C for every 100-meter rise in elevation. This explains why high mountains may have tropical jungles at their base but permanent ice and snow at their peaks. In general, a city at a high elevation will have a cooler climate than one at a low elevation.

MOUNTAIN RANGES

Mountain ranges serve as barriers to outbreaks of cold air. In this way, the Alps protect the Mediterranean coast and the Himalayas protect India's lowlands from cold, polar air. The side of a mountain range facing the prevailing winds tends to have a cool, moist climate, while the other side of the mountain range has a warm, dry climate. This happens because air blown over mountain ranges is forced to rise and cool. This causes condensation and forms clouds and precipitation. By the time the air reaches the top of the mountains, it has lost much of its moisture. When the air descends on the other side of the mountains, it is warmed by compression. The air is warm and dry, and precipitation is less likely. This explains why Tillamook, a city west of the Cascade Mountains in Oregon, has a cool, wet climate, while Bend, Oregon, on the other side of the Cascades, has a warm, dry climate. See Figure 12.19.

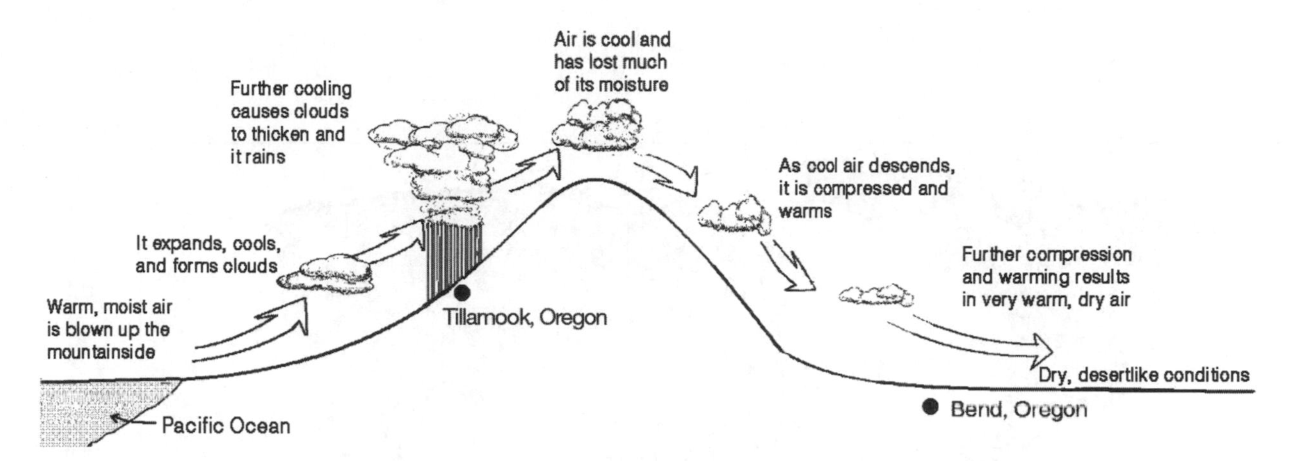

Figure 12.19
How mountain ranges affect climate.

VEGETATION

Vegetation affects the water cycle by influencing the processes of transpiration and surface runoff, which, in turn, influence rainfall. Vegetation also influences temperature. Vegetation reflects less insolation back into space than bare surfaces, which tends to warm the climate. However, this effect is small compared with the cooling effect when vegetation absorbs carbon dioxide from the air and thereby decreases the greenhouse effect. In the ***greenhouse effect***, carbon dioxide in the air lets sunlight pass through to Earth's surface and warm it but blocks heat rays from escaping. Thus, deforesting the land tends to warm the climate.

TYPES OF CLIMATES

Climates are classified by moisture, temperature, and vegetation patterns. Several systems are in use today. Table 12.3 summarizes terms often used to describe different patterns of moisture, temperature, and vegetation. Figure 12.20 shows the major climates of the world.

Table 12.3 Terms Used to Describe Climate Patterns

Moisture	Temperature	Vegetation
Arid	Polar	Desert
Semiarid	Subpolar	Grassland, steppe, taiga
Subhumid	Subtropical	Deciduous forest Coniferous forest
Humid	Tropical	Rain forest

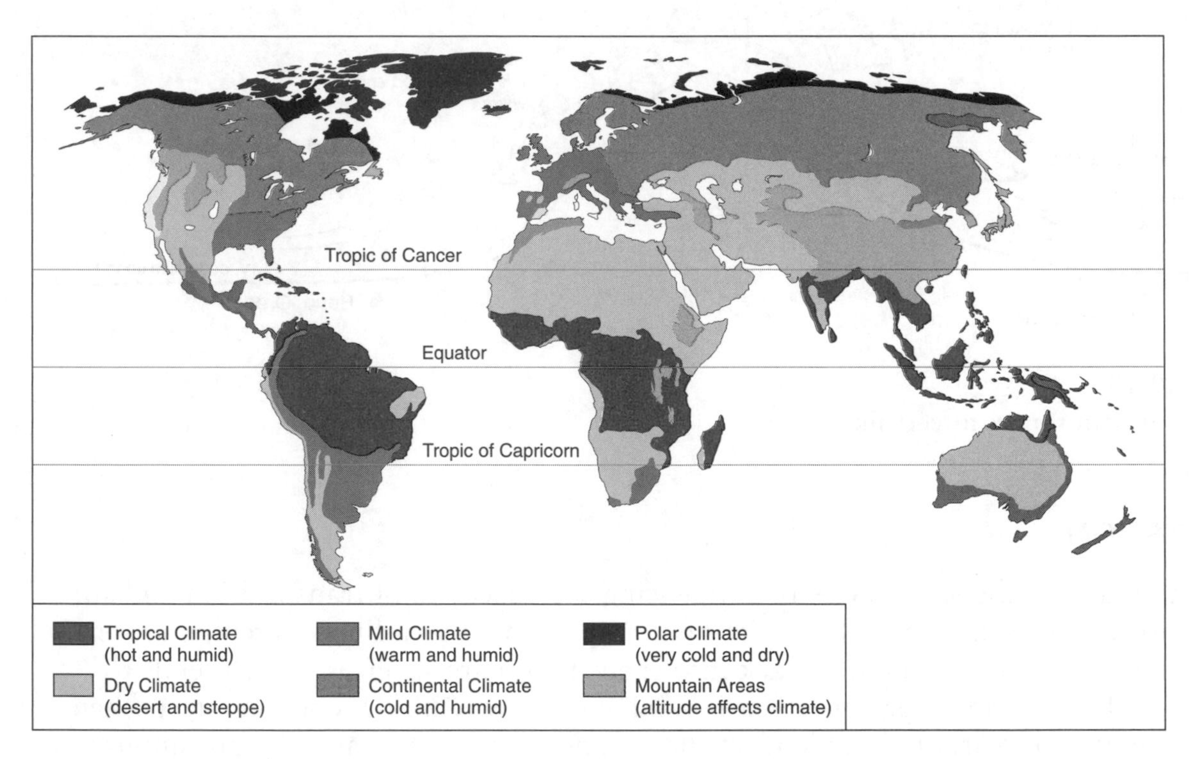

Figure 12.20
Major climates of the world.

Practice Review Questions

1. The major cause of Earth's weather is

1. changes in the distance between Earth and the Sun
2. continental drift and seafloor spreading
3. unequal heating of the atmosphere
4. tidal forces due to the Moon's changing distance from Earth

2. Energy is transferred from the Sun to Earth mainly by

1. radiation
2. molecular collisions
3. density currents
4. red shifts

3. In which diagram does the sunlight cause the most heating of Earth's surface?

1. 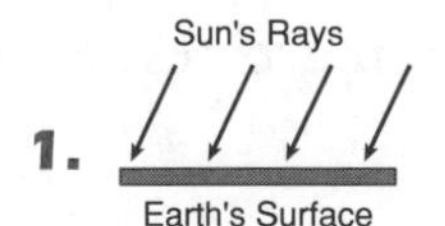

3. 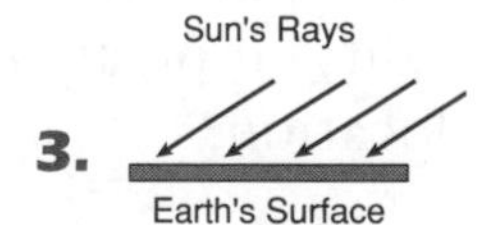

2. 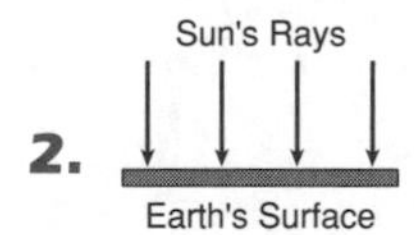

4. 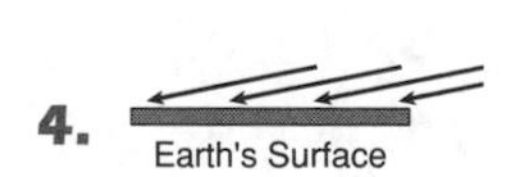

4. When sunlight strikes a flat, snow-covered field at a low angle, most of the energy will be

1. absorbed by the snow
2. reflected by the snow
3. radiated by the snow
4. conducted by the snow

5. The change from water vapor to liquid water is called

1. evaporation
2. transpiration
3. precipitation
4. condensation

Note that question 6 has only three choices.

6. As warm, moist air moves into a region, the air pressure in the region will generally

1. increase
2. decrease
3. remain the same

7. Surface winds on Earth are mainly caused by differences in

1. distances from the Sun during the year
2. ocean wave heights during the tidal cycle
3. rotational speeds of Earth's surface at various latitudes
4. air density due to unequal heating of Earth's surface

8. Which factor most directly affects the wind speed between two locations?

1. air pressure
2. cloud cover
3. Coriolis force
4. time of day

GO ON

9. Which cross section shows the most likely direction of the surface winds that will develop over a land area next to the ocean on a hot, sunny afternoon?

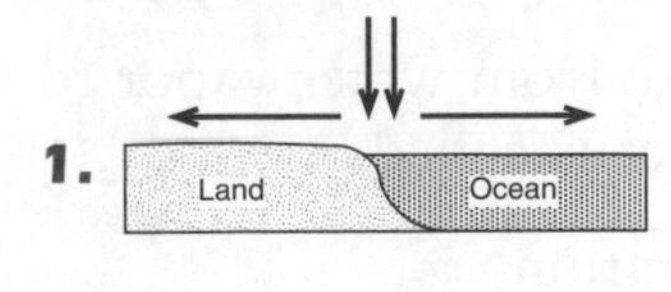

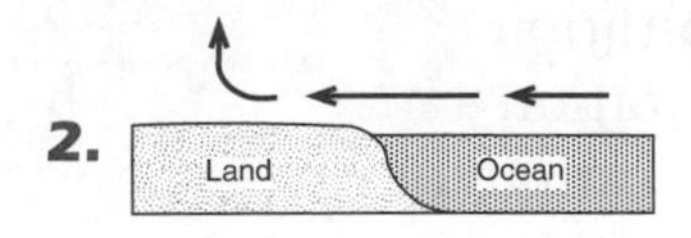

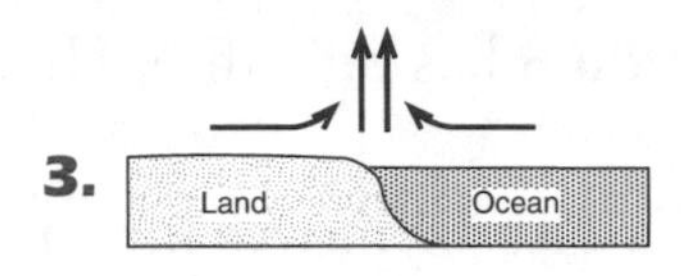

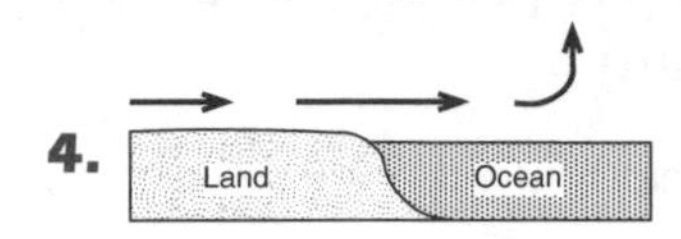

10. Winds appear to curve to the right in the Northern Hemisphere. This curving to the right is caused by Earth's

1. rotation
2. revolution
3. size
4. shape

11. Which substance is a form of precipitation?

1. hail
2. frost
3. fog
4. dew

Questions 12–14 refer to the following diagrams, which show station models with weather data collected at four different times during one day at a location in New York State.

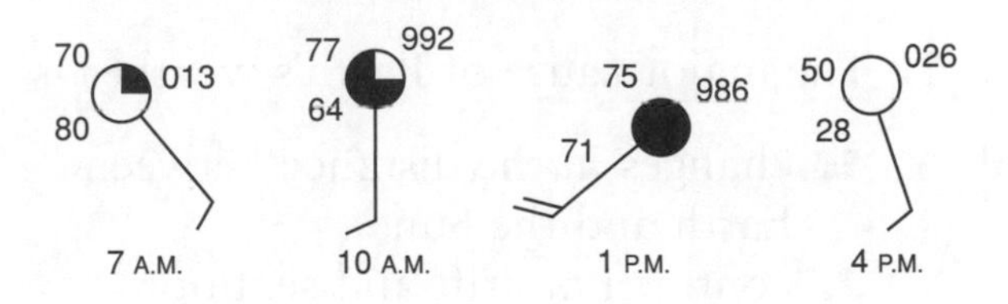

12. What was the air pressure at 4 P.M.?

1. 1,026 mb
2. 260 mb
3. 1,002.6 mb
4. 26.0 mb

13. At which time of day was the greatest wind speed recorded?

1. 7 A.M.
2. 10 A.M.
3. 1 P.M.
4. 4 P.M.

14. At what time of day was the probability of precipitation greatest at this location?

1. 7 A.M.
2. 10 A.M.
3. 1 P.M.
4. 4 P.M.

15. Present-day weather predictions are based mainly on

1. ocean currents
2. cloud height
3. air mass movements
4. land and sea breezes

GO ON ➡

Questions 16–18 refer to the weather map below showing the location of a front and the air mass influencing its movement:

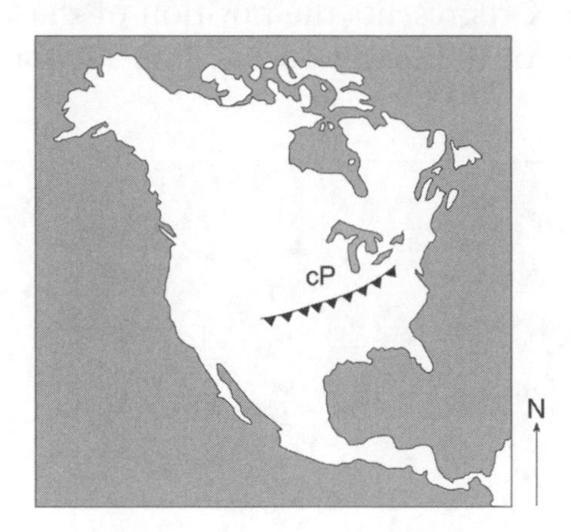

16. Which region is the most likely source of the air mass labeled **cP** on the map?

1. Gulf of Mexico
2. central Canada
3. North Atlantic Ocean
4. southwestern United States

17. The **cP** air mass is identified on the basis of its temperature and

1. wind direction
2. windspeed
3. moisture content
4. cloud cover

18. What type of front is shown in this weather map?

1. warm
2. cold
3. occluded
4. stationary

19. The diagram below shows how prevailing winds have caused different climates on the windward and leeward sides of a mountain range.

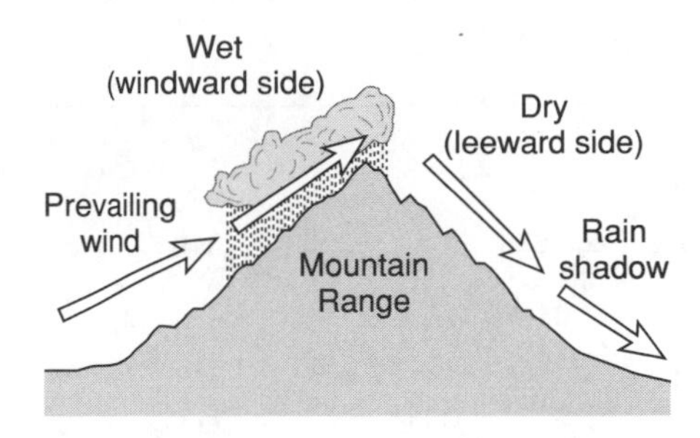

Why does the windward side of this mountain have a wet climate?

1. Rising air expands and warms, causing water droplets to evaporate.
2. Rising air expands and cools, causing water vapor to condense.
3. Rising air compresses and warms, causing water vapor to condense.
4. Rising air compresses and cools, causing water droplets to evaporate.

20. The average temperature, amount of sunlight, and amount of rainfall or snowfall for many years would determine an area's

1. climate
2. weather
3. earthquake risk
4. rate of crustal uplift

GO ON ➡

CONSTRUCTED-RESPONSE QUESTIONS

Questions 21–23 refer to the weather map below, which shows partial weather station data for several cities in eastern North America.

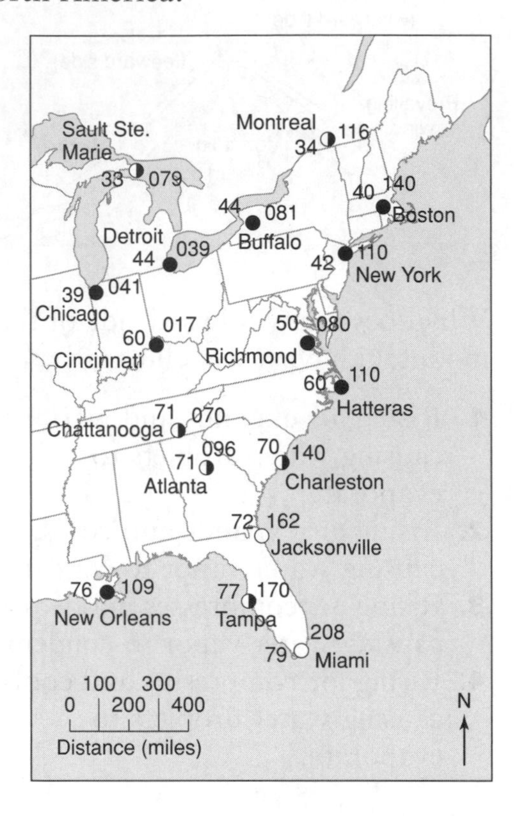

21. On the given weather map, draw isotherms every 10°F, starting with 40°F and ending with 70°F. (*Isotherms must extend to the edges of the map.*)

22. State the general relationship between air temperature and latitude for locations shown on this map.

23. State the actual air pressure, in millibars, shown at Miami, Florida, on the given weather map.

Questions 24–27 refer to the map below, which represents a satellite image of Hurricane Gilbert in the Gulf of Mexico. Each X represents the position of the center of the storm on the date shown.

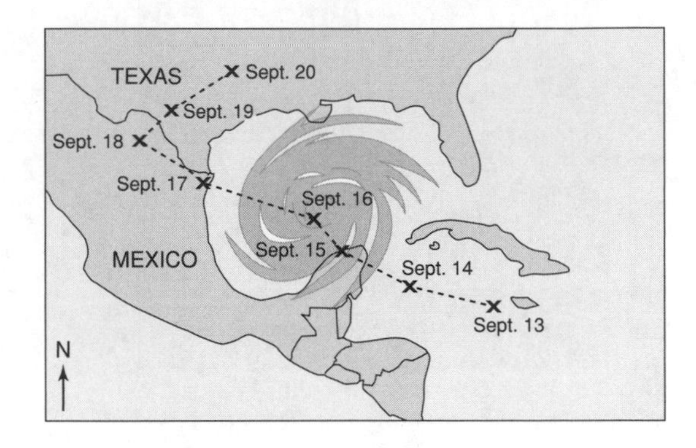

24. State the general direction of Hurricane Gilbert's track from September 13 through September 18.

25. State *one* reason Hurricane Gilbert weakened between September 16 and September 18 in the given map.

26. State *two* dangerous conditions that could cause loss of human life and property as the hurricane strikes the coast.

27. Describe one emergency preparation humans could take to avoid a problem caused by one of these dangerous conditions.

Chapter 13

Chemistry

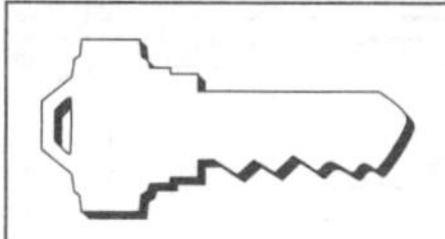

Key Idea: Matter is made up of particles whose properties determine the observable characteristics of matter and its reactivity.

UNDERSTANDING MATTER

Chemistry is the science dealing with the composition, structure, and properties of matter; the changes in matter; and the energy involved in changes in matter. What is matter? *Matter* is anything that has mass and takes up space. A block of wood, a diamond, a puddle of water, air—all are examples of the millions of different kinds of matter found in the world. Although these things are very different, they all share two characteristics: They all have mass and they all take up space.

MASS AND WEIGHT

The *mass* of an object is the amount of matter it contains. A boulder has a large mass. A grain of sand has a small mass. In general, the more mass an object has, the heavier it feels. However, mass should not be confused with weight. Although the two terms are often used interchangeably, they are two different concepts.

Weight is the amount of the force of gravity between one object and another object such as Earth. Gravity is a force of attraction that exists between all matter. The force of gravity between two objects depends on how much matter they contain and how far apart they are. The more matter an object contains, the stronger its gravity. The farther apart the centers of two objects, the weaker the force of gravity between them.

The mass of an object does not vary. It is the same everywhere in the universe. On the other hand, the weight of an object varies with its distance from the center of Earth or any other planet or body on which it is measured. You weigh slightly more at sea level than you do on top of Mount Everest because you are closer to Earth's center at sea level. However, your mass is the same in both places. Weight also varies with the mass of the planet or body on which it is measured. On the Moon, you weigh about one-sixth what you weigh on Earth because the Moon has much less mass than Earth. That does not mean that your body would be made up of one-sixth as much flesh, bone, and blood! Your body would contain the same amount of matter on the Moon as it does on Earth, that is, your mass would be the same.

MEASURING MASS AND WEIGHT

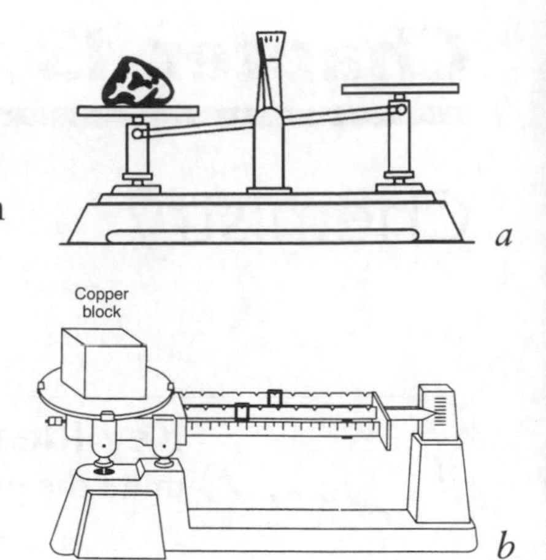

Figure 13.1
Some types of balances used to measure mass. (*a*) Pan balance. (*b*) Triple beam balance.

Mass can be measured with a ***balance***—an instrument that compares the mass of two different objects by using a seesaw principle. See Figure 13.1. One object is placed onto each pan of the balance. If their masses are equal, Earth's gravity will attract them equally and the pans will balance out. If one object has more mass than the other, the attraction between Earth and the object with more mass will be greater than the attraction between Earth and the object with less mass. The pan with more mass in it will fall, and the pan with less mass in it will rise.

To measure mass, an object of unknown mass is placed onto one pan of the balance. Then objects of known mass are added to another pan or to a beam until the two masses are balanced. Adding the masses of the known objects gives the mass of the unknown object. In chemistry, the most widely used units of mass are the kilogram, gram, and milligram (1 kilogram = 1,000 grams; 1 gram = 1,000 milligrams). What makes a balance so useful is that it works the same in weak gravity and strong gravity. If you placed equal masses onto a balance on the Moon, the pans would still balance.

Weight is measured with a spring ***scale***, such as a grocery scale. When you place objects into the pan of a grocery scale, gravity pulls the objects toward the center of Earth, causing a spring to stretch. If you did this on the Moon, its weaker force of gravity would pull less on the objects. Therefore, the spring would not stretch as much, and the scale would show a lesser weight for the same objects.

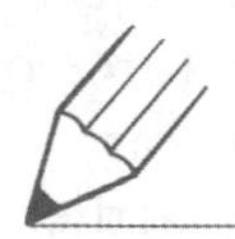

TRY THIS

1. The weight of an object
   1. is the quantity of matter it contains
   2. refers to its size
   3. is basically the same quantity as mass but is expressed in different units
   4. is the force with which it is attracted to Earth
2. On Earth, a box of cookies with a weight of 1 pound has a mass of 454 grams. If the box of cookies is taken to the Moon, which measurement would remain the same?

VOLUME

The ***volume*** of an object is the amount of space it takes up. A basketball has more volume than a baseball because the basketball takes up more space. The volume of a cube or rectangle can be obtained by multiplying its length by its width by its height. The unit of volume is the cube of distance, or $(\text{distance})^3$. For example, the volume of a cube measuring 3 centimeters on a side is

$$\textbf{volume of cube} = \textbf{length} \times \textbf{width} \times \textbf{height} = 3\,\text{cm} \times 3\,\text{cm} \times 3\,\text{cm} = 27\,\text{cm}^3$$

Other formulas can be used to calculate the volume of regular shapes, such as spheres, cones, and cylinders.

MEASURING VOLUME

The volume of a liquid is measured with a ***graduated cylinder*** (also called a graduate). This instrument is marked with *graduations*, or lines, to show volume. It is used like a kitchen measuring cup. Liquid volume is usually measured in a metric unit called the liter. Since 1 liter = 1,000 cubic centimeters and 1 liter = 1,000 milliliters, 1 milliliter = 1 cubic centimeter ($1\,\text{mL} = 1\,\text{cm}^3$).

Most liquids form a curved surface, or ***meniscus***, in a graduate. The volume is read at the lowest point of the meniscus. See Figure 13.2*a*. It can also be used to measure the volume of an irregular solid by the volume of liquid the solid displaces. See Figure 13.2*b*. Other kinds of graduated glassware can be used to measure the volume of a gas in a similar way.

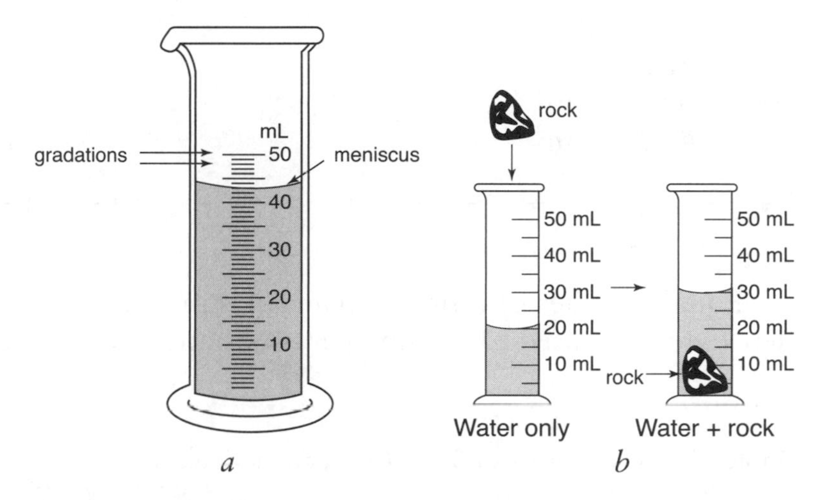

Figure 13.2
Graduated cylinder (*a*) showing volume at the meniscus, (*b*) determining volume by displacement (volume of rock = 10 mL).

DENSITY

Density is a measure of how closely the matter of a given substance is packed into a given volume. In other words, density is the mass (m) per unit volume (V) of a substance. The official unit of density is kilograms per cubic meter. However, in chemistry, you will normally deal with small samples of matter, so density is given in units of grams per cubic centimeter (g/cm^3) or grams per milliliter (g/mL).

$$\text{density} = \frac{\text{mass}}{\text{volume}} \quad \textit{or} \quad d = \frac{m}{V}$$

TRY THIS

The diagram below shows a mineral sample and a graduated cylinder containing the amount of water shown.

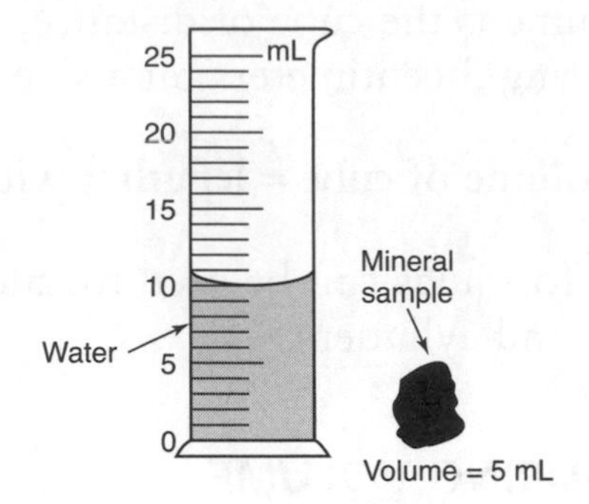

1. If the mineral sample is put into the graduated cylinder, the new water level reading will be
   1. 5 mL
   2. 12 mL
   3. 15 mL
   4. 19 mL
2. If the mineral sample had a mass of 15 grams, its density would be
   1. 0.3 g/cm^3
   2. 3.0 g/cm^3
   3. 15 g/cm^3
   4. 75 g/cm^3

Equal volumes of different pure substances usually have different masses. Therefore, density can be used to tell one pure substance from another. See Table 13.1.

Table 13.1 The Density of Some Common Substances

Solids	Density (g/cm^3)	Liquids	Density (g/cm^3)	Gases	Density (g/cm^3)
Balsa wood	0.13	Gasoline	0.68	Hydrogen	0.00009
Pine	0.45	Kerosene	0.82	Nitrogen	0.0012
Ice	0.92	Liquid water	1.0	Air	0.0018
Sugar	1.6	Mercury	13.6		
Glass	2.6				
Iron	7.8				
Lead	11				
Gold	19				

TYPES OF MATTER

A ***pure substance***, which we will simply call a ***substance***, is one of the many types of matter found in the universe. Pure substances have the same composition and properties throughout. Pure substances can be classified as either *elements* or *compounds*.

An ***element*** is one of the 110 or so basic building blocks of all matter. In other words, all matter is composed of one or more elements. Elements are made of very small particles called ***atoms***. The atoms of an element are like all other atoms of that element but are different from the atoms of other elements. (Atoms will be discussed in more detail later in this chapter.) Examples of elements include iron, copper, aluminum, sulfur, carbon, oxygen, helium, and neon. Scientists have discovered 92 elements that occur naturally on Earth and have made very small amounts of 18 others in the laboratory. All living and nonliving materials are made of these elements or combinations of these elements.

A ***compound*** is a form of matter in which of two or more elements are joined together in a fixed ratio, or *chemically combined*. In a compound, atoms of two or more elements join together in well-defined groups, or ***molecules***. The links that holds the atoms together are called ***chemical bonds***. Chemical bonds contain chemical energy, which can be released if the links are broken.

When linked by chemical bonds, each group of atoms forms a tiny particle. For example, water is a compound of the elements hydrogen and oxygen joined together in a fixed 2:1 ratio. It consists of molecules made of two hydrogen atoms linked to one oxygen atom. All compounds are composed of atoms that are grouped together in different arrangements. More than 6 million compounds of the 110 known elements have been identified. Each week, thousands of new compounds are added to the list. The food, air, and water you ingest; the clothes you wear; and the many other products you use likely contain more than 50,000 different compounds.

Mixtures are made of two or more substances that are not chemically combined with each other and are capable of being separated. Mixtures do not have the same composition and properties throughout. The common rock granite is an example of a mixture. It is made of several mineral compounds, each with its own properties—white quartz, pink feldspar, black mica, and so on. Therefore, granite varies in composition and does not have the same color throughout. Granite also varies in hardness, density, and many other properties. If you break a piece of granite into tiny pieces, you can separate the grains of the different minerals from one another.

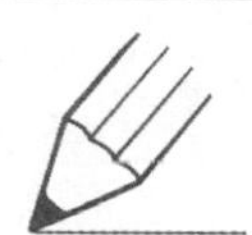

TRY THIS

1. In one or more complete sentences, explain the difference between a compound and a mixture.
2. In what way is a compound like a mixture?

PHASES OF MATTER

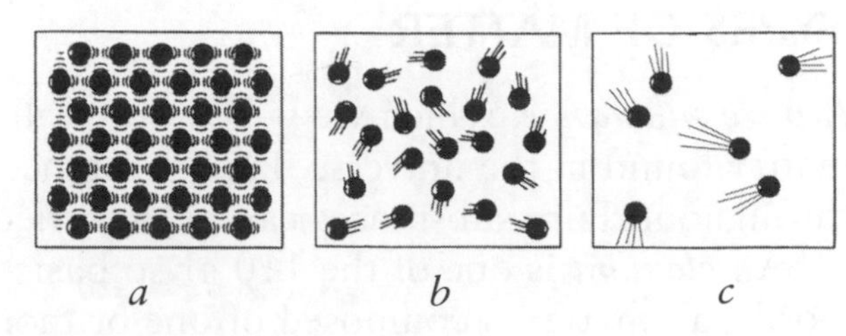

Figure 13.3
The spacing and motion of particles in the three phases of matter (*a*) solid, (*b*) liquid, (*c*) gas.

Matter may exist in three ***physical states***, or ***phases***: *solid*, *liquid*, and *gas*. The phase of a sample of matter is the form in which you find it under a given set of conditions. Each phase has certain characteristics. See Figure 13.3.

Solids have a definite shape and a definite volume. When a small sample of a solid is placed into a container, it does not take the shape or the volume of the container. The particles in a solid are closely locked in position and can only vibrate.

Liquids have a definite volume but do not have a definite or rigid shape. The particles in a liquid are held together loosely and can slide past one another. When a liquid is poured into a container, its particles can flow or move freely enough to take the shape of the bottom and sides of the container. However, they do not move freely enough to fill the entire volume of the container.

Gases have no fixed shape or volume. The particles in a gas are so weakly attracted to one another that they are free of one another except during collisions. In a gas, particles are free to move in any direction. When placed into a closed container, the particles in a gas move about freely, filling the container completely and assuming its shape.

TRY THIS

Use check marks to complete the following chart:

Phases of Matter

Property	Solid	Liquid	Gas
Takes up space			
Has mass			
Takes the shape of its container			
Fills container			

THE KINETIC MOLECULAR THEORY

The explanation of solids, liquids, and gases you just read is based on an idea known as the ***kinetic theory of matter***. The word *kinetic* refers to motion and comes from the same root word as cinema (*moving* pictures). The kinetic theory is based on three simple ideas:

1. Matter is not continuous but is made of small particles separated by empty space.
2. The particles in matter are in perpetual motion.
3. The energy of the moving particles is heat.

What makes it such a powerful theory is that it is able to explain why solids, liquids, and gases behave as they do. It also helps explain how matter acts in solutions. Over time, it led to the next advance in chemistry—atomic theory.

According to the kinetic theory, what we perceive as heat is actually the motion of atoms or molecules. The more heat a sample of matter contains, the faster its particles move. This, in turn, affects the spacing of the particles. The faster the particles are moving, the harder they collide. The harder they collide, the farther apart they rebound after the collision and the more widely spaced they become—and the matter expands. When cooled, the reverse happens. The particles lose energy and move more slowly. They rebound less vigorously and become more closely spaced—and the matter contracts.

What we sense as temperature is really an indication of how fast the particles in a sample of matter are moving. In fact, we measure temperature by the expansion and contraction of matter. A *thermometer* is nothing more than a rigid tube in which a liquid expands or contracts. Since expansion and contraction occur due to changes in the speed at which the particles in matter are moving, you might think of a thermometer as a particle speedometer.

The kinetic theory also explains pressure in terms of particle motion. ***Pressure*** is the amount of force exerted on a given area. The pressure that a gas exerts on a surface is the force of its particles colliding with the surface. Pressure is affected by the speed of the particles (temperature of the matter). The faster the particles are moving, the more force they exert when they collide with a surface. Pressure is also affected by the spacing of the particles. The more closely spaced the particles, the greater the number of particles that collide with a surface and the more total force they exert.

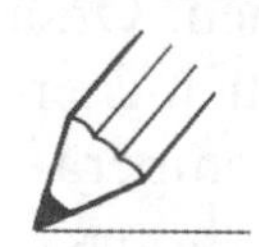

TRY THIS

1. List five solids, five liquids, and two gases found in the home.
2. The temperature of a sample of gas in a closed container is raised. Why does the pressure exerted by the gas on the walls of the container also increase?

CHANGES IN PHASE

By changing temperature, pressure, or both, we can change a substance from one phase to another. At normal air pressure, water is a solid at any temperature below 0°C, a liquid between 0°C and 100°C, and a gas at any temperature above 100°C. Changing the phase of a substance does not change its composition. Water is a compound of two atoms of hydrogen joined to one atom of oxygen—no matter its phase. Changing phase changes only the speed and spacing of the particles.

In ice, the particles are so close together that the force of attraction that exists between those particles keeps them locked in position. They can only vibrate. Heating ice causes its particles to vibrate more rapidly. Eventually, their motion becomes great enough to overcome the forces of attraction that hold them together, and the ice melts. ***Melting*** is the change from the solid phase to the liquid phase. The temperature at which a solid melts is called its ***melting point***. Different substances are made of different particles. Different particles are held together with different forces of attraction. Therefore, different substances have different melting points. The melting point of a substance is barely affected by pressure, except under extreme conditions. The change from the liquid phase to the solid phase is called ***freezing***. The temperature at which freezing occurs is called the ***freezing point***. The freezing point and melting point of a solid are always the same.

In a liquid, the particles are still close enough to be attracted to one another but not close enough to be held in a rigid structure. The particles are free to move around and can slide past one another. The change from the liquid phase to the gaseous phase is called ***vaporization***. In this process, fast-moving particles near the surface of the liquid break free and become a gas. A sample of water at room temperature and pressure is always undergoing vaporization. Over time, all of the liquid water will change to water vapor. In ***boiling***, the temperature of the liquid rises high enough that *all* of the particles move fast enough to break free of attractive forces and become a gas. During boiling, small bubbles form throughout the liquid, rise to the surface, and release their vapor. The temperature at which boiling occurs *at normal pressure* is called the ***boiling point***.

Unlike the melting point, the boiling point is greatly affected by pressure. At normal pressure, water boils at 100°C. As long as water is boiling, its temperature does not rise above 100°C. However, the higher the pressure above the liquid, the more difficult it is for a particle in the liquid state to enter the gaseous state. The constant battering of the liquid surface by gas particles above it beats back particles trying to leave the liquid. Therefore, if water is placed into a pressure cooker, the pressure increases and the water will boil at a temperature *higher* than 100°C. Since the water gets hotter in a pressure cooker, food cooks faster, which is why it is a handy device to have in the kitchen. On a mountaintop, the air pressure is lower than normal. Particles do not need as much energy to push their way into the gas above the liquid. Therefore, water boils at a temperature lower than 100°C. On a high mountaintop, hard-boiling an egg could take hours.

Figure 13.4 shows a *heating curve*, or graph showing the changes in temperature over time as a sample of matter is heated steadily and changes phases. First, it shows ice being heated and melted to form liquid water. Then the graph shows this water being heated until it boiled and formed water vapor. Finally, the graph indicates the temperature as the water vapor was heated.

The line on the graph shows temperature changes versus time. It is labeled to show what was happening to the water during different time periods. Note the two flat areas, or plateaus, on

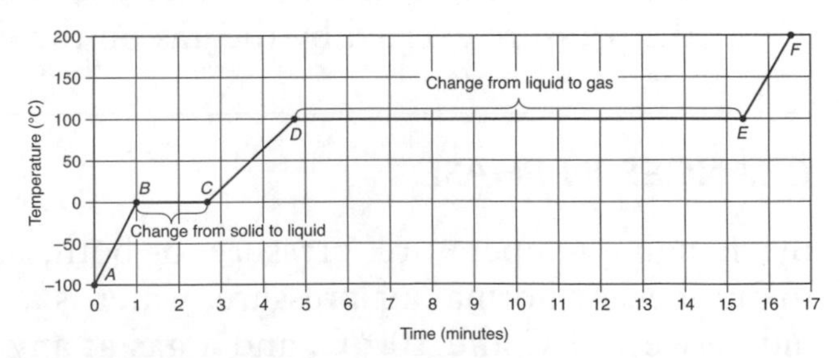

Figure 13.4
The heating curve of water.

the graph. These show that while the water was changing phase, it did not change temperature. How could heating the water not cause its temperature to change?

During melting, heat energy added to a solid is used to overcome attractive forces, *not* to increase the speed of the particles. Therefore, the temperature of a solid stays the same while it is changing into a liquid. Ice has a melting point of 0°C. A mixture of melting ice and water will remain at 0°C until all the ice has melted.

During vaporization, heat energy added to a liquid is again used to overcome attractive forces, *not* to increase the speed of the particles. Therefore, the temperature of a liquid stays the same while it is changing from a liquid to a gas. A sample of boiling water will remain at 100°C until all the water has vaporized.

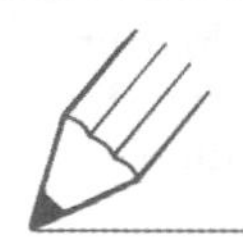

TRY THIS

1. Why is an ice cube at 0°C more effective at cooling a drink than the same mass of water at 0°C?
2. A student is unhappy with the rate at which eggs cook in a pot of boiling water. Would the eggs cook faster if the student turned up the flame?

PHYSICAL PROPERTIES

Each element and compound has a unique set of properties that makes it different from all other elements and compounds. A ***property*** is a characteristic or quality that distinguishes or identifies something. Although two substances may have several properties in common, no two are alike in all of their properties. ***Physical properties*** can be observed or measured without changing the chemical composition of a sample of matter. Some examples of physical properties are listed in Table 13.2.

CHEMICAL PROPERTIES

Chemical properties can be observed or measured during a change in the chemical composition of an element or compound. For example, if paper is ignited, it bursts into flame and burns. In the process, the chemical composition of the paper changes. Energy is released, and the paper is converted into water, carbon dioxide, and maybe some ash. The ability to burn in air is a chemical property. If electricity is passed through water, the water breaks down into hydrogen gas and oxygen gas. In the process, the chemical composition of the water is changed. The ability of a compound to be broken down into its elements by electricity is another example of a chemical property.

CHANGES IN MATTER

You have already read about many ways in which matter can change. Changes in matter can be classified as *physical* or *chemical* based on whether they alter the chemical composition of the matter.

Table 13.2 Some Physical Properties of Matter

Property	Description
Color	The appearance due to the dominant wavelength of light it emits, usually described in terms of hue (particular color), lightness and darkness, and intensity
Odor	The way in which a sample of matter is perceived by the sense of smell
Phase of matter at room temperature	Whether a sample of matter is a solid, liquid, or gas at room temperature (about 20°C)
Heat conductivity	The rate at which heat travels through a sample of matter by conduction
Electrical conductivity	The rate at which electricity is transferred through a sample of matter
Density	The concentration of matter, measured by the mass per unit volume
Solubility	The amount of one substance (solute) that will dissolve in and just saturate a given amount of another substance (solvent) at a given temperature, often given as grams of solute per 100 grams of solvent
Hardness	The ability of one substance to scratch or indent another
Boiling point	The temperature at which a substance boils under *normal atmospheric pressure*
Freezing point	The temperature at which a substance changes from a liquid to a solid; the freezing point and melting point of a solid are always the same

PHYSICAL CHANGES

A ***physical change*** results in a change in one or more physical properties of a sample of matter without a change in its chemical composition. In other words, no new substances form as a result of the change.

CHANGES IN SIZE AND SHAPE

Changing the size and shape of a sample of matter, such as crushing ice or crumpling aluminum foil, are physical changes. Crushed ice and crumpled aluminum foil are still the same substances—water and aluminum. Crushing, grinding, tearing, and cutting simply change the size or shape of a sample of matter without changing the matter into a new substance.

CHANGES IN PHASE

All changes in phase are physical changes. Ice, liquid water, and water vapor may have different physical properties, but they all have the same chemical composition.

MIXTURES

Simply mixing two or more substances together is making a physical change. If you mix sand and iron filings, no new substance is formed. The iron is still iron, the sand is still sand, and both still have their original properties. The particles of the two types of matter are just intermingled.

Separating Mixtures

Differences in the physical properties of the substances in a mixture can be used to separate the parts of a mixture. For example, a mixture of sand and water could be poured through a filter with holes small enough to keep the sand from passing through but large enough to allow the water to pass through and collect in a container below the filter. The sand could also be separated from the mixture simply by allowing the water to evaporate. Some of the methods used to separate the substances in a mixture by physical means are given in Table 13.3.

Table 13.3 **Methods of Separating the Parts of a Mixture by Physical Means**

Technique	Description
Filtration	The process of separating a mixture of a liquid and a solid by straining it through a porous material. The liquid passes through the openings, the solid particles are trapped because they are too large to pass through. The liquid that passes through is the *filtrate*, the solid that is trapped by the filter is the *residue*, e.g., separating pasta from water in a kitchen colander or coffee grinds from coffee.
Sieving	The process of separating a mixture of solids that have different-sized particles by passing the mixture through a screen. Particles smaller than the openings in the screen pass through, larger particles do not, e.g., separating pebbles mixed in garden soil by shaking the soil through a screen.
Evaporation	Allowing a liquid to evaporate and leave behind any solid that was mixed with it, e.g., letting the water evaporate from seawater to recover the salt.
Distillation	The process of heating a mixture of two liquids with different boiling points until some of its ingredients vaporize and then cooling the vapor to recover it in liquid form by condensation, e.g., separating alcohol from wine.
Recrystallization	The process of separating substances with different solubilities from a solution. The solution is heated and then allowed to cool. The substance with the lower solubility crystallizes out of the solution first and is separated by filtration.
Centrifuging	Separation of substances with different densities by spinning containers at a high speed around a central axis. Centrifugal force causes the densest substance to settle to the bottom of the container.
Decanting	The process of pouring off the liquid and leaving a solid that has settled out of a mixture of a solid and a liquid, e.g., pouring off iced tea into a clean cup after sand gets into it at the beach and settles to the bottom.
Magnetic separation	Using a magnet to remove a magnetic material from a mixture of a magnetic and a nonmagnetic material, e.g., using a magnet to remove iron nails that are mixed in a pile of sawdust.

Solutions: A Special Kind of Mixture

A special kind of mixture forms when you make iced tea from instant tea mix. To make this mixture, you simply add instant tea mix to the water and stir. The tea mix is soluble in water. ***Soluble*** means "able to be dissolved." When a substance ***dissolves***, its particles spread out evenly in another substance. The result is a ***solution***, a mixture that is a uniform blend of the two substances and has the same composition and properties throughout. All parts of the solution have the same physical and chemical properties, such as color, odor, taste, or acidity. For example, all of the solution made from instant tea mix is brown in color, tastes sweet, smells like tea, and is slightly acidic. Not all mixtures are the same throughout. A box of raisin bran is a mixture, but every bowl does not have exactly the same number of raisins or bran flakes. Therefore, raisin bran is not a solution.

In a solution, the substance present in the largest amount is called the ***solvent***. The substance present in the smallest amount is called the ***solute***. When a small scoop of instant tea mix is dissolved in a pitcher of water, the tea mix is the solute and the water is the solvent. Similarly, when a small amount of carbon dioxide is dissolved in water to make seltzer, the carbon dioxide is the solute and the water is the solvent. One usually thinks of a solution as a solid dissolved in a liquid. As shown in Table 13.4, however, solutions can involve mixtures of substances in any of the three phases of matter.

Table 13.4 Solutions of Substances in Various Phases

	Phase of Matter	
Solution	**Solute**	**Solvent**
Seltzer	Gas carbon dioxide	Liquid water
Rubbing alcohol	Liquid isopropyl alcohol	Liquid water
Karo syrup	Solid sucrose (sugar)	Liquid water
Fog	Liquid water droplets	Gas air
Air	Gas oxygen and other gases	Gas nitrogen
Brass	Solid copper	Solid zinc

Solubility

In some solutions, the relative amounts of solute and solvent can be varied greatly. For example, a solution of water and alcohol can range from almost pure alcohol to almost pure water. However, only a certain amount of salt will dissolve in a given amount of water at a specific temperature. A solution in which the amount of solute is small compared with the amount of solvent is said to be ***dilute***. As more solute is added, the solution becomes more concentrated. A ***concentrated*** solution has a relatively large amount of solute compared with the amount of solvent. When the maximum amount of solute has dissolved in a solvent at a given temperature, the solution is ***saturated***.

Solubility is the maximum amount of a solute that will dissolve in a given amount of solvent at a given temperature. In other words, it is the amount of solute that will saturate a solution at a given temperature. Increasing the temperature of the solvent can increase the solubility of many solutes. For example, more sugar will dissolve in a cup of hot water than will dissolve in a cup of cold water. Figure 13.5 shows how the solubility of several substances changes with the temperature of the solvent.

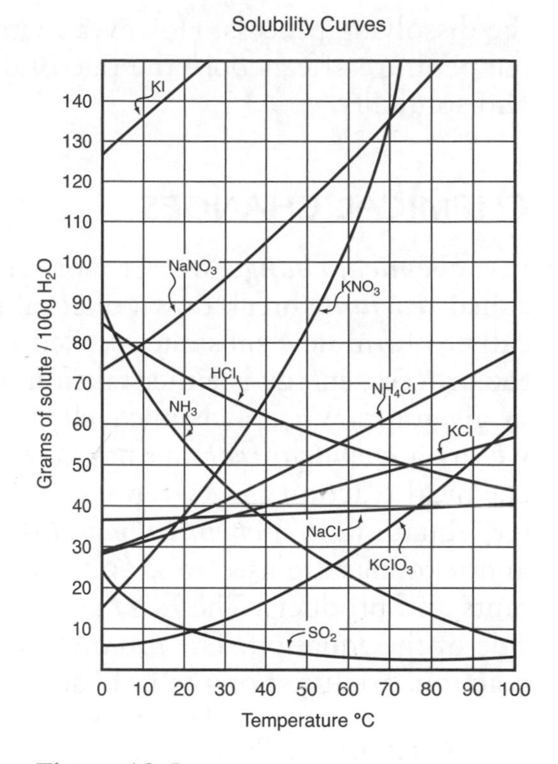

Figure 13.5
The effect of temperature on the solubility of various substances in water.
Key:
SO_2—sulfur dioxide
$KClO_3$—potassium chlorate
NaCl—sodium chloride (table salt)
KCl—potassium chloride
NH_4Cl—ammonium chloride
NH_3—ammonia
HCl—hydrochloric acid
$NaNO_3$—sodium nitrate
KNO_3—potassium nitrate
KI—potassium iodide

Rate of Dissolving

Have you ever stirred a cup of tea to speed up the rate at which a spoonful of sugar dissolves? Stirring is one of three ways to increase the amount of solute that a solvent dissolves per second, or the ***rate of dissolving***.

Stirring or shaking: This increases the rate at which solute and solvent particles are spread out through the solution. It also brings fresh solvent into contact with solute that has not yet dissolved.

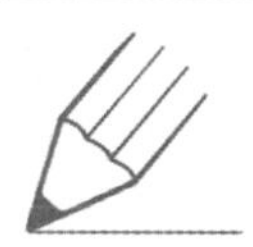

TRY THIS
Based your answers to the following questions on the graph in Figure 13.5.

1. At what temperature would equal masses of hydrochloric acid (HCl) and potassium chloride (KCl) dissolve in a 100-gram sample of water?
2. Name two compounds that decrease in solubility as temperature is increased.

Crushing or grinding: Breaking the solute into smaller pieces exposes more surface area to the solvent at the same time. See Figure 13.6. This allows the solvent to dissolve and spread out more solute molecules at the same time. Therefore, the smaller the solute particles, the more rapidly they dissolve.

Increasing the temperature makes particles move faster. At higher temperatures, solute and solvent particles collide more frequently and with greater impact. Therefore, the solute dissolves and spreads out more rapidly.

Note that neither stirring and shaking nor crushing and grinding have any effect on the total *amount* of solute that dissolves at a given temperature. They simply speed up the dissolving process. However, increasing temperature affects *both* the rate of dissolving and solubility.

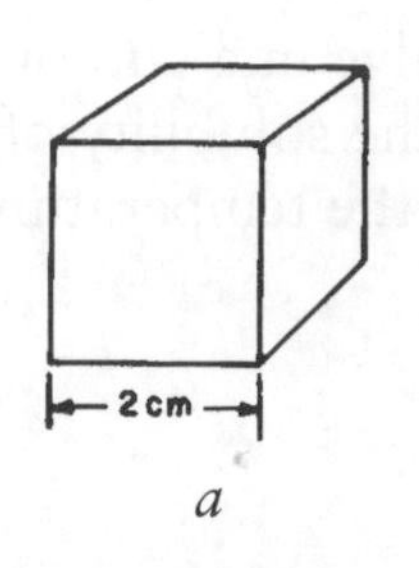

a

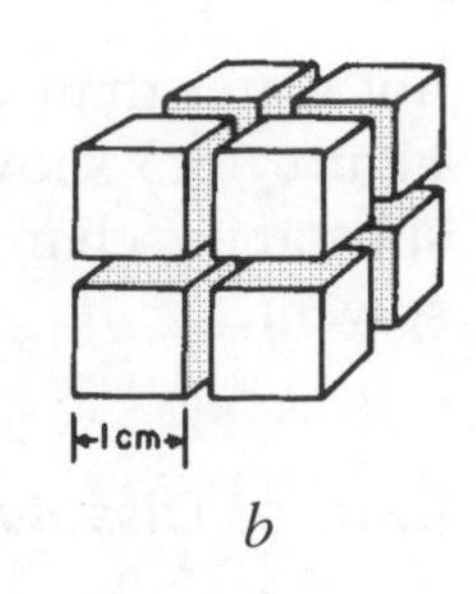

b

Figure 13.6
Breaking a solute into smaller pieces increases its surface area. (*a*) This cube has a surface area of 24 square centimeters. (*b*) Each of these eight small cubes has a surface area of 6 square centimeters, for a total of 48 square centimeters.

CHEMICAL CHANGES

In a ***chemical change***, one or more substances called *reactants* break down or combine chemically to form new substances called *products*. In the process, energy is either taken in and used or given off. When a chemical change occurs, we say a ***chemical reaction*** has taken place. A chemical reaction can be represented by a chemical equation. In a ***chemical equation***, numbers and formulas are used to keep track of the reactants and products. The ***reactants***, or substances you start with, are written on the left side of the equation. The ***products***, or substances that are formed during the chemical reaction, are listed on the right side.

Reactants → Products

Suppose, for example, you burned a sample of the element carbon in the presence of pure oxygen gas. The two elements would combine to form a compound called carbon dioxide—a colorless, odorless gas. In the process, energy is given off. This is an example of a chemical reaction in which two elements combine to form a compound.

It is the main chemical reaction that occurs when you burn coal. Coal is mostly carbon. When coal is burned, the carbon combines with the oxygen in the air to produce carbon dioxide. In most chemical reactions, a small amount of energy is needed to get the reaction started. The energy needed to start a reaction is called activation energy. When you use a match to ignite a piece of coal, you are adding activation energy. However, once the reaction is started, heat is given off. The amount of heat given off during the reaction is much greater than what was put in to get it started, which explains why coal is used as a fuel. The equation for the burning of coal would be written as

Reactants	⟶	Products
carbon + oxygen	⟶	**carbon dioxide + energy**
element + element	⟶	**compound**

A chemist would read this equation as "carbon plus oxygen yields carbon dioxide plus energy."

Breaking down a compound into its elements is also a chemical change. If you pass enough electricity through pure water, it will break down into oxygen gas and hydrogen gas. This is an example of a chemical reaction in which a compound is decomposed into its elements. In the process, energy is taken in and used to break apart the water molecules. The equation for the decomposition of water would be written as

Reactant	→	Products
water + energy	→	hydrogen + oxygen
compound	→	element + element

A chemist would read this equation as "water plus energy yields hydrogen plus oxygen."

In both of the above examples, the new substances produced by the chemical change have physical and chemical properties that differ from those of the original substances that reacted. See Figure 13.7. During a chemical reaction, the atoms are rearranged to form new substances. This involves breaking the chemical bonds that join atoms together as molecules or forming new chemical bonds between atoms. The properties of the new substances are different from those of the original substances because the original substances no longer exist. The energy taken in or given off during a chemical reaction involves making or breaking chemical bonds.

	Reactants: **Coal** (carbon) +	**Oxygen** →	Products: **Carbon dioxide**
Physical Properties	Black solid Density 3.51 g/cm^3 Freezing point > 3,550°C Boiling point > 4,827°C Insoluble in water Conducts electricity	Colorless, odorless gas Density 0.0014 g/cm^3 Freezing point −219°C Boiling point −183°C Soluble in water	Colorless, odorless gas Density 0.0019 g/cm^3 Sublimes (changes directly from solid to gas) at −78.5°C Soluble in water
Chemical Properties	Burns	Does not burn Supports burning of other elements or compounds	Does not burn Does not support burning of other elements or compounds

Figure 13.7 Chemical Reaction: Burning (oxidation) of coal.

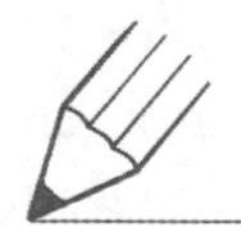

TRY THIS

1. A tree is chopped down, cut into pieces, and then burned. Which of the following shows the correct order of the changes that have occurred?
   1. physical, physical, physical
   2. physical, physical, chemical
   3. physical, chemical, chemical
   4. chemical, chemical, chemical
2. Which is a chemical change?
   1. Element 1 is hammered into a thin sheet.
   2. Element 2 is heated and turns into a liquid.
   3. Element 3 turns a greenish color as it sits in air.
   4. Element 4 is ground into a fine, slippery powder.

FACTORS AFFECTING CHEMICAL REACTION RATES

If you have ever made a campfire, you know that a big log burns slowly. If you take the same log and chop it into small pieces, it burns much faster. Clearly, chemical reactions such as burning do not always occur at the same rate. Chemists who study reaction rates have found that a number of factors affect the speed at which reactions take place. To explain how and why these factors change reaction rates, they have proposed the ***collision theory of reaction rates***. According to this theory, atoms and molecules are in constant motion. If they collide with enough force, a chemical reaction can take place. Thus, factors that affect collisions between atoms and molecules affect the rate at which they react.

Increasing Temperature Increases Reaction Rate

Increasing temperature increases the speed at which atoms and molecules are moving. Faster moving molecules are more likely to have enough energy to cause a reaction when they collide. When you cook food, you increase the rate at which certain reactions occur. In boiling water, egg white may take several minutes to change from a slimy liquid to a white solid. On a hot frying pan, it may happen almost immediately. When you put food into the refrigerator, you slow down the rate of chemical reactions that cause food to spoil. As a general rule, the reaction rate doubles for every 10°C increase in temperature.

Increasing Concentration Increases Reaction Rate

If only 10 students are in the gym, you are less likely to bump into someone when running than if 100 students were in the gym. In a similar way, as atoms and molecules become more concentrated, they are more likely to collide and react.

Increasing Surface Area Increases Reaction Rate

In order for atoms and molecules to react, they must come into contact with one another. Breaking a substance into smaller pieces creates more surface area along which it can come into contact with other substances. Increasing the surface area increases reaction rate by exposing more atoms or molecules to other reactants.

Increasing Light May Increase Reaction Rate

Light is a form of energy. Some substances can absorb enough energy from light so that when their particles do collide with others, a reaction can take place. For example, a mixture of hydrogen gas and chlorine gas will not react at room temperature if kept in the dark. If exposed to sunlight, though, the chlorine absorbs energy and may react with the hydrogen explosively!

Catalysts Increase Reaction Rate

A catalyst is a substance that can increase the rate of reaction without being used up in the reaction. Therefore, it can be recovered and used over and over again. For example, solid margarine is made by reacting hydrogen with liquid oils, such as soybean oil. This reaction takes place in the presence of finely divided nickel, but the nickel does not become part of the margarine. The nickel acts as a catalyst. Catalysts work in many different ways. However, most chemists think that they allow reactions to happen in a series of steps that require less energy. Your body is filled with catalysts that help keep you alive and well. Burning (reacting with oxygen) can release the energy in carbohydrates. Normally, your body temperature is about 37°C. This is too cool for colliding molecules of carbohydrates and oxygen to react and release energy fast enough to be useful. However, the catalysts in your body allow you to obtain energy from food by helping reactions to occur fast enough to produce useful amounts of energy even at low temperatures.

TELLING PHYSICAL FROM CHEMICAL CHANGES

Since chemical changes result in new substances, they almost always result in one or more physical changes. This sometimes makes it difficult to tell whether a physical change or a chemical change has taken place. In general, though, one can tell that a chemical change has taken place if one or more of the following physical changes occur.

- The reactants burn or burst into flame.
- A large amount of heat or light is given off.
- Odors and/or colors change.
- Bubbles of gas form.
- A clear solution turns cloudy, and an insoluble solid (called a *precipitate*) settles out of the solution.

Table 13.5 gives some examples of chemical reactions and the physical and chemical changes that accompany them.

TRY THIS

Suppose you have two test tubes. In each test tube is a strip of copper. A different liquid is added to each test tube. What kinds of results would show that a chemical change took place in one test tube but not the other?

Table 13.5 **Some Examples of Chemical Reactions and the Physical and Chemical Changes That Accompany Them**

Chemical Change	Physical Changes That Accompany It
Milk turns sour. A new substance, lactic acid, is produced.	Change in odor. Milk may form clumps that settle out of the milk.
Iron rusts (combines with oxygen). A new substance, iron oxide, is produced.	Magnetic iron changes to nonmagnetic iron oxide. Colorless oxygen and silvery iron change to reddish brown color. Density changes from 7.8 to 5.1 g/cm^3.
Mix a solution of silver nitrate in water with a solution of sodium chloride (salt) in water.	When the two colorless solutions are mixed, they become cloudy. A white solid, silver chloride, that is not soluble in water forms and settles out of the solution.
Mix vinegar and baking soda.	Mixture fizzes, and bubbles of gas are given off.
Burn magnesium in air.	The silvery, metallic magnesium burns, giving off heat and a brilliant white light. It is converted into magnesium oxide, a white, powdery solid.
Mix a solution of hydrochloric acid and a solution of sodium hydroxide.	No obvious visible changes are seen, but the solution gets very hot.

CONSERVATION OF MATTER

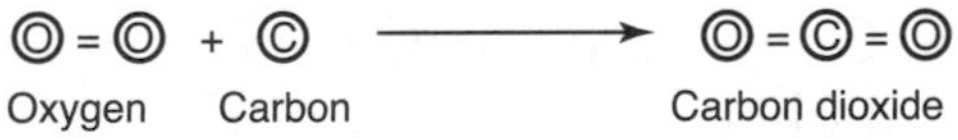

Figure 13.8
The burning of carbon.

When a physical or chemical change takes place, no new atoms are created and none are destroyed. All of the atoms that were present before the change are still present after the change. What has changed is the way the atoms are arranged. Physical changes merely change the spacing and/or motion of atoms or molecules. When 10 grams of ice melt, 10 grams of water are the result. Chemical reactions change only the way atoms are joined together. The total mass of the reactants equals the total mass of the products.

Figure 13.8 shows what happens when carbon is burned. A chemical reaction takes place in which carbon combines with oxygen to form carbon dioxide. How many carbon atoms are present before the reaction? How many oxygen atoms are present before the reaction? Compare this with the number of carbon and oxygen atoms after the reaction. There are two oxygen atoms and one carbon atom before and after the reaction. What has changed? Before the reaction, two oxygen atoms were bonded to one another. After the reaction, two oxygen atoms were bonded to a carbon atom. Since all of the matter that was present before the change is still present after the change, the mass remains the same. This is an example of the ***law of conservation of matter***. It states that during an ordinary chemical reaction, matter cannot be created or destroyed. In chemical reactions, the total mass of the reactants equals the total mass of the products.

TRY THIS

A glass of water with ice cubes in it has a mass of 300 grams. What will the mass be immediately after the ice has melted? Explain your answer.

THE STRUCTURE OF MATTER

The idea that matter is made of tiny particles, or atomism, dates back to the time of the Greek philosopher Democritus. He suggested that matter was made of tiny particles that could not be divided into anything smaller. He called these particles *atomos* from the Greek word for "indivisible." Democritus did not arrive at this idea by experimenting but by deductive reasoning. Atomism is not an obvious idea, for, to our senses, most matter appears to be all one piece. A gold coin or a drop of water does not look like it is made of particles. Neither does a sandy beach when you look at it from a distance. Only when we look at it close up can we see that it is made of countless small grains of sand. Perhaps, Democritus thought, gold or water is made of particles too small to see. However, he had no evidence that atoms did exist, so his idea was not widely accepted.

ATOMS

The idea of atoms came into favor again when the English chemist Robert Boyle experimented with gases. Boyle used the idea of atoms to explain why gases did not change mass when compressed. Boyle also suggested that elements were made of atoms. He reasoned that if elements were the simplest substances that made up the universe, you should not be able to break down an element into anything simpler or make an element by combining simpler substances. As soon as a substance was broken down into something simpler, it could no longer be considered an element. Therefore, the smallest particle of an element is one that cannot be broken down into anything smaller—an ***atom***. For quite a while, chemists could never be sure if a substance was an element. They could not be sure if some new way to break down matter would be developed that would make it possible to break apart a substance that had previously been unbreakable. For example, chemists found they could make water by burning pure hydrogen in pure oxygen. Since water could be formed from two substances, they realized that water was not an element.

To make it easier to keep track of elements, the Swedish chemist Jöns Jakob Berzelius suggested using a ***chemical symbol*** for each one. The symbols were the first letter of their Latin name. If other elements started with the same letter, another letter (usually the second) from the name was added. Thus, the symbol for fluorine is F. The symbol for iron, whose Latin name is *ferrum*, is Fe. Hydrogen is H, helium is He, and so on. This system made sense and was so easy to use that we still use it today to represent the atoms of the different elements.

MOLECULES

Of course, not all substances found in nature are elements. Most substances are made of two or more elements joined together. As chemists experimented with compounds, they began to study more than just their properties. They began to measure the amount of each substance in a chemical reaction to keep track of what was used or produced. By 1789, the French chemist Antoine Laurent Lavoisier had collected enough evidence to propose the law of conservation of matter. Based on this law, chemists began to measure the mass of each of the substances that made up a compound. For example, the French chemist Joseph Louis Proust studied a compound called copper carbonate. He found that it was made of the elements copper, carbon, and oxygen always in the same ratio by weight—5 parts copper, 4 parts oxygen, and 1 part carbon. This was true of every sample of copper carbonate he tested. Proust found that other compounds were also made of elements present in fixed ratios. It seemed as if elements would join together only in fixed ratios, a discovery that became known as the ***law of fixed proportions***.

Proust's work gave further support to the idea of atoms. Suppose copper is made of tiny copper atoms, carbon of tiny carbon atoms, and oxygen of tiny oxygen atoms. Now suppose that copper carbonate is made of a copper atom, a carbon atom, and an oxygen atom joined together in a tight group. Such a group of atoms is called a ***molecule*** (from Latin for "a small mass"). What if it just so happened that a carbon atom had a mass of 1, a copper atom had a mass of 5, and an oxygen atom had a mass of 4? Then, breaking apart a copper carbonate molecule would give you a mass ratio of 1 part carbon to 5 parts copper to 4 parts oxygen. Of course, if the ratios were 1 part carbon to 5.5 parts copper to 3.5 parts oxygen, the idea would not work, because you would have to work with fractions of atoms. *However, this never happens!*

In 1803, the English chemist John Dalton proposed the modern atomic theory. Based on the law of fixed proportions, it seemed clear that

1. Each element is made of atoms that have the same mass.
2. Different elements are made of atoms of different masses.
3. Compounds form by the joining together of atoms into molecules.

From the law of fixed proportions, working out the relative masses of different atoms is possible. For example, Dalton found that in forming water, 1 part hydrogen by weight combines with 8 parts oxygen by weight. He believed that water was made of molecules that contained 1 hydrogen atom and 1 oxygen atom. Therefore, he reasoned that oxygen atoms had 8 times the mass of hydrogen atoms. This does not tell us the actual mass of each atom; simply that one is 8 times heavier than the other. Dalton decided to use the hydrogen atom as a reference and assigned it a mass of 1 because it was the lightest element yet discovered. On the basis of the mass of hydrogen = 1, an oxygen atom would have a mass of 8.

At about the same time, though, water was first being broken down into hydrogen and oxygen by an electric current. When this was done, it was found that breaking down water always yielded 2 parts hydrogen to 1 part oxygen *by volume*. Soon, it was shown that a water molecule consisted of 2 atoms of hydrogen joined to 1 atom of oxygen. However, if water had 2 hydrogen atoms and not 1, then the mass ratio of a hydrogen atom to an oxygen atom was really $^1/_2$ part hydrogen to 8 parts oxygen, or a ratio of 1:16. So if you want to set the atomic weight of a hydrogen atom at 1, then the atomic weight of oxygen must equal 16. By using a system in which the weight of a hydrogen atom is 1, chemists worked out the atomic weight of element after element.

THE PERIODIC TABLE

By the 1860s, more than 60 elements had been identified. They displayed a wide variety of properties. Some were gases at room temperature, a very few were liquids, and most were solids. Some were colored, others were not. Some were metals, others nonmetals. Some metals were heavy, others were light. Some elements were very reactive, while others would not react at all.

In trying to make sense out of the elements, chemists looked for some underlying order. One logical approach was to arrange the elements by atomic weight. This did not seem to help much until the Russian chemist Dmitri Mendeleev came up with a system of arranging the elements by both their weight *and* their properties.

First, Mendeleev listed the elements in order of their atomic weight. Then he arranged them in a table of rows and columns so that elements with similar properties fell in the same column. As you go through the table, properties of a certain kind repeat after a fixed number of columns, or periodically. See Table 13.6. Therefore, Mendeleev's table became known as a ***periodic table.***

Table 13.6 **Some Elements Arranged by Weight**

Element*	Atomic Mass
Hydrogen	1.0079
Helium	4.0026
Lithium	6.941
Beryllium	9.012
Boron	10.81
Carbon	12.011
Nitrogen	14.0067
Oxygen	15.999
Fluorine	18.998
Neon	20.179
Sodium	22.989
Magnesium	24.305
Aluminum	26.982
Silicon	28.086
Phosphorus	30.974
Sulfur	32.06
Chlorine	35.45
Potassium	39.09
Argon	39.95
Calcium	40.08

* The noble gases (helium, neon, argon) are printed in bold. Note that they reappear at periodic intervals in the list.

The problem Mendeleev faced was that the list of elements was incomplete. When he arranged the known elements so that those with similar properties fell into the same column, he had to leave gaps. He believed, though, that these gaps were elements that had not yet been discovered. In 1871, he predicted the properties of the missing elements by comparing them with the properties of known elements in the same column above and below the gap. Within 15 years, three of the missing elements had been discovered and their properties closely matched those predicted by Mendeleev. Clearly, Mendeleev's table worked, but no one knew exactly why. That had to wait until the twentieth century and the discovery of subatomic particles and atomic structure. Over time, all of the missing elements were found. Today, the table has been expanded as new elements have been discovered or created in the laboratory.

A modern periodic table of the elements is shown in Figure 13.9. It consists of horizontal rows called ***periods*** and vertical columns called ***groups***. Periods are identified by

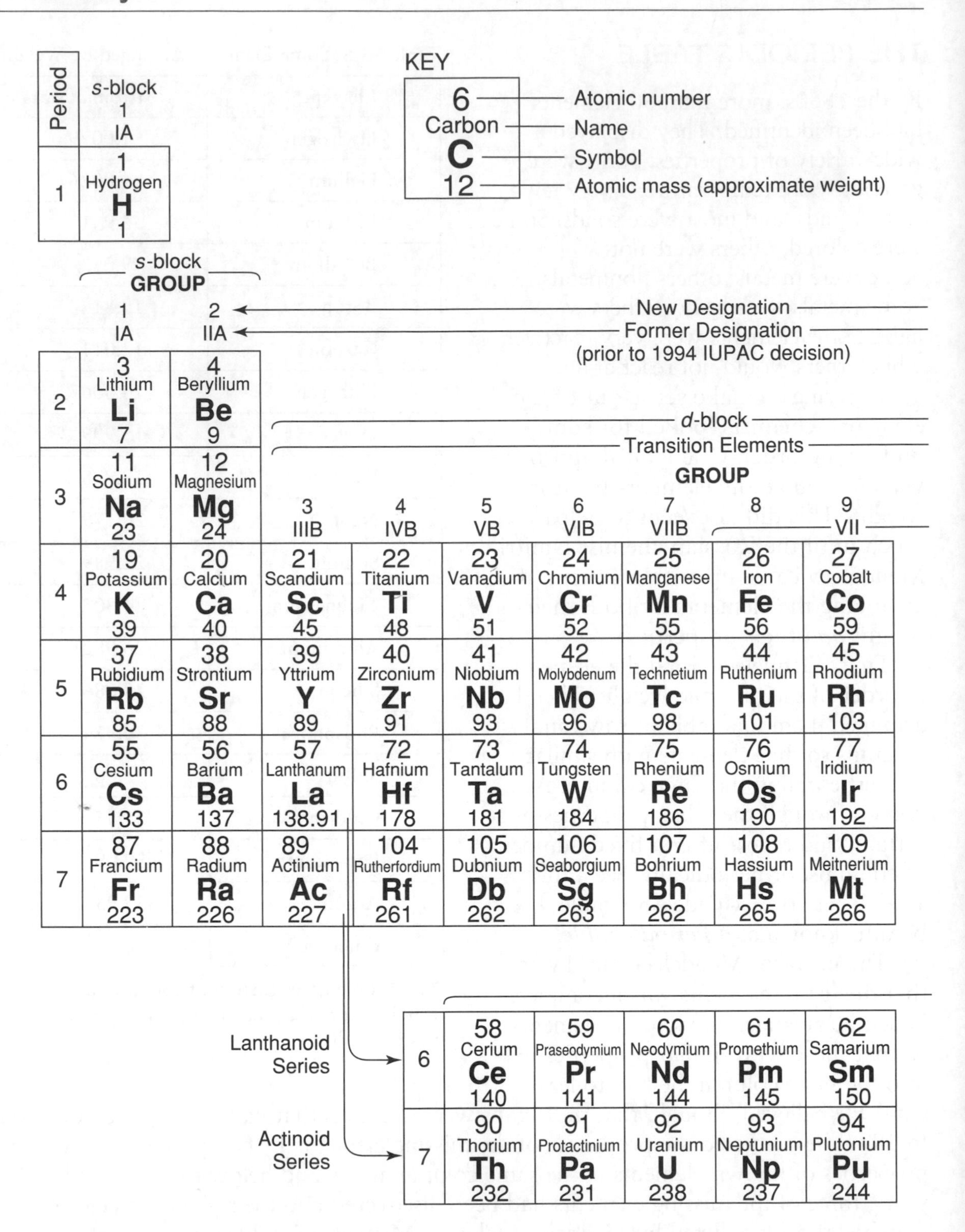

Figure 13.9
The periodic table of the elements.

s-block

p-block

GROUP

10	11 IB	12 IIB	13 IIIA	14 IVA	15 VA	16 VIA	17 VIIA	18 O
								2 Helium **He** 4
			5 Boron **B** 11	6 Carbon **C** 12	7 Nitrogen **N** 14	8 Oxygen **O** 16	9 Fluorine **F** 17	10 Neon **Ne** 20
			13 Aluminum **Al** 27	14 Silicon **Si** 28	15 Phosphorous **P** 31	16 Sulfur **S** 32	17 Chlorine **Cl** 35	18 Argon **Ar** 40
28 Nickel **Ni** 59	29 Copper **Cu** 64	30 Zinc **Zn** 65	31 Gallium **Ga** 70	32 Germanium **Ge** 73	33 Arsenic **As** 75	34 Selenium **Se** 79	35 Bromine **Br** 80	36 Krypton **Kr** 84
46 Palladium **Pd** 106	47 Silver **Ag** 108	48 Cadmium **Cd** 112	49 Indium **In** 115	50 Tin **Sn** 119	51 Antimony **Sb** 122	52 Tellurium **Te** 128	53 Iodine **I** 127	54 Xenon **Xe** 131
78 Platinum **Pt** 195	79 Gold **Au** 197	80 Mercury **Hg** 201	81 Thalium **Tl** 204	82 Lead **Pb** 207	83 Bismuth **Bi** 209	84 Polonium **Po** 209	85 Astatine **At** 210	86 Radon **Rn** 222
110 Ununnilium **Uun** 269	111 Unununium **Uuu** 272	112 Ununbium **Uub** 277						

f-block

63 Europium **Eu** 152	64 Gadolinium **Gd** 157	65 Terbium **Tb** 159	66 Dysprosium **Dy** 163	67 Holmium **Ho** 165	68 Erbium **Er** 167	69 Thulium **Tm** 169	70 Ytterbium **Yb** 173	71 Lutetium **Lu** 175
95 Americium **Am** 243	96 Curium **Cm** 247	97 Berkelium **Bk** 247	98 Californium **Cf** 251	99 Einsteinium **Es** 252	100 Fermium **Fm** 257	101 Mendelevium **Md** 258	102 Nobelium **No** 259	103 Lawrencium **Lr** 262

numbers (e.g., 1, 2). Groups are identified by Roman numerals followed by the letter A or B (e.g., IA, IIIB). The elements in each group have many similar properties. Since members of a family often share common traits, elements that share similar properties are often referred to as a family. We will consider several of the more well-known families of elements.

TRY THIS

List the names and symbols of the elements present in each of the following household substances:

Substance	Formula	Elements It Contains
Ammonia	NH_3	
Baking soda	$NaHCO_3$	
Hydrogen peroxide	H_2O_2	
Natural gas	CH_4	
Sugar	$C_{12}H_{22}O_{11}$	
Table salt	NaCl	

Group VIIIA is made up of the elements helium (He), neon (Ne), argon (Ar), krypton (Kr), xenon (Xe), and radon (Ra). All are gases at room temperature, have low melting and boiling points, and almost never react with other elements or compounds. For this reason, they are called the ***noble gases***. Until 1962, it was thought that none of them ever took part in chemical reactions. Since then, it has been discovered that some of them will react with fluorine and certain other elements.

Group VIIA is made up of fluorine (F), chlorine (Cl), bromine (Br), Iodine (I), and astatine (At). These elements are known as the ***halogens*** (from Greek for "salt formers"). One member, chlorine, is responsible for the family name. Chlorine reacts with sodium to form ordinary table salt. The others react with sodium to form compounds very similar to salt. All of the halogens are nonmetals that react with metals to form compounds that are called *salts*.

Group IA contains lithium (Li), sodium (Na), potassium (K), rubidium (Rb), cesium (Cs), and francium (Fr). These elements are known as the ***alkali metals***. All are soft, silvery solids; easily melted; and very reactive. Sodium is so reactive that it will explode if mixed with water! This family gets its name from the word *alkali*, an Arabic phrase for "ash." Already in ancient times, the ashes of certain plants (known as soda ash and potash) were used in soap making and glass making. Since the metals sodium and potassium were first isolated from these substances, they were called alkali metals.

On the periodic table, the properties of elements change in a cyclic pattern from left to right. Elements to the far left in a period are highly reactive ***metals***. As you move to the right, the metals become less reactive, and the elements become less metallic and more nonmetallic. The elements to the right of the zigzag line are ***nonmetals***. Table 13.7 contrasts some of the physical and chemical properties of metals and nonmetals.

Since the zigzag line marks the boundary between metals and nonmetals, not surprisingly, some elements along the line have some metallic *and* some nonmetallic properties. Eight elements—B, Si, Ge, As, Sb, Te, and Po—have both metallic and nonmetallic properties and are called ***metalloids***. As you move to the right among the nonmetals, reactivity increases. When you reach the column farthest to the right, the noble gases, the elements are nonreactive. As you move to the next row, this pattern repeats itself. This repeating pattern of properties is the reason for the name *periodic* table.

Table 13.7 **Some Physical and Chemical Properties of Metals and Nonmetals**

	Metals	Nonmetals
Physical Properties	Most have a shiny, lustrous surface.	Most have a dull surface that does not have a metallic luster.
Physical Properties	Some are malleable, meaning they can be hammered or rolled into thin sheets without breaking.	Most are brittle in their solid state and will break into pieces if hammered or rolled.
Physical Properties	Most are ductile, meaning they can be pulled out into wires without breaking.	Most are brittle and are therefore not ductile in their solid state.
Physical Properties	All are fairly good conductors of heat and electricity.	Except for carbon, all are poor conductors of heat and electricity.
Chemical Properties	Most do not easily combine with other metals to form compounds.	Except for the noble gases, most can combine with each other to form compounds.
Chemical Properties	Most tend to combine with nonmetals to form compounds.	Most tend to combine with metals to form compounds.

Practice Review Questions

1. Mass is best defined as a measure of:

1. the amount of space an object occupies.
2. the length of an object.
3. the amount of matter an object contains.
4. Length × width × height

2.* The masses of three blocks were compared.

Which one of the following statements about the mass of the three blocks is true?

1. Block *A* has the most mass.
2. Block *B* has the least mass.
3. Block *C* has the most mass.
4. All three blocks have the same mass

* Reproduced from TIMSS 2003 Science Released Items for Grade 8, TIMSS International Study Center, Boston College, MA

3. The diagram below shows a mineral sample and a graduated cylinder containing some water.

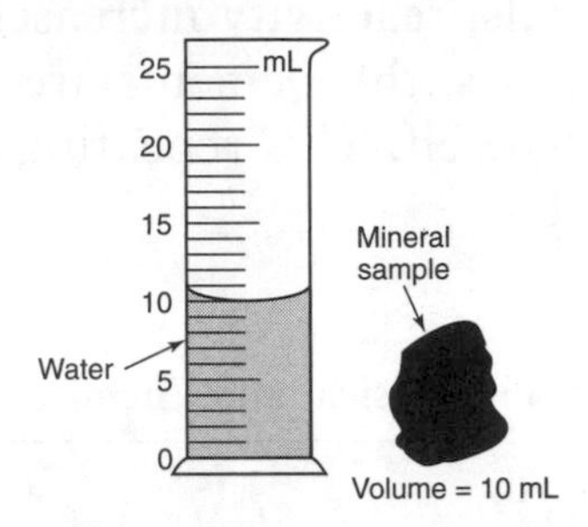

What will be the water level in the graduated cylinder after the mineral sample is placed in the water?

1. 5 ml
2. 10 ml
3. 15 ml
4. 20 ml

4. The diagram below shows the dimensions of a metal block that has a mass of 1,500 grams.

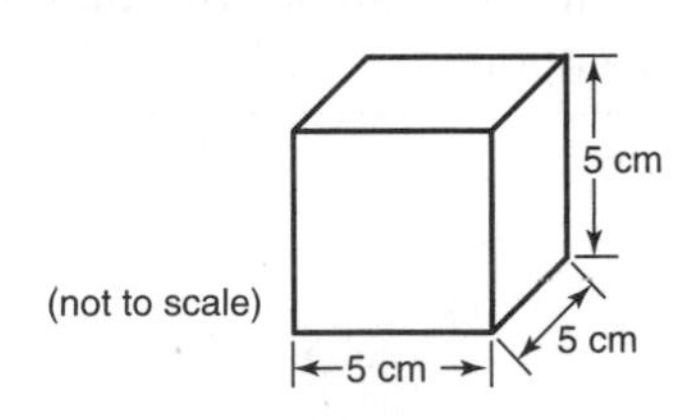

What is the density of the metal?

1. 12 g/cm^3
2. 60 g/cm^3
3. 100 g/cm^3
4. 1500 g/cm^3

5. The chart below shows the density of some common liquids.

Liquid	Density (g/cm^3)
Water	1.0
Olive oil	0.9
Seawater	1.03
Gasoline	0.66
Glycerine	1.26
Turpentine	0.85

If olive oil, water, gasoline, and glycerin are put into a test tube, in what order will they form layers from bottom to top?

1. glycerin, water, olive oil, and gasoline
2. water, olive oil, gasoline, glycerin
3. olive oil, water, gasoline, glycerin
4. gasoline, glycerin, olive oil, water

6. The diagram below shows a cylinder containing four different liquids and their densities.

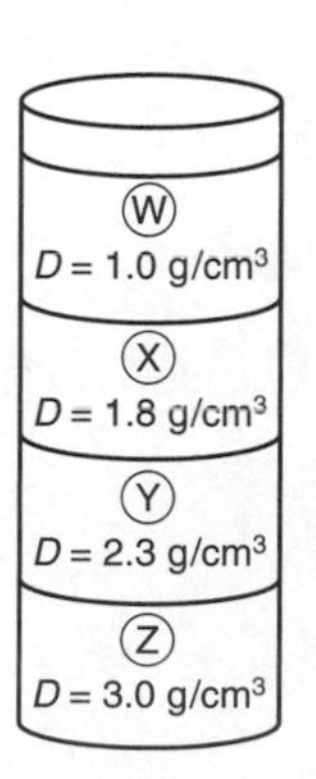

A piece of solid quartz with a density of 2.7 g/cm^3 is placed on the surface of liquid *W*. When the quartz is released, it will pass through

1. *W*, *X*, *Y*, and *Z*
2. *W*, *X*, and *Y*, but not *Z*
3. *W*, but not *X*, *Y*, or *Z*
4. *W* and *X*, but not *Y* or *Z*

7. All the atoms that make up an element are

1. alike, but different from those of other elements
2. always spaced the same distance apart
3. different, but have the same mass as atoms of other elements
4. always moving at the same speed

8.* If you took all the atoms out of a chair, what would be left?

1. The chair would still be there, but it would weigh less.
2. The chair would be exactly the same as it was before.
3. There would be nothing left of the chair.
4. Only a pool of liquid would be left on the floor.

GO ON ➡

9. The diagram below shows a model of a water molecule.

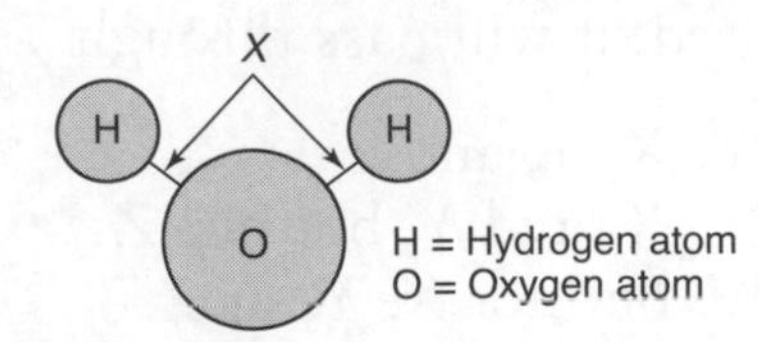

What do the lines labeled *X* represent?

1. gravitational forces
2. light energy
3. chemical bonds
4. magnetic fields

10. The chart below shows some common substances and their chemical formulas.

Substance	Formula
Carbon	C
Hydrochloric acid	HCl
Nitrogen	N_2
Nitric acid	HNO_3
Salt	NaCl
Water	H_2O

Based on the information in the table, which substances are elements?

1. vinegar and hydrochloric acid
2. salt and water
3. carbon and nitrogen
4. nitrogen and nitric acid

11. Which materials would be most useful if you wanted to separate a mixture of salt and sand?

1. triple-beam balance and graduated cylinder
2. filter paper and water
3. battery and voltmeter
4. magnet and beaker

12. As an ice cube melts, its molecules

1. absorb heat energy and move farther apart
2. absorb heat energy and move closer together
3. release heat energy and move farther apart
4. release heat energy and move closer together

13. Letters *A* through *D* in the diagram represent four processes that can occur when water changes phase.

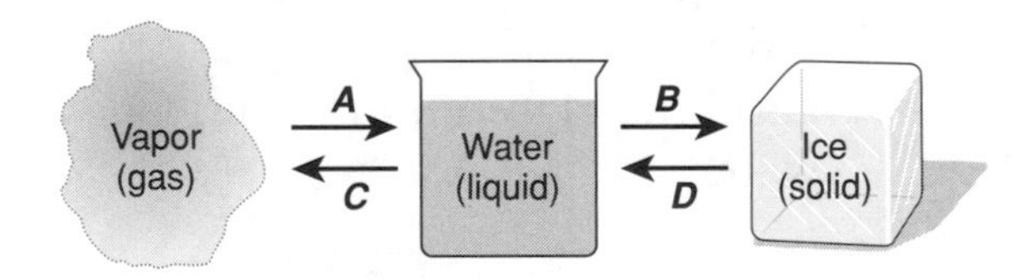

Which process requires the addition of the greatest amount of heat energy?

1. *A*
2. *B*
3. *C*
4. *D*

GO ON

14. The diagrams below show sealed containers that contain only one substance. Which diagram best represents the molecules of a substance in the gaseous state?

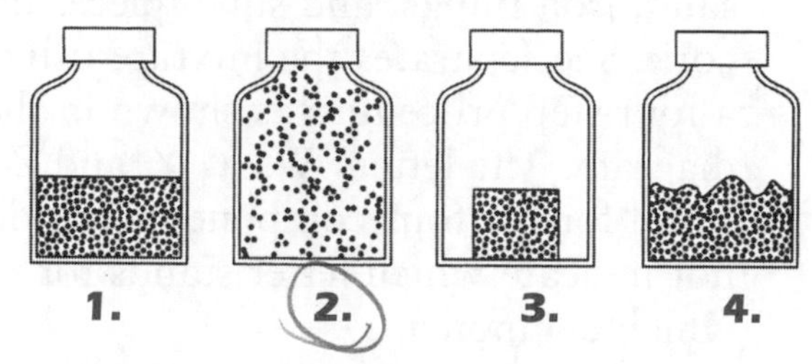

15. Ethanol has a freezing point of –114.6°C and a boiling point of 78.4°C. At which of the following temperatures would ethanol be in its liquid state?

1. –80°C
2. 80°C
3. –150°C
4. 150°C

Questions 16 and 17 refer to the graph below, which shows the results of an experiment in which a 200-gram sample of ice at –50°C was placed in an open beaker and heated for 70 minutes while being constantly stirred.

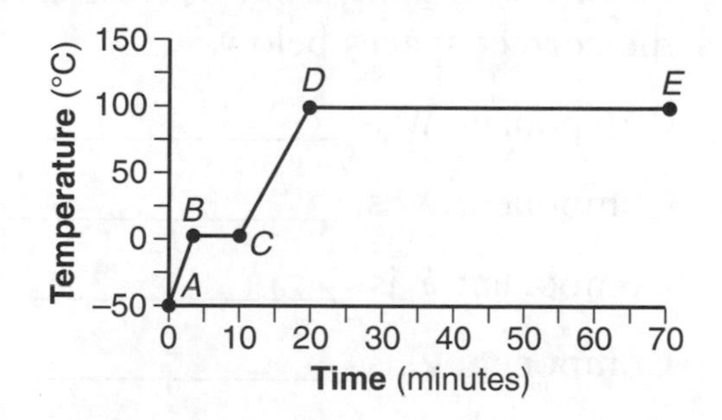

16. Which change occurred between point *A* and point *B*?

1. Water condensed.
2. Water froze.
3. Ice melted.
4. Ice warmed.

17. The greatest amount of energy was absorbed by the water between points

1. *A* and *B*
2. *B* and *C*
3. *C* and *D*
4. *D* and *E*

18. The chart below shows the solubility of two substances at various temperatures.

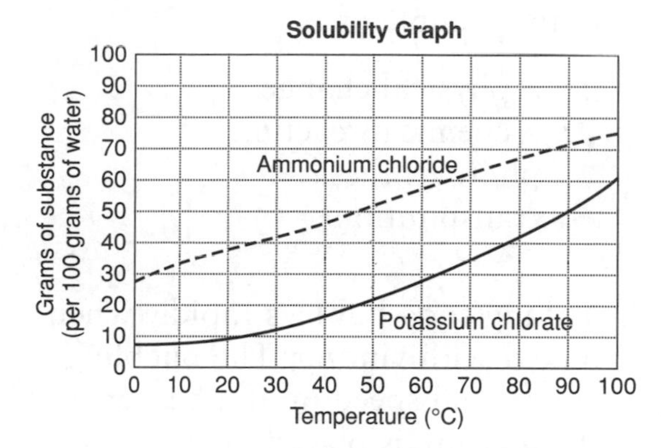

Compared to potassium chlorate, how many more grams of ammonium chloride will dissolve in 100 grams of water at 50°C?

1. 20
2. 30
3. 40
4. 50

19.* Which is a chemical change?

1. Element 1 is polished to form a smooth surface.
2. Element 2 is heated and evaporates.
3. Element 3 develops a white, powdery surface after standing in air.
4. Element 4 is separated from a mixture by filtration.

GO ON

20. A physical change results in

1. formation of a new substance
2. no change in chemical properties
3. a change in color
4. an increase in mass

21. When hydrochloric acid (HCl) is added to a test tube containing the mineral calcite ($CaCO_3$), calcium chloride and water are formed and carbon dioxide gas is released. This is an example of

1. a physical change
2. a chemical reaction
3. photosynthesis
4. transpiration

22. Baking soda bubbles rapidly when mixed with vinegar. The bubbling action will speed up if a beaker of the mixture is placed in

1. ice
2. water at room temperature
3. cold water
4. hot water

23. The modern periodic table shows

1. elements arranged in alphabetical order
2. elements arranged in order of increasing atomic number
3. elements arranged in order of increasing density
4. elements listed in the order in which they were discovered

CONSTRUCTED-RESPONSE QUESTIONS

24.* Teresa is given a mixture of salt, sand, iron filings, and small pieces of cork. She separates the mixture using a four-step procedure as shown in the diagram. The letters *W*, *X*, *Y*, and *Z* stand for the four components but do not indicate which letter stands for which component.

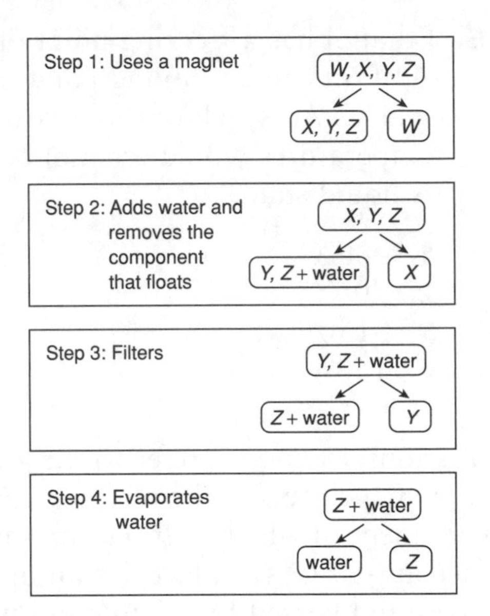

Identify what each component is by writing *salt*, *sand*, *iron*, or *cork* in the correct spaces below.

Component *W* is iron.

Component *X* is cork.

Component *Y* is sand.

Component *Z* is salt.

GO ON

25. The data table below shows the volume and mass of three different samples of the mineral pyrite.

Pyrite

Sample	Volume (cm³)	Mass (g)
A	2.5	12.5
B	6.0	30.0
C	20.0	100.0

a. On the grid below, plot the data (volume and mass) for the three samples and connect the points with a line.

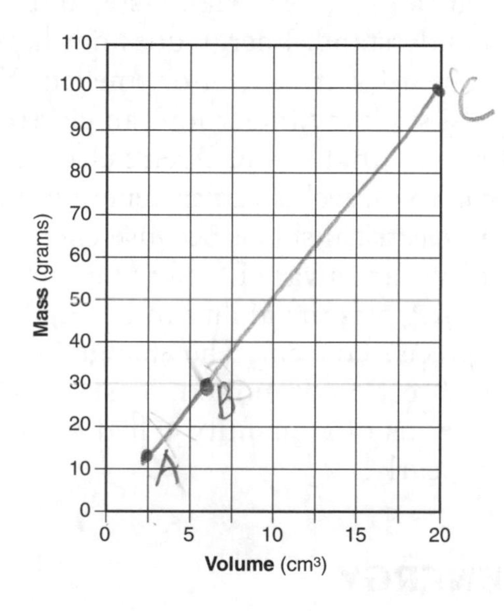

b. State the mass of a 10.0-cm³ sample of pyrite.

26. Complete the chart below by placing a check in the boxes corresponding to the properties of each of the three phases of matter.

Phases of Matter

Property			
Has a definite shape	✓		
Has a definite volume	✓	✓	
Takes the shape of its container		✓	✓

27. Although water consists of hydrogen, which explodes, and oxygen which supports burning, water can be used to put out fires. Give a brief explanation of this statement.

Both of those elements together creates a new item which has different propeties. There are many elements like this which have different properties than their components.

Chapter 14

Energy

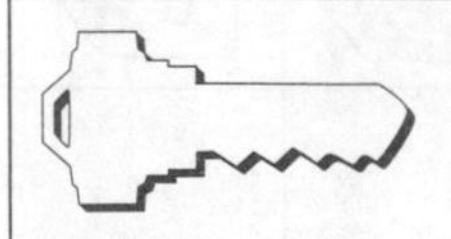

Key Idea: Energy exists in many forms, and when these forms change, energy is conserved.

The concept of energy is one of the most important ideas in science. Together, matter and energy are what make up the universe. Matter is the substance of the universe, and energy is what moves matter. The concept of matter is fairly simple to understand. Matter is stuff; it has mass and takes up space. You can see, hear, feel, taste, and smell matter. The concept of energy is more difficult to understand. Energy does not have mass, does not take up space, and cannot be seen, heard, felt, tasted, or smelled. You might say, "Wait a minute, don't I feel electricity if I stick my finger into an electrical socket and get a shock? Don't I feel heat when I touch a hot stove? Aren't electricity and heat forms of energy?" Yes, they are! However, what you feel is not the energy; it is the *effects* of the energy on matter. You feel heat or an electrical shock because energy causes changes in the matter making up your finger. Since we are aware of energy by its *effects*, we define energy by what it can *do*. What we call *energy* is the ability to bring about changes in matter. However, change is not a very precise concept. The change may be in the speed at which a body of matter is moving, its direction of motion, its size, its shape, and so on. So what we will do is define change in terms of a quantity called *work* and then see how it allows us to describe energy more clearly.

WORK AND ENERGY

When energy acts upon matter, it does so in the form of a force. A ***force*** is any influence that changes the speed of a body of matter, its direction of motion, or both. The quantity called *work* is a measure of the *amount* of change that a force brings about when it acts upon something. ***Work*** is defined as the amount of force exerted onto an object multiplied by the distance the object moves in the direction of the force. In equation form:

work = force × distance the object moves in the direction of the force *or*
work = force × distance *or*
$w = f \times d$

When you lift a box waist high, work is done on the box. You exert an *upward force*, and the box *moves upward* to waist height. Notice that two things happen when work is done: (1) a force is exerted and (2) something is moved in the direction of that force. Increasing either of these increases work. The higher the box is lifted or the heavier the box, the more work is done. To lift a heavier box, you must exert a greater upward force. If you lift the box overhead, the box moves a greater distance.

However, a force does not always cause something to move. If a force does not cause something to move in the direction of the force, no work is done on it. For example, you can push against a wall for hours and get really tired. If the wall does not move any distance in the direction you push, though, no work is done on the wall. *You* may get tired because of biological work done on your *muscles* when they contract and relax, but that work was not done on the *wall*.

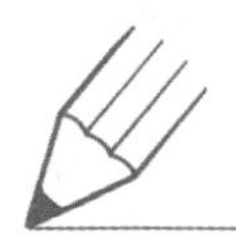

TRY THIS

1. What two things are necessary for work to be done?
2. How much work is being done to hold up the Empire State Building?

Now we will use the idea of work to explain the concept of energy. When you lift a box, the work done comes from the human arm. Where does the work of the human arm come from? You might say that the work, or something equivalent to work, could be stored in the human body and that this "stored work" could be called upon when needed to produce real, physical work. At first, it might seem that this stored work has something to do with life, because living things seem filled with the ability to do work, while dead things just seem to lie still and not do work. However, thinking that only living things can do work is clearly wrong. The wind is not alive, yet it can move ships. Running water is not alive, yet it can turn waterwheels. In both cases, a force is being exerted through a distance. Thus, work can clearly be stored in both living and nonliving things.

In 1807, the English scientist Joseph Young suggested that this stored work be called energy, from Greek words meaning "work within." The term *energy* is now widely used to describe anything that can be converted to work. In other words, ***energy*** is what something has if it can do work. When we say something has energy, we mean it is able to do work. The more energy something has, the more work it can do.

On the other hand, when we do work *on* something, we have added to it an amount of energy equal to the work done. For example, if you lift a box, it becomes capable of doing work that it could not do when resting on the floor. If you let go of the box, it will fall and can do work on something else as it falls. In a sense, when work is being done, stored work, or energy, is being transferred from one thing to another.

POTENTIAL AND KINETIC ENERGY

The first phenomenon that was clearly recognized as energy was motion itself. Work involves motion since an object has to be moved through a distance. So, it was not surprising that motion could do work. *Still* air does not move a ship, but *moving* air does. *Still* water does not turn a waterwheel, but *moving* water does. Thus, it is not the air or water that contain energy but the *motion* of the air or water. In fact, any object that moves contains energy because if it collides with another object, it can set that object's mass into motion. Thus, it will do work on that object; it will exert a force on that object that will move its mass through a distance. The energy of motion is called ***kinetic energy***, from a Greek word meaning "motion."

Now, suppose you throw an object up into the air. The object is moving and, therefore, has kinetic energy. However, as it climbs upward, it slows down because of Earth's gravity and eventually stops moving upward. You might suppose that its kinetic energy disappeared and it could no longer do any work. However, after the object reaches maximum height, it begins to fall again. It falls faster and faster, gaining more and more kinetic energy. When it hits the ground, it has all of the kinetic energy that it started with.

The kinetic energy was not really lost on the way up, it was simply stored in some form other than motion. As it fell, the stored energy changed back to kinetic energy of the falling object. Work had to be done to lift the object to a certain height, even if it stopped moving when it reached that height. The work must be stored in the form of an energy that the object contains because of its position. As the object rises, kinetic energy is stored little by little as "energy of position." At maximum height, all of the kinetic energy has become energy of position. When the object falls, the energy of position is converted back into kinetic energy. Since the energy of position has the potential to change back to kinetic energy, the energy an object has because of its position is called ***potential energy***. A box held above the floor has gravitational potential energy because it can do work on something else as it falls. A nail held near a magnet has magnetic potential energy because it can do work as it moves toward the magnet. A stretched-out spring has elastic potential energy because it can do work on something else as it recoils.

FORMS OF ENERGY

From everyday experiences, it is clear that kinetic and potential energy can take many forms. For example, if you drill a hole into a piece of metal, the metal gets hot. Clearly, the kinetic energy of the moving drill gives rise to heat. On the other hand, heat can be turned into kinetic energy. Each day, the heat of the Sun lifts huge amounts of water vapor high up into the air. The work done on the water vapor to lift it is stored as potential energy. When the water falls back down as rain, its potential energy is changed back into kinetic energy. Since the energy of the falling water can be traced back to the Sun's heat, heat must be a form of energy too. If that is true, then any phenomenon that gives rise to heat must also be a form of energy. An electric current can heat a wire, so electricity is a form of energy. A piece of wood has the potential to give rise to heat if it burns, so it must contain a form of potential energy (which we call chemical energy). There are

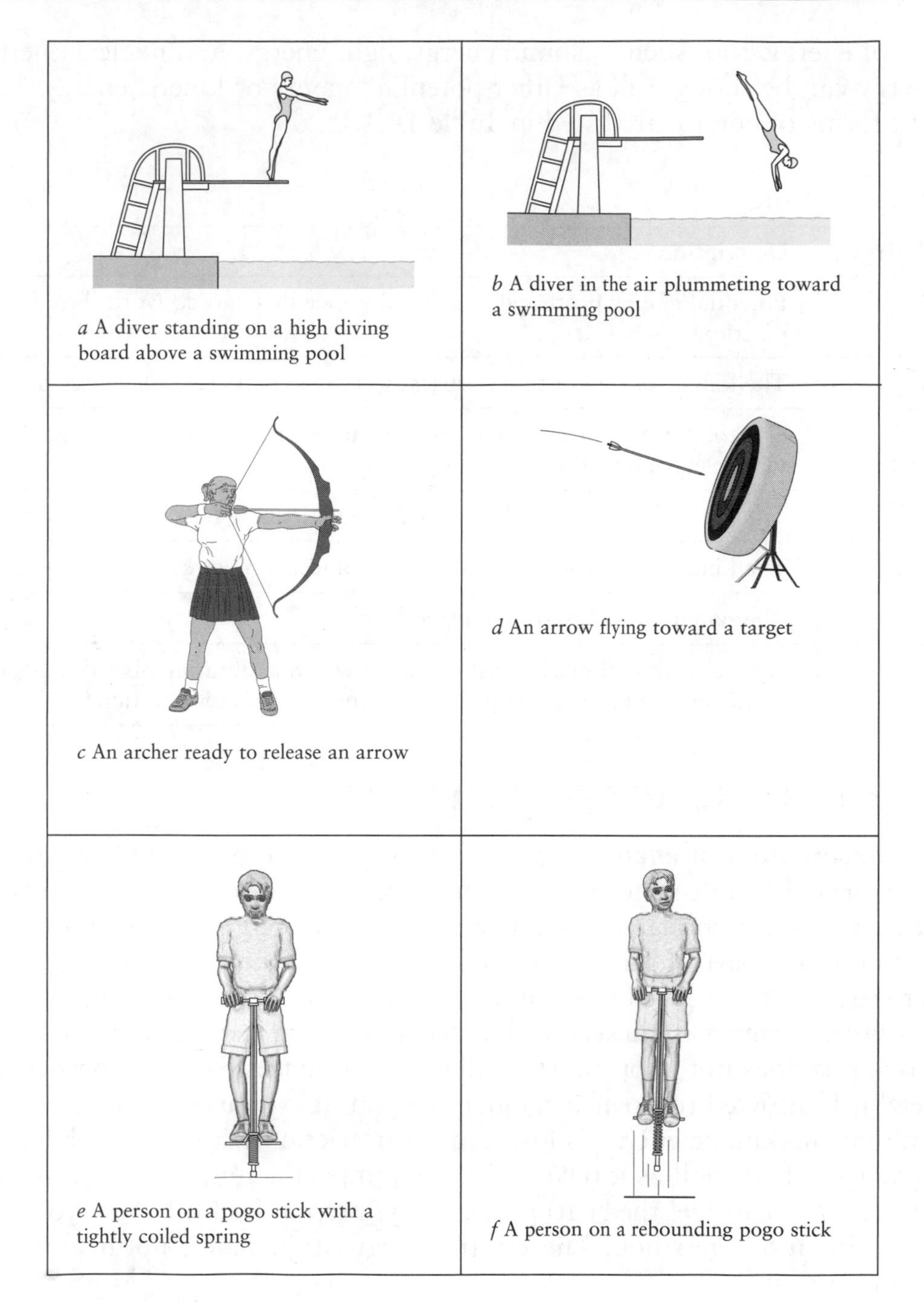

Figure 14.1
Objects with potential or kinetic energy.

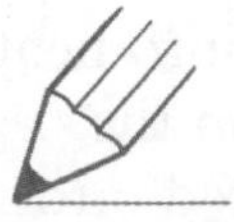

TRY THIS

In each of the examples shown in Figure 14.1, decide whether the object has kinetic energy or potential energy.

other forms of energy, too, such as sound energy, light energy, and nuclear energy. Each form of energy can be thought of as either potential energy or kinetic energy. Some of the different forms of energy are listed in Table 14.1.

Table 14.1 **Forms of Energy**

Form of Energy	Description
Chemical	Potential energy possessed by any substance that can do work through chemical reactions.
Electrical	The kinetic energy of tiny electrically charged particles called electrons.
Heat	The total kinetic energy of all the tiny atoms and molecules of which a body of matter is composed.
Light	The motion of invisible electric and magnetic force fields that surround matter.
Mechanical	The kinetic energy with which moving objects do work.
Nuclear	The potential energy stored inside atoms.
Sound	A special kind of mechanical energy in which a vibrating object's kinetic energy is transferred through matter in the form of waves of vibrations.

THE LAW OF CONSERVATION OF ENERGY

The *law of conservation of energy* states that energy cannot be created or destroyed. It may be transformed from one form to one or more other forms of energy, but the *total* amount of energy never changes. The transformation of energy from one form to another is an important part of this law. If we look at only one form of energy, it does not always seem to be conserved. For example, if you drop a basketball from a given height, it should (if *kinetic energy* is conserved) bounce and return exactly to its original height. As you know, this does not happen. The ball always bounces to somewhat less than its original height. If allowed to bounce again and again, it eventually comes to rest. It might seem that the kinetic energy is lost. However, friction with the air, the ground, and within the matter of the ball as it flexes with each impact converts some of the kinetic energy to heat. (You can feel the heat produced by internal friction when you repeatedly bend a paper clip and it gets hot.) The kinetic energy of the ball is not lost; it is simply transformed into heat.

ENERGY TRANSFORMATIONS

In fact, most activities of everyday life involve one form of energy being transformed into another. In every transformation, though, some of the energy is converted into heat. For example, a lightbulb converts electrical energy into light energy. However, in the process, not all of the electricity is transformed into light. Quite a bit is converted into heat, as you probably know if you have ever tried to unscrew a lightbulb right after turning it off.

In some systems, one form of energy may be changed into *many* forms of energy. For example, an automobile engine is a system designed to transform the chemical energy of gasoline into the mechanical energy of a car. When gasoline explodes in a cylinder, some of the chemical energy is transformed into sound energy and released into the environment—the roar of the engine. Some is converted into light energy, which is absorbed by the cylinder walls and transformed into heat. However, most of the chemical energy is converted into heat. So much, in fact, that the engine needs a cooling system to carry off the heat and release it into the environment or the engine's matter might actually become hot enough to melt! Only a fraction of the heat is transformed into mechanical energy by expanding gases in the cylinders. Even then, a lot of the mechanical energy is transformed back into heat as the engine parts rub against one another. The amount of chemical energy in gasoline that actually ends up as mechanical energy of the car is very small.

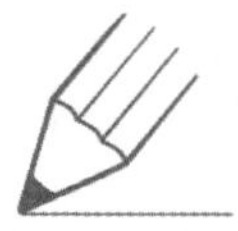

TRY THIS

For each of the common devices shown in this diagram, identify the starting form of energy and the form into which it is transformed by the device. For each device, identify two additional forms of energy into which the starting energy is transformed.

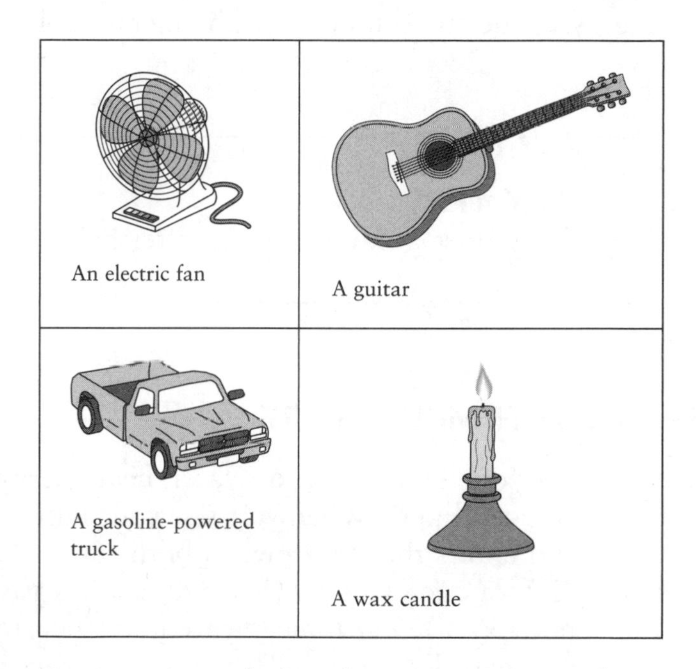

Some common devices that convert energy from one form to another.

Some systems transform energy with less loss of heat than others. The amount of energy a system transforms into a desired form of energy compared with the total amount of energy put into the system is known as the system's ***efficiency***. For example, diesel engines transform more heat energy into mechanical energy than do gasoline engines. Less energy is lost to the environment as heat. Therefore, diesel engines are said to be more efficient than gasoline engines.

HEAT

We have seen that in the process of transforming energy from one form to another, some energy is always converted to heat. What exactly is heat? You have already learned that matter is composed of atoms and molecules that are in constant motion. Each atom or molecule is a tiny mass. When a mass is in motion, it has kinetic energy. Therefore, each tiny atom or molecule in a sample of matter has kinetic energy. The total kinetic energy of all the particles that make up a sample of matter is called its ***heat energy***.*

Heat energy is involved in many everyday activities, from cooking your food to warming your hands. The atoms and molecules of every substance are constantly jiggling or moving around in some way. The greater the kinetic energy of the molecules in a substance, the hotter it feels. When you drill a hole into a piece of metal, the metal gets warm. This is because the moving drill collides with some of the molecules of metal, causing them to move faster. Later, as the molecules slow down by giving off some of their energy to the surrounding air, the metal cools.

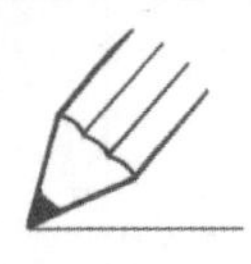

TRY THIS
Why does a nail get warm after being hammered into a piece of wood?

HEAT AND TEMPERATURE

When you put your hand into hot water, heat energy enters your hand because the water is hotter than your hand. When you put your hand into ice water, heat energy leaves your hand and enters the ice water. In both cases, the heat energy was transferred from a hotter body into a cooler body. Heat will always pass from a body at a higher temperature into a body at a lower temperature until their temperatures are the same. However, that does not mean that heat will always pass from a body with more heat energy into a body with less heat energy! Just because something feels hot does not mean it contains more heat energy than something that feels cold.

Suppose you are at the beach and leave a bottle of water out in the Sun so now it is hot. Even though the ocean is cooler, it is made of many, many more molecules than the bottle of hot water. When added up, the total energy of the molecules in the ocean is far greater than that of the few molecules in the bottle of hot water. Therefore, the ocean

---

*The author respectfully points out that what the New York State core curriculum refers to as heat energy should more correctly be called *thermal* energy and that what is customarily called *heat* is thermal energy in transit.

contains far more heat energy than the bottle of hot water. Does this mean that because the ocean has more heat energy, heat will pass from the ocean into the bottle? No! If you put the bottle of hot water into the cool ocean, heat does not pass from the ocean into the bottle. Instead, heat passes from the hot bottle into the cooler surrounding water until the two are at the same temperature. In doing so, heat moves from the body with less heat energy into the body with more heat energy. The important factor was *temperature*, not the amount of heat energy. Heat never moves of itself from a cooler body into a hotter body.

Now consider another example. Suppose you put a small amount of water and a large amount of water at the same temperature into two identical pots on a hot stove for the same length of time. At the end of this time, the small amount of water will have increased in temperature more than the large amount of water. Both pots received the same amount of heat, but one increased in temperature more than the other. Why is this? The same amount of heat was distributed among a different number of molecules. Think of it this way. If you divide $1,000 among 1,000 people, each person gets $1. If you divide $1,000 among 10 people, each person gets $100. In the sample with fewer molecules, each molecule received a greater share of the heat energy. The more energy each molecule received, the faster it moved, and the higher the temperature of the substance.

From these examples, it should be clear that heat and temperature are two different things. ***Heat energy*** is a measure of the *total* kinetic energy of all the particles in a body. ***Temperature*** is a measure of the *average* kinetic energy of each particle in a body. Each molecule in a body at a high temperature has, on average, a large kinetic energy. Each molecule in a cooler body has, on average, a small kinetic energy. In other words, how hot or cold something feels depends on how fast its molecules are moving. Heat energy, on the other hand, depends not on only how fast the molecules are moving but also on the total *number* of molecules. So, even if its molecules are moving slower, a body made of huge numbers of molecules can contain more total heat energy than a body made of just a few molecules.

So why does heat transfer only from something at a higher temperature to something at a lower temperature? Because when a fast-moving particle and a slow-moving particle collide, the slow-moving particle always speeds up and the fast-moving particle always slows down. There is no way for the two to collide in such a way that the slow particle moves slower and the fast particle moves faster. Therefore, even if slow-moving molecules *outnumber* fast-moving molecules, they will speed up if they come into contact with fast-moving molecules. Thus, the slow molecules speed up and the fast molecules slow down, until, eventually, all of the molecules are moving at the same speed and the two bodies are at the same temperature.

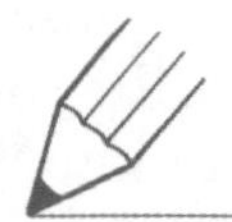

TRY THIS

1. When water absorbs heat, the speed of the water molecules' movement
   1. increases
   2. decreases
   3. remains the same

2. Two glasses of water are placed into a room that has an air temperature of 25°C, as shown in the diagram below.
   After 1 hour, what will happen to the water temperatures?

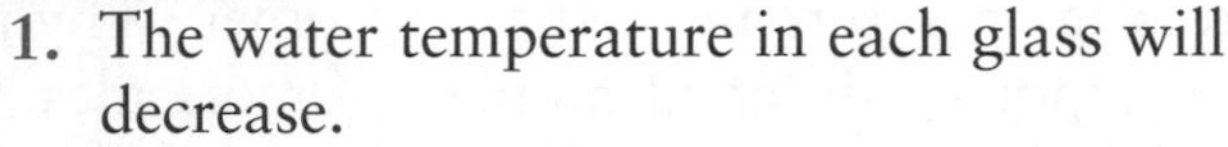

   1. The water temperature in each glass will decrease.
   2. The water temperature in each glass will increase.

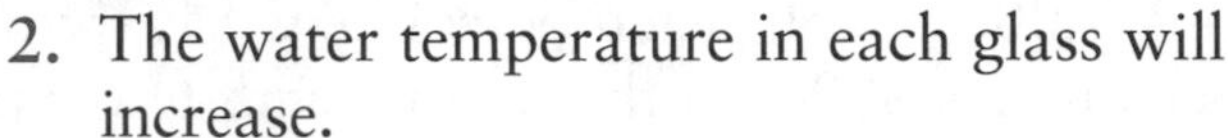

   3. The water temperature will decrease in glass *A* and increase in glass *B*.
   4. The water temperature will increase in glass *A* and decrease in glass *B*.

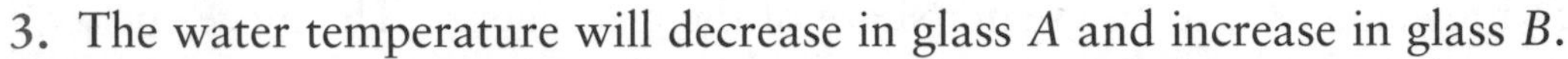

3. Using water to cool a nuclear reactor increases the temperature of the water. Which body of water would probably be most affected when the warmer water from the reactor is returned to it?
   1. an ocean
   2. a small lake
   3. a large river
   4. a sea

MEASURING TEMPERATURE

Nearly all substances expand when their temperature is raised and shrink when their temperature is lowered. This happens because when the temperature of a substance is raised, its molecules move faster. Collisions between faster-moving particles force them to move farther apart, resulting in expansion of the substance. (Water is an exception, expanding when changing to ice.)

A ***thermometer*** is an instrument that measures temperature by the expansion and contraction of a liquid, usually mercury or colored alcohol. When a thermometer is placed into water that is changing phase, the liquid inside it stops expanding for a time because water does not change temperature while changing phase. Thus, the freezing point and boiling point of water make convenient reference points for measuring expansion of the liquid in a thermometer. The thermometer is placed into freezing water, and the height of the liquid is marked. Then it is placed into boiling water, and the height of the liquid is marked. The space between the two marks is divided into 100 equal parts, called ***degrees***. A value of 0°C is assigned to the freezing point and 100°C to the boiling point. A thermometer marked off in this way used to be called a centigrade thermometer (from the Latin *centi* = 100, *gradus* = step). It is now called the ***Celsius*** thermometer in honor of the Swedish scientist Anders Celsius who first invented it. Scientists worldwide use the Celsius thermometer.

The German physicist Gabriel Daniel Fahrenheit invented the ***Fahrenheit*** thermometer, in which the space between the freezing and boiling point marks is divided into 180 equal parts. A value of 32°F is assigned to the freezing point and 212°F to the boiling point of water. The Fahrenheit scale is used mainly in English-speaking countries.

Arithmetic formulas convert one temperature scale to the other, but you will not be asked to do so on the intermediate-level science exam. Besides, many thermometers list both scales side by side. You can convert from one scale to another simply by reading the height of the liquid on the corresponding scale. See Figure 14.2.

Theoretically, there is no upper limit to temperature since there is no upper limit to speed. On the other hand, there is a lower limit to temperature since particles cannot move any slower than not to move at all. Theoretically, when a particle stops moving, it has reached its coldest possible temperature, or ***absolute zero***. This lower limit to temperature occurs at 273.16° *below* zero on the Celsius scale and 459.69° *below* zero on the Fahrenheit scale. The ***Kelvin*** temperature scale assigns a value of 0 K to absolute zero and is divided into degrees that are the same size as those on the Celsius scale. Therefore, water's freezing point is 273.16 K and its boiling point is 373.16 K on the Kelvin scale.

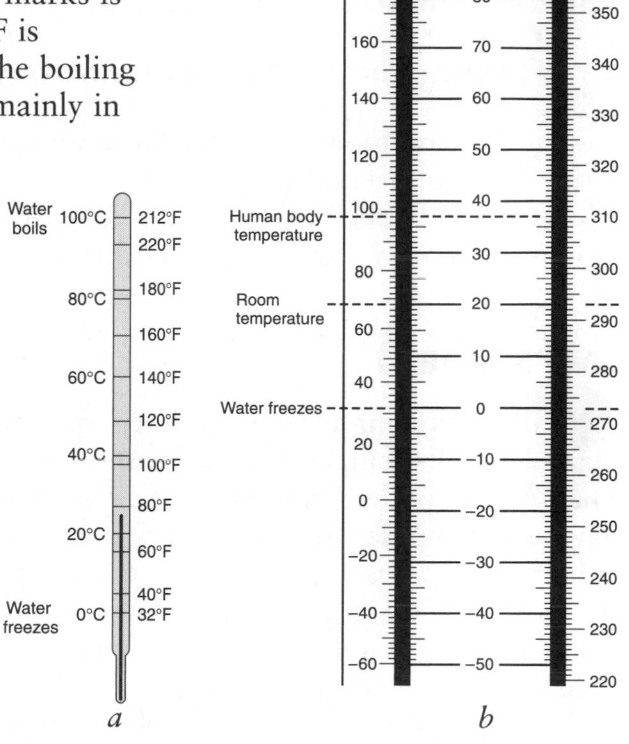

Figure 14.2
(*a*) A thermometer with Celsius and Fahrenheit scales printed side by side. (*b*) Temperature scales.

TRY THIS

1. Refer to Figure 14.2 to answer the following questions. Which two events occur at the same temperature?
   1. Water boils and freezes.
   2. Water boils and ice melts.
   3. Water freezes and ice melts.
   4. Ice melts and water reaches its greatest density.
2. The temperature outside is 85°F. What is the Celsius equivalent of this temperature?

MEASURING HEAT

When heat is transferred from a hot body to a cold body, a change in temperature occurs. The more heat that is transferred from one body to another, the greater the temperature change. Therefore, heat is measured by how much of a temperature change it produces. The most commonly used unit of heat energy is the ***calorie***, the amount of heat needed to change the temperature of 1 gram of water by 1°C.

1 calorie of heat will change the temperature of 1 gram of water 1°C

Another common unit is the kilocalorie, the amount of heat needed to change the temperature of 1 *kilogram* of water by 1°C. Since 1 kilogram equals 1,000 grams, 1 kilocalorie equals 1,000 calories. The Calorie (note the capital C) used to measure the energy content of foods is actually a kilocalorie.

LATENT HEAT

Typically, when matter gains heat energy, its temperature increases. When matter loses heat energy, its temperature decreases. However, in the last chapter you learned that when matter is changing phase, it absorbs or gives off heat energy with no resulting change in temperature. The energy is used to overcome forces of attraction rather than to increase the speed at which molecules of matter are moving. Since the heat gained or lost during a phase change does not show up as a temperature change, it is called hidden heat, or ***latent heat.***

Energy is *absorbed* when a solid changes to a liquid and when a liquid changes to a gas. Energy is *released* when a gas changes to a liquid and when a liquid changes to a solid. The quantity of heat absorbed when water changes from a solid to a liquid or released when water changes from a liquid to a solid, called the *latent heat of fusion*, amounts to 80 calories per gram. The quantity of heat absorbed when water changes from a liquid to a gas or released when water changes from a gas to a liquid, called the *latent heat of vaporization*, amounts to 540 calories per gram. See Figure 14.3.

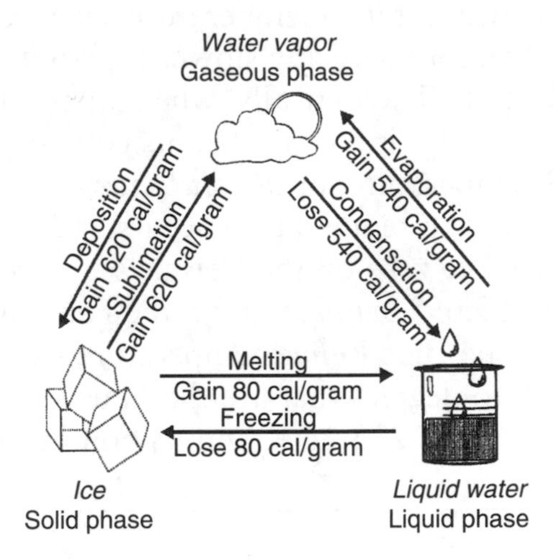

Figure 14.3
Energy gain and loss during phase changes.

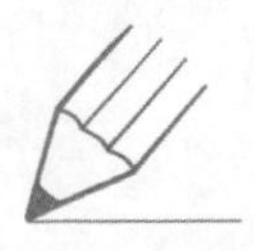

TRY THIS

Why does the evaporation of sweat cool the surface of your skin?

HEAT TRANSFER

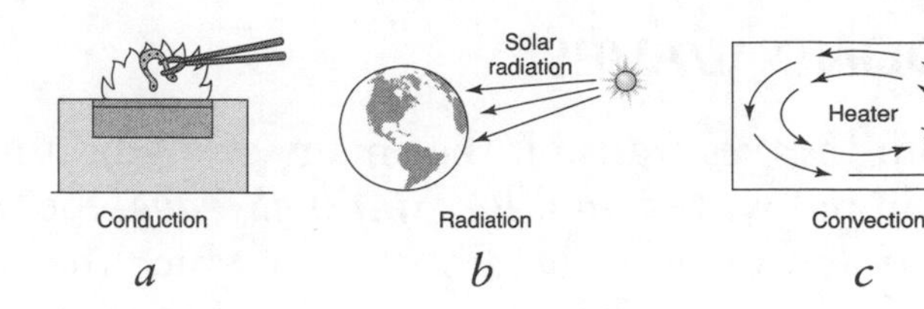

Figure 14.4
Mechanisms of heat transfer.

Heat can be transferred from one place to another in three ways: conduction, radiation, and convection. See Figure 14.4.

In ***conduction***, heat is transferred by means of collisions between rapidly moving molecules at the hot end of a body of matter and the slower molecules at the cold end. When faster molecules collide with slower molecules, some of the kinetic energy of the faster molecules passes to the slower molecules. As a result of successive collisions, there is a flow of heat through the body of matter. Solids, liquids, and gases all conduct heat. Gases are the poorest conductors of heat because their molecules are spaced farthest apart and collide less frequently than those in solids and liquids.

In ***convection***, a volume of hot fluid (gas or liquid) moves from one region to another, carrying heat energy along with it. Hot fluids are less dense than cold fluids. For example, when a pan of water is heated on a stove, the hot water on the bottom of the pan expands slightly and becomes less dense. The hot water rises to the surface, while colder, denser water sinks to take its place at the bottom. These rising and sinking motions form a circular pattern called a ***convection current.***

In ***radiation***, energy is carried by the electromagnetic waves emitted by every object. As atoms and molecules vibrate, they send out waves of energy that travel through space. The higher the temperature of an object, the greater the rate at which it radiates energy. When you sit in front of a campfire, you are warmed by heat energy radiated to you by the fire.

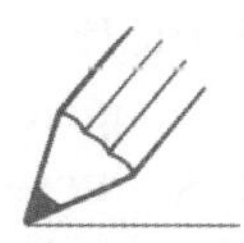

TRY THIS

1. In one or more complete sentences, explain why a piece of metal feels colder to the touch than a piece of wood when you are outdoors in the winter.
2. How would Earth be different if heat were not able to travel by radiation?
3. Why can you bring your finger close to the *side* of a candle flame without being burned, but if you place your finger the same distance *above* the flame, you will be badly burned?

SOUND

Sound is a form of energy produced by vibrating objects. When an object vibrates, it moves back and forth rapidly. The back and forth motion causes particles of matter around the object to be alternately squeezed together and spread apart. Continued vibrations produce successive layers of squeezed together and spread-out matter that travel outward from the source, or ***sound waves.***

SOUND WAVES

Sound waves transmit the energy of a vibrating object through matter in the form of longitudinal waves. In a ***longitudinal wave***, the individual particles of a medium vibrate back and forth in the direction in which the waves travel. See Figure 14.5. In air, sound waves spread out from a vibrating object in the form of a series of pressure fluctuations.

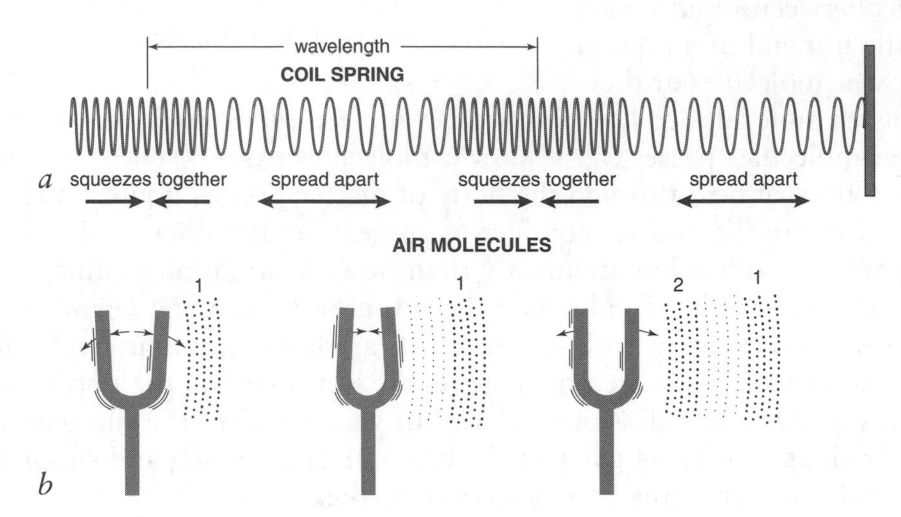

Figure 14.5
Longitudinal waves (*a*) in a coil spring and (*b*) in air around a tuning fork.

Consider a tuning fork. When a tuning fork is struck, its prongs vibrate rapidly. When a prong moves outward, the air molecules next to it are pushed closer together. This forms a region of high pressure that pushes outward in front of the prong. The prong then moves inward, thereby increasing the volume available to nearby air molecules. Air molecules spread out to fill the space and form a region of low pressure right behind the high-pressure region. The continued vibrations of the prong send out successive layers of air that are squeezed together and spread apart. When these changes in pressure come into contact with your eardrum, they cause it to vibrate with the same frequency. This is what causes the sensation of sound.

THE SPEED OF SOUND

Sound waves can travel through solids, liquids, and gases. However, they cannot travel through a vacuum because there is no matter to be squeezed together or spread apart. In air, at sea level and 0°C, sound waves travel at 331 meters per second, or 1,096 feet per second. In general, the denser and stiffer the material, the faster the sound waves travel through it. See Table 14.2. This makes sense, because

Table 14.2 The Speed of Sound in Various Substances

Material	Speed (meters per second)
Air	331
Aluminum	5,000
Carbon dioxide gas	259
Water	1,486
Wood (oak)	3,850

density and stiffness depend on how tightly particles are packed and coupled together. The closer and more tightly coupled the particles, the more immediately they respond to one another's movements.

The speed of sound is also affected by temperature. Raising temperature speeds up the particles that make up matter. This, in turn, increases the speed at which pressure changes pass through the matter. Therefore, the speed of sound increases with temperature.

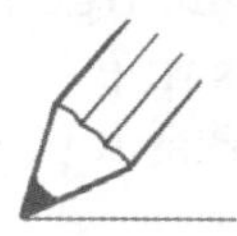

TRY THIS

Base your answers to the following questions on Table 14.2, which shows the speed of sound in different materials.

1. How much faster does sound travel through aluminum than through wood?
   1. 1,050 meters per second
   2. 1,150 meters per second
   3. 2,364 meters per second
   4. 8,850 meters per second
2. Which conclusion about the speed of sound in different materials is best made from the data table?
   1. Sound travels fastest through gases.
   2. Sound travels fastest through solids.
   3. Sound travels slowest through solids.
   4. Sound travels slowest through liquids.

ELECTRICAL ENERGY

Electrical energy is the ability to do work by virtue of the forces of the attraction and repulsion between electric charges. ***Static electricity*** involves forces exerted on matter because of an imbalance of electric charge. ***Current electricity*** refers to the flow of electric charge through matter.

STATIC ELECTRICITY

Have you ever had a hair-raising experience when you pulled off a sweater on a dry winter day? Have you ever made a balloon stick to the wall after rubbing it against your hair? These things happen because of static electricity. In order to understand static electricity, you first need to recall some of the basics about atoms.

All matter is made up of atoms. Inside an atom are protons, electrons, and neutrons. The protons have a ***positive electric charge,*** the electrons have a ***negative electric charge,*** and the neutrons have no electric charge. Therefore, all matter is made up of electrical charges. Opposite charges attract each other (positive and negative attract). Like charges repel each other (positive repels positive and negative repels negative). Most of the time, positive and negative charges are balanced in an object, which makes that object *neutral*. What exactly is charge? We can only say that it is charge, just as mass is, well, mass.

Static electricity is the result of an imbalance between negative and positive charges in an object. If you rub certain substances against one another, you can transfer negative charges, or electrons, from one substance to the other. These charges can build up on the surface of an object until they find a way to be released, or discharged.

For example, when you take off your sweater, electrons are transferred from the sweater to your hair. Your hair then has a surplus of electrons and is negatively charged. Remember that like charges repel each other. Because individual hairs have the same charge, they try to get as far away from each other as possible, creating an interesting hairdo!

When you rub a balloon against your hair, you are adding extra electrons to the surface of the balloon. The balloon becomes more negatively charged than the wall. The more positively charged wall attracts the more negatively charged balloon. When the two come into contact, the balloon will stick to the wall because opposite charges attract.

If your shoe rubs against a carpet as you walk across a room, your body collects extra electrons. The electrons cling to your body until they can be released. As you reach out and touch a metal doorknob, you get a shock. Do not be scared by it. The shock is only the extra electrons being released from you to the doorknob.

Another interesting observation is that once a material has been electrically charged, it attracts uncharged materials and may either attract or repel charged materials *without touching them!* The forces of attraction and repulsion due to electric charge extend beyond the object! The region of space around a charged object in which these forces exist is called an *electric field*. Although an electric field is invisible, we know it exists because we can see its effects. The field itself can exert forces of attraction or repulsion on matter, even if the charged object does not come into contact with the matter. See Figure 14.6.

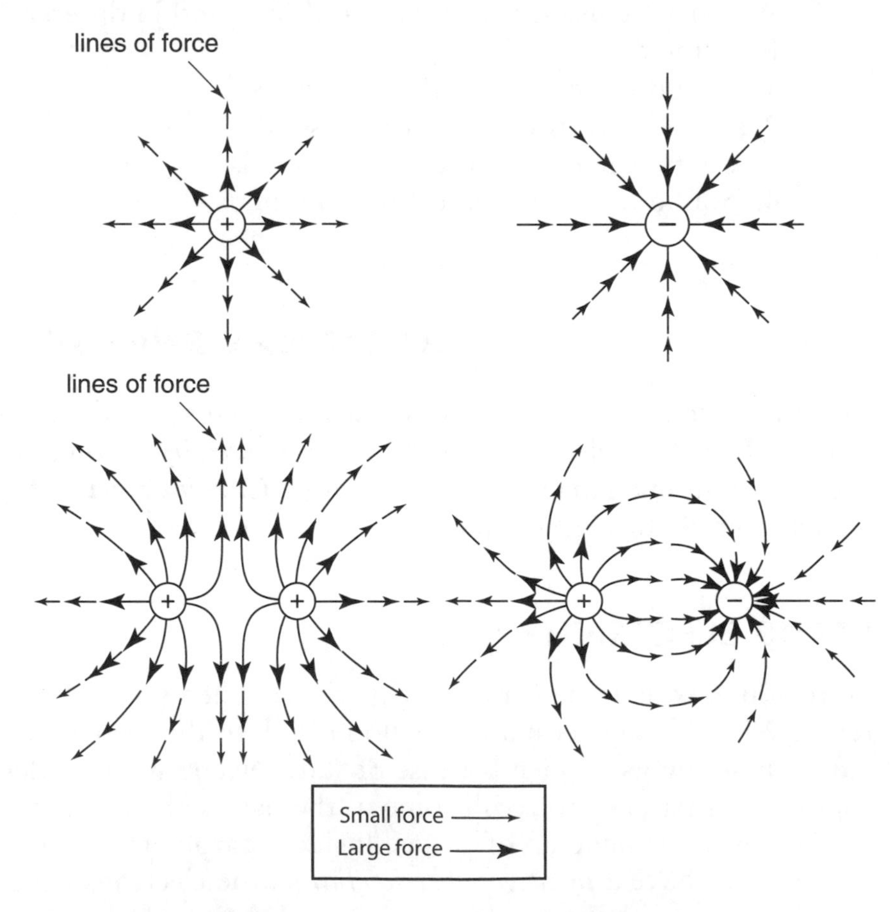

Figure 14.6
(*a*) Electric fields around some charged particles. Arrows indicate the direction in which force is exerted. (*b*) Electric fields when like and unlike charges are brought close to one another.

CURRENT ELECTRICITY

An *electric current* is a flow of charge from one place to another. Almost all substances fall into one of two categories: conductors and insulators. ***Conductors*** are substances through which charge flows easily. ***Insulators*** are substances through which charge can flow only with great difficulty. Metals and many liquids and gases whose particles are charged are good conductors. Nonmetals, liquids, and gases whose particles are electrically neutral are insulators.

In order to produce a flow of electric charge, you must have a separation of positive and negative charges. ***Electric cells***, such as the D cells in a flashlight, use chemical reactions to separate positive and negative charges. Electric power plants use complex systems to make charge flow through the wires that connect to your home. The two prongs on the plugs of all your household appliances connect to the positively and negatively charged conductors in the power outlet. Electric charge will flow through a conductor if it forms a continuous pathway, or ***closed circuit***, from matter that is positively charged to matter that is negatively charged. However, if a break occurs in the conducting pathway, it is called an *incomplete* or ***open circuit***, and electric charge will not flow. Never touch both conductors in a power outlet at the same time, or *you* could become part of the circuit. Electric charge could then flow through your body and give you a severe shock, which could be fatal.

A ***switch*** is a device that allows one to control the flow of charge through a circuit by opening and closing the circuit. A device that uses flowing electric charge to do work is called a *load*. Lightbulbs, televisions, CD players, and the like are all loads. In a lightbulb, charge is made to flow through a thin wire, called a filament. The moving charges exert forces onto the molecules in the wire that cause them to vibrate more rapidly, and the wire gets so hot that it glows. If too much electricity flows through the wires in your house, they too can heat up and cause a fire. Therefore, household circuits are protected by *fuses* or *circuit breakers* that open the circuit if the wires get too hot.

A load can be connected to an electric circuit in several ways. In a ***series circuit***, electric charge can travel along only one pathway through the circuit. See Figure 14.7*a*. Some Christmas lights are wired in a series circuit. If one of the lights burns out, the conducting pathway is broken and electric charge stops flowing. Then all of the lights go out at once. In a ***parallel circuit***, the electric charge can

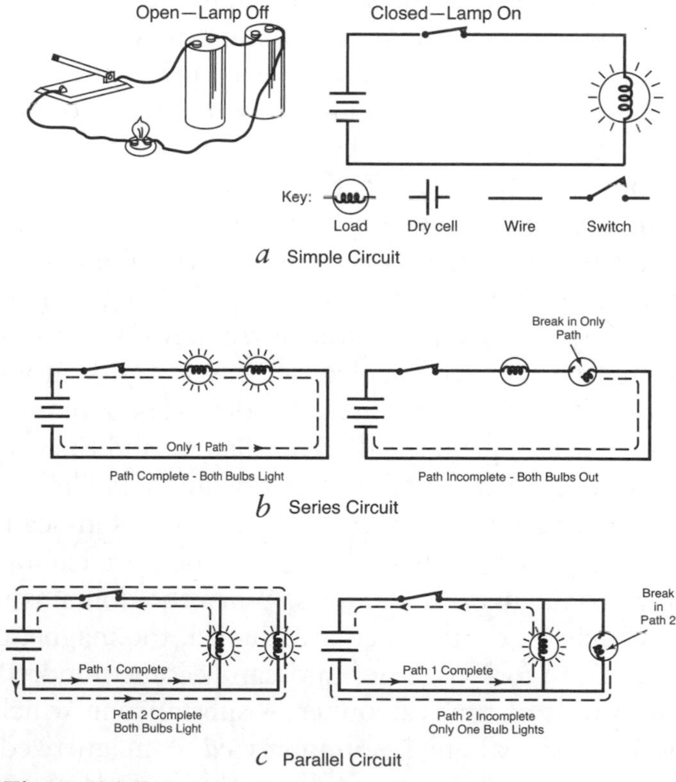

Figure 14.7
Electric circuits. (*a*) A picture of a simple circuit and the same circuit shown using symbols. (*b*) A series circuit. (*c*) A parallel circuit.

flow through the circuit along more than one pathway. See Figure 14.7*b*. Christmas lights wired in a parallel circuit do not all go out at once. If one light burns out, electric charge can flow along one of the other paths that is still unbroken. See Figure 14.7*c*.

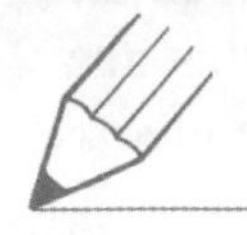

TRY THIS
The diagram below shows an incomplete electrical circuit.

1. Which object, when connected to the wires at position *X*, would complete the circuit and light the bulb?
   1. wooden matchstick
   2. glass test tube
   3. iron nail
   4. rubber hose

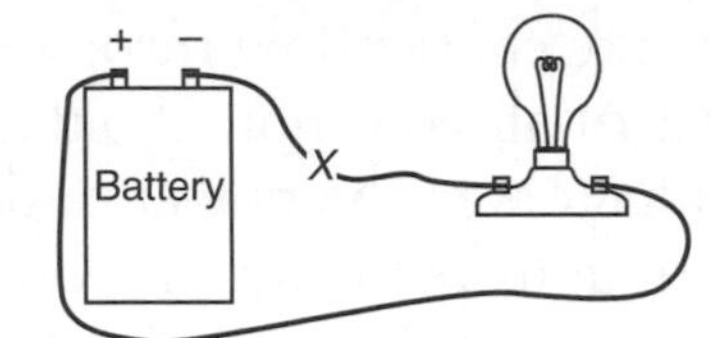

2. By using the symbols shown in Figure 14.7, draw a diagram of a circuit with two dry cells, one switch, and three lamps. In your circuit, what will happen to the lamps if one of them burns out?

MAGNETISM

Electric charges that are in motion exert forces onto one another that are very different from when they are at rest. If we put two wires through which electric charge is flowing in the same direction next to each other, the wires are attracted to one another. If electric charge is flowing through the two wires in opposite directions, the wires repel. See Figure 14.8. These forces disappear if we stop the flow of charge. The forces that come into existence when electric currents interact are called ***magnetic forces***. Magnetic forces can cause matter to move and thus have the ability to do work. Therefore, you might think that magnetism is another form of energy. Since all magnetic forces can be traced back to *moving electric charges*, however, magnetism is really a type of electrical energy.

Magnetism can also result from electric charges moving inside atoms. In certain types of matter, the moving electric charges inside atoms cause their magnetic forces to align. The atoms lock together in a group, called a ***domain***, and their magnetic forces work together. A domain acts like a tiny magnet. Substances that contain domains are called *ferromagnetic*. Iron, nickel, and cobalt are examples of ferromagnetic substances. When the domains in a ferromagnetic substance are aligned, the magnetic forces all work together and can become hundreds or thousands of times stronger. A substance in which the domains are aligned is ***magnetized***. A magnetized object is called a ***magnet***. If an object made of a pure ferromagnetic substance is brought near a magnet, its domains are forced to align in such a way that it is attracted to the magnet. The substance then becomes

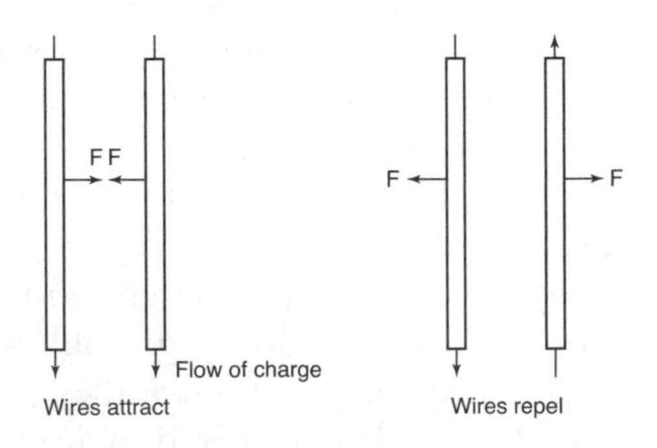

Figure 14.8
Wires carrying electric charge attract or repel depending on the direction in which the charge flows.

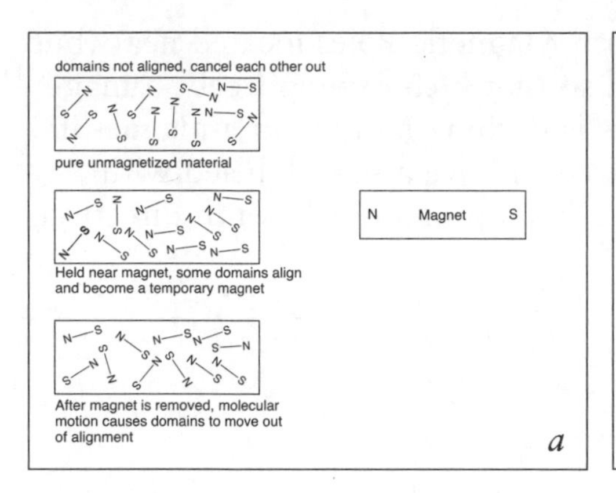

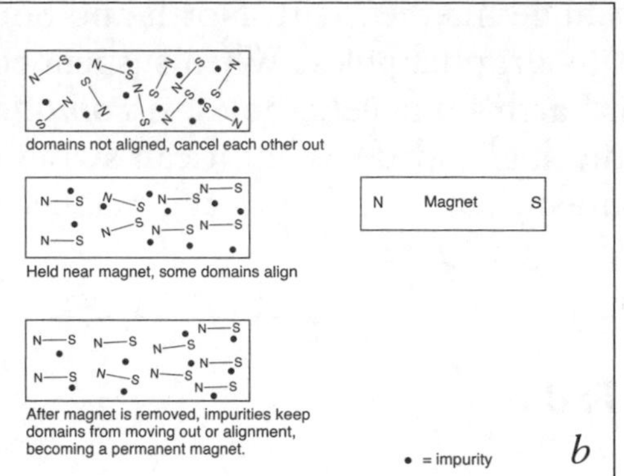

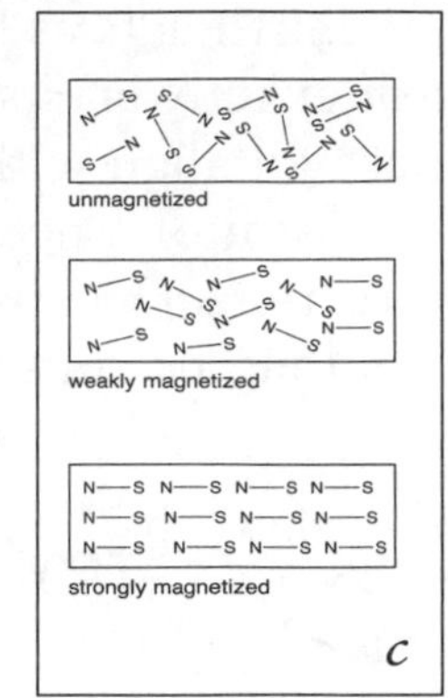

Figure 14.9
Magnetic domains in a temporary magnet (*a*) and a permanent magnet (*b*). As domains become more aligned, magnetic strength increases (*c*).

a temporary magnet. It is called a temporary magnet because once the magnet is removed; normal molecular motion in the substance causes the domains to shift out of alignment, and it no longer acts as a magnet. If the ferromagnetic substance is impure, as in the case of steel (a mixture of iron and carbon), once the domains are forced into alignment, the impurities keep the domains from moving out of alignment. So, once the object is magnetized, it stays magnetized and is called a permanent magnet. See Figure 14.9.

A magnet attracts certain materials and either attracts or repels other magnets *without direct contact*. The forces of magnetic attraction and repulsion extend *beyond the object!* The region of space around a magnet in which magnetic forces exist is called a ***magnetic field.*** If we sprinkle iron filings (iron is ferromagnetic) around a magnet, the magnetic forces in its field cause the bits of iron to align in a pattern as shown in Figure 14.10. The lines in the pattern are called magnetic ***lines of force.***

LAW OF MAGNETIC POLES

As with electric charges, the fact that magnetic forces can attract *or* repel suggests that there are two types of magnetic forces. When suspended from a string, magnets align in a north-south direction, pointing roughly in the direction of Earth's North and South Poles. Therefore, the ends that point toward Earth's north or south are called a magnet's ***poles.*** The magnetic forces exerted by a magnet are strongest at its poles.

The end of a magnet that points north is called its north-seeking pole, or simply its ***north pole*** (labeled N in a diagram of a magnet). The end that points south is called a south-seeking pole, or ***south pole*** (labeled S). If we experiment with several magnets, we find that north poles repel north poles, south poles repel south poles, and north poles and south poles attract one another. These observations are summarized in the ***law of magnetic poles***:

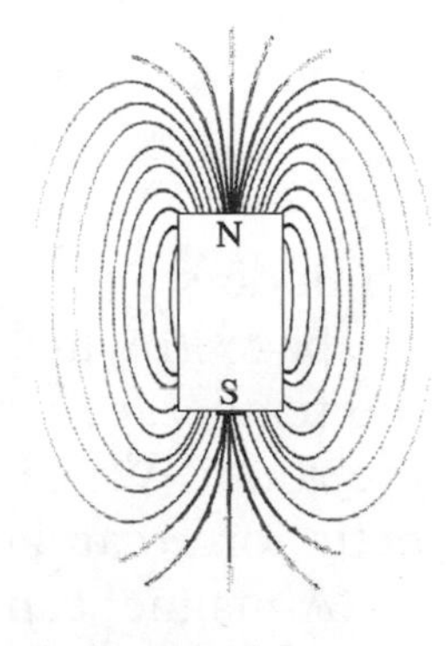

Figure 14.10
Magnetic field surrounding a magnet.

Like poles repel, unlike poles attract.

Earth acts as a huge magnet, with North and South Magnetic Poles located near (but not exactly at) its geographic poles. When suspended so that it can swing freely, a magnet will align with Earth's magnetic poles. A ***compass*** is nothing more than a magnet in the form of a needle suspended so that it can swing freely above a scale labeled with directions. The tendency of a freely moving magnet to line up in a north-south direction has long been used to navigate.

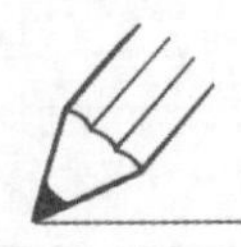

TRY THIS

1. Two bars of iron are the same size and shape. One of the bars is a strong magnet and the other is a weak magnet. With the aid of a diagram, explain how this is possible.
2. What conditions must be present on Mars to make possible the use of a magnetic compass?

ELECTROMAGNETS

You learned earlier that wires carrying an electric current exert magnetic forces. If a wire is formed into a coil, the magnetic forces surrounding the wire line up and work together, causing the force to become stronger. If a ferromagnetic material is placed inside the coil, the magnetic force of the coil forces the material's domains to align and become a magnet. These types of magnets are called ***electromagnets*** because the magnetism is produced by a flow of electricity. See Figure 14.11.

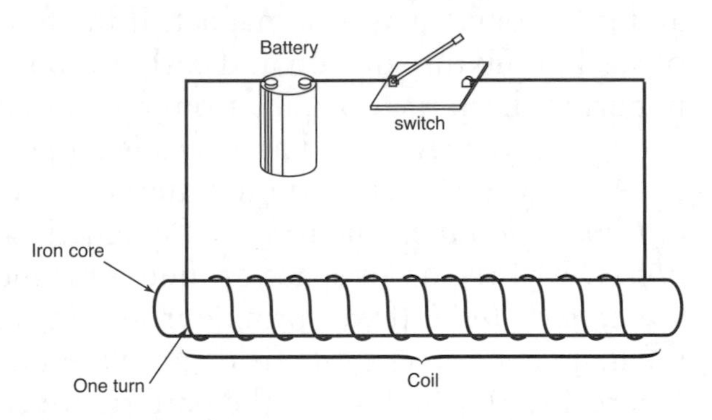

Figure 14.11
An electromagnet.

LIGHT

To understand light energy, you must first understand what a field is. A *field* is a region of space in which a measurable quantity is present at every point. Thus, we can speak of the temperature field in a room because a temperature can be measured at every point in the room. We can also speak of a magnetic field surrounding a magnet because a magnetic force can be measured at every point in the space around it.

A magnet can make a compass needle move from a distance because the magnet is surrounded by an invisible magnetic field. If you move the magnet, the magnetic field will move, and the moving magnetic field will cause the compass needle to move. In much the same way, a balloon charged with static electricity will attract your hair from a distance because it is surrounded by an invisible electric field. If you move the balloon, the invisible electric field moves with it, and you can feel the effect of the moving field on your hair. Such fields extend outward infinitely in all directions from their sources, but they weaken with distance from the source.

Every atom of matter has electric charges in it and is surrounded by an electric field. Whenever electric charges move, a magnetic force is produced, and the particle is then also surrounded by a magnetic field. Since the atoms and molecules of all matter are in constant motion, these two fields—an electric field and a magnetic field—exist simultaneously around all matter. Together, these fields are called an ***electromagnetic field*** and extend outward infinitely in all directions around the particle.

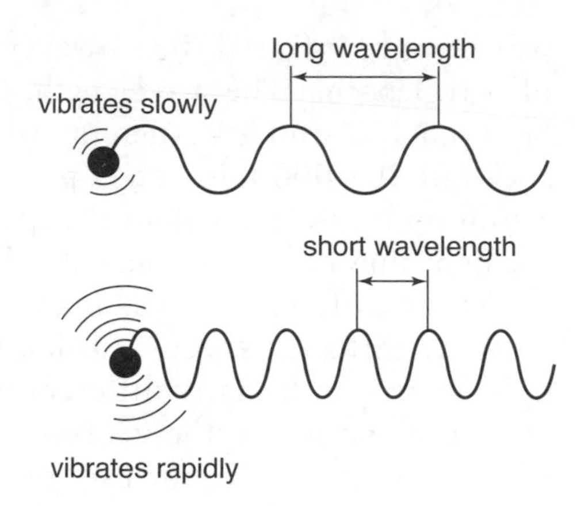

Figure 14.12
The motion of a particle creates waves in an invisible electromagnetic field.

When a particle moves back and forth, its electromagnetic field moves along with it. Like a ripple in a pond, this movement spreads out through the field in the form of a light wave. Light waves travel in straight lines. That is why objects that block light waves cast shadows.

The way that the wave moves through the electromagnetic field depends on the way in which the particle moves. The faster the particle vibrates, the shorter the ***wavelength***, or distance between ripples in the electromagnetic field. Different speeds of vibration produce waves of different wavelengths. The number of waves that pass a given point in a given amount of time is called the ***frequency*** of the wave. See Figure 14.12.

A moving wave contains energy. It can exert forces onto matter with which it interacts. Since an electromagnetic field does not require a medium to exist, it can extend through space. Disturbances in electromagnetic fields around particles can travel through space as they move outward, or *radiate*, through the field. In this way, energy is transferred from the Sun to Earth without the existence of a physical medium between the two.

THE ELECTROMAGNETIC SPECTRUM

Electromagnetic waves are classified by their wavelength, ranging from short waves, such as X rays, to long waves, such as radio waves. See Figure 14.13. The electromagnetic spectrum is a continuum in which electromagnetic waves are arranged in order, from longest to shortest

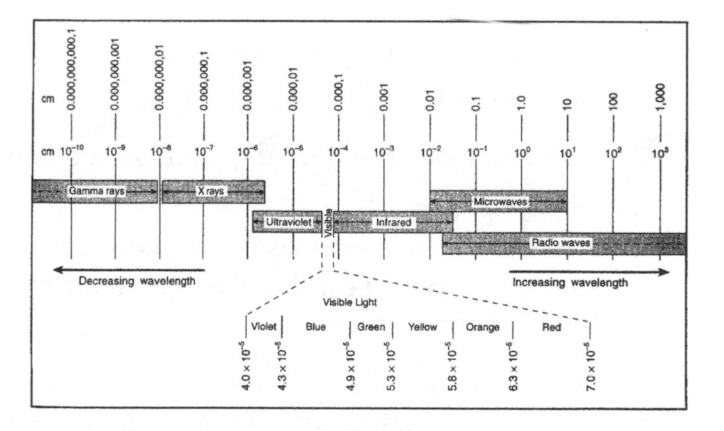

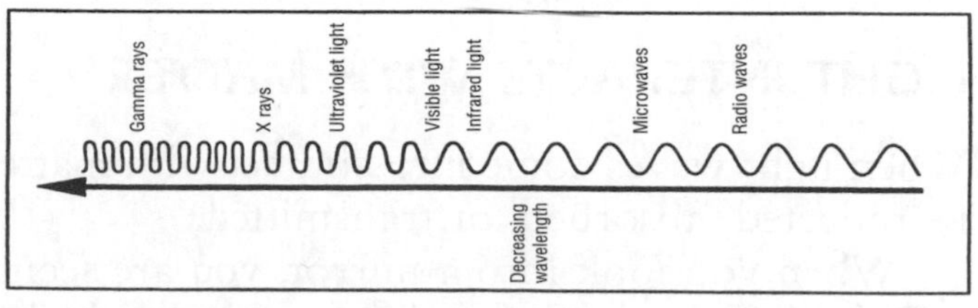

Figure 14.13
The electromagnetic spectrum.

wavelength. Infrared (heat) waves and visible light waves fall roughly in the middle range of wavelengths. The wavelength, or distance between ripples of visible light waves, falls between 10^{-6} and 10^{-7} meters. All electromagnetic radiation travels at the incredible speed of 300,000 kilometers per second (about 186,000 miles per second). This is about 1 million times faster than the speed of sound (331 meters per second), which is why we see lightning before we hear the thunder it creates.

Many different waves emanate from the Sun because it contains particles moving at many different speeds. Ordinary white light is actually made of a mixture of visible light waves with many different wavelengths. However, most of the waves coming from the Sun are in the visible-light and infrared range. Our eyes can detect light in the visible range. To our eye, the different wavelengths and frequencies of visible light appear as the colors of the visible spectrum. Light with the shortest wavelength appears violet, and light with the longest wavelength appears red. When ordinary white light strikes an object, the color you see is actually the wavelength of light that is reflected by the object. The object absorbed the other visible wavelengths in ordinary white light.

Earth's magnetic field and the Van Allen belt of charged particles in the upper atmosphere deflect many of the waves with short wavelengths. This is fortunate since most short-wavelength radiation is harmful to living things. For example, X rays and ultraviolet rays can damage cellular material and have been linked to genetic mutations and cancer.

TRY THIS

1. Light travels in the form of __________ waves.
2. Light travels in __________ lines.
3. The color of an object depends on the wavelength of the light __________ by its surface.
4. List the following types of electromagnetic radiation in order from longest to shortest wavelength: blue visible light, microwaves, gamma rays, orange visible light, ultraviolet rays.

LIGHT INTERACTS WITH MATTER

When light waves come into contact with matter, three things can happen. The light may be reflected, absorbed, or transmitted.

When you look into a mirror, you are seeing light that is bounced back, or ***reflected***, from its surface. The angle at which incoming light strikes a surface is called the ***angle of incidence***, and the angle as it bounces off is called the ***angle of reflection***.

See Figure 14.14*a*. When light is reflected, it bounces off the surface at the same angle at which it struck the surface.

The angle of incidence equals the angle of reflection.

The smoother the surface, the more closely the reflected light will match the incoming light and the more accurate the reflection. Rough materials have many tiny surfaces all at different angles, so the incoming light is *scattered*, or bounces off in many different directions. See Figure 14.14*b*. This is why you do not see your image when you look at a wall.

Your car gets hot when the Sun shines onto it because light was ***absorbed***. Its energy was transferred to the molecules of your car, causing them to move faster, and was thus converted to heat. Different substances absorb light to varying degrees. Dark-colored substances *absorb* more light and convert it to heat energy. Light-colored objects *reflect* more light, so less is converted to heat energy. That is why people usually wear light-colored clothing during the summer. Rough surfaces tend to absorb more light than smooth ones, because light may bounce off a rough surface several times before leaving it, with some energy being absorbed during each bounce. See Figure 14.14 again.

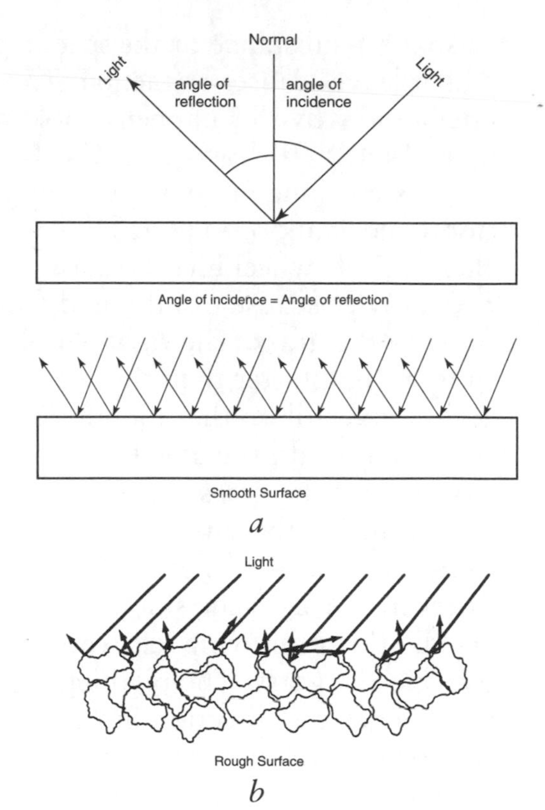

Figure 14.14
Light (*a*) reflecting from a smooth surface and (*b*) scattering from a rough surface.

You can see something through glass because light is transmitted, or passes through. Substances differ in their ability to transmit light. *Transparent* substances, such as window glass, allow almost all of the light that reaches them to pass through. *Translucent* substances, such as wax paper, allow some but not all of the light that reaches them to pass through. They also scatter some of the light so that images are not transmitted clearly. *Opaque* substances, such as concrete or steel, do not allow any light to pass through them. See Figure 14.15.

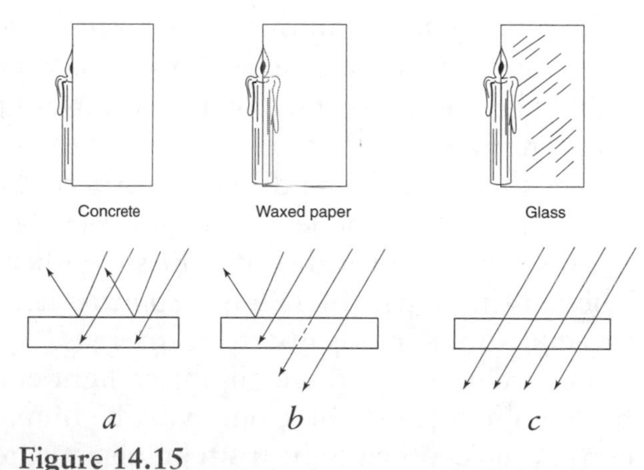

Figure 14.15
(*a*) Opaque, (*b*) translucent, and (*c*) transparent substances.

REFRACTION AND LENSES

Light is bent, or ***refracted***, whenever it passes at an angle from one transparent substance (such as air) into another (such as glass.) The slowing down of light when it enters a denser transparent substance or the speeding up as it leaves causes refraction. To understand why this happens, look at Figure 14.16*a*. It shows two wheels on an axle rolling at an angle from a smooth sidewalk onto a grass lawn. When the left wheel hits the grass, it slows down because of increased friction with the grass. The right wheel is now rolling faster than the left because it is still on the smooth sidewalk. As a result, the axle tends to pivot toward the grass. This is because during the same time, the faster-moving right wheel travels a greater distance on the sidewalk than the slower-moving left wheel travels in the grass. The pivoting action bends the direction in which the axle is traveling toward the normal, the dotted line drawn perpendicular to the boundary between the grass and the sidewalk.

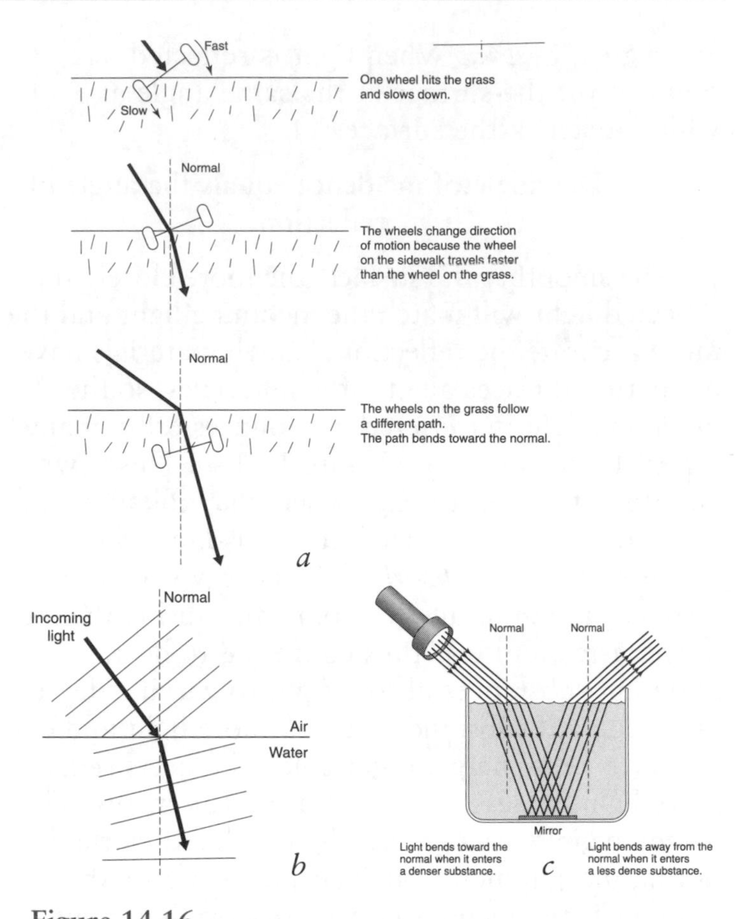

Figure 14.16
(*a*) Axle changes direction of travel. (*b*) Light bends toward the normal when going from air to water. (*c*) Light going from air to water bends toward normal, light going from water to air bends away from normal.

A similar situation occurs when light strikes the surface of a transparent substance at an angle. See Figure 14.16*b*. When the speed of light decreases by going from one transparent substance into another, the light bends toward the normal. When the speed of light increases by going from one transparent substance into another, the light bends away from the normal. See Figure 14.16*c*.

A *lens* is a curved piece of glass that bends, or refracts, light. The curved surface of a lens causes light to strike it at different angles at different places on the surface. This causes the light hitting different parts of the lens to bend at different angles. The lens in Figure 14.17*a* is a ***convex lens***. It causes light to bend so it comes together, or *converges*. The lens in Figure 14.17*b* is a ***concave lens***. It causes light to bend so it spreads apart, or *diverges*.

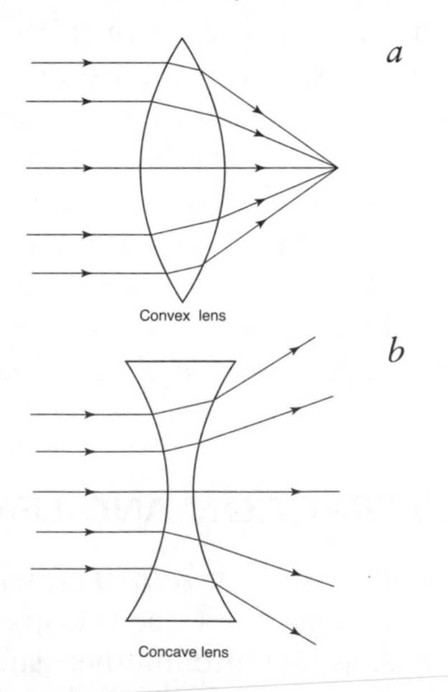

Figure 14.17
(*a*) Convex and (*b*) concave lenses.

In order for us to see an object, light emitted or reflected by the object must enter our eye. The human eye contains a convex lens. When light from an object enters our eye, the lens bends the light so that it forms an image of the object. How does the eye form an image? Place an object, like a flagpole, at some distance from a lens.

Whether it gives off its own light or reflects light from another source, every point on the object sends out light in all directions. A point at the tip of the flagpole reflects light in all directions. The lens in your eye bends the light coming from the top of the flagpole so that it intersects at one point, or converges, on the other side of the lens. This is the image of the point at the tip of the flagpole. We can follow the same process with the light coming from the flag, the middle of the flagpole, the bottom of the flagpole, and every point in between. In this way, point by point, an image of the entire object is built up on the other side of the lens. See Figure 14.18. The human eye forms an image on the retina, a layer of nerve cells at the back of the eye. Messages from the nerve cells in the retina are sent via the optic nerve to the brain, where they are interpreted. We then see the object.

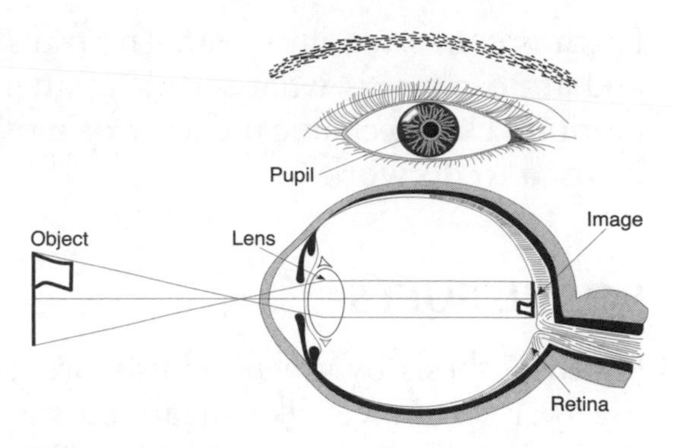

Figure 14.18
How the eye forms an image.

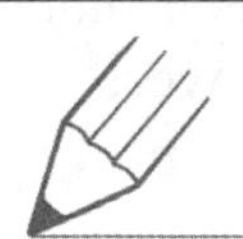

TRY THIS

1. Materials through which light cannot pass are __________.
2. Green glass transmits only __________ light.
3. Light entering a medium perpendicular to the surface is not __________.
4. Match the following terms with the corresponding letters in the illustration: incident light ray, reflected light ray, normal, angle of incidence, angle of reflection.

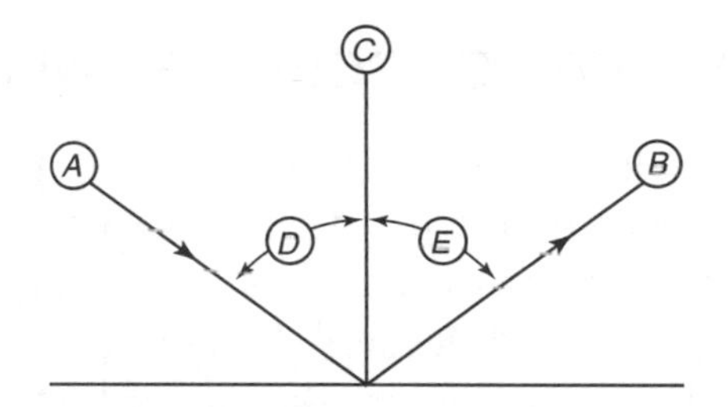

ENERGY SOURCES

Humans can take advantage of many sources of energy on Earth. Running water, ocean currents, and winds all have mechanical energy. Many substances on Earth contain chemical energy that can be released through chemical reactions. The atoms of all the matter of which Earth is made contain nuclear energy. However, almost all sources of energy on Earth can be traced back to the Sun.

SOLAR ENERGY

The Sun is Earth's primary source of energy. The energy contained in the electromagnetic waves that emanate from the Sun is called ***solar energy***. Much of the solar energy striking

Earth is converted into heat. The transfer of heat among the lithosphere, hydrosphere, and atmosphere is what sets the matter in running water, ocean currents, and winds into motion. The mechanical energy of moving water and wind derived from solar energy can be used to do work.

FOSSIL FUELS

Photosynthesis by green plants stores solar energy in the chemical bonds that hold together substances like sugar and starch. When animals eat the plants, this energy becomes part of their bodies, too. When plants and animals die, that energy is still present in the chemicals that make up their remains. Over time, the remains of living things have accumulated on Earth and have been changed to energy-rich substances such as coal, petroleum, and natural gas. A substance that can be used as an energy source is called a ***fuel***. Coal, petroleum, and natural gas are called ***fossil fuels*** because they formed from the remains of once-living things and can be burned to obtain heat energy.

Coal is a black rock made mostly of carbon that formed from decomposed plant remains that were deeply buried and changed by heat and pressure. ***Petroleum*** is a black, oily liquid that formed from the decomposed remains of marine organisms that were buried and changed by heat and pressure. ***Natural gas*** is a mixture of gaseous compounds that formed along with petroleum but separately from the liquid.

OTHER SOURCES OF ENERGY

Geothermal energy is the heat energy within Earth. Some of this heat was trapped inside Earth when it formed. Some of it comes from radioactive elements in Earth's rocks. Great quantities of this heat escape from Earth's interior in places like volcanoes, hot springs, and geysers. Scientists are beginning to find ways of tapping into these sources of heat. In Reykjavík, Iceland, steam and hot water from geysers and hot springs is used to heat homes. It is also used to heat greenhouses where food can be grown.

Nuclear energy is the energy contained within the atoms that make up matter. Nuclear energy can be released by nuclear reactions, or forcing atoms to split apart or combine to form new atoms. Nuclear reactions release energy without the combustion products of burning fuels. However, the radioactivity of nuclear fuels and the products of nuclear reactions pose other risks, which may last for thousands of years.

GENERATING ELECTRICITY

Electrical energy has become essential to our society. What makes electricity so useful? Electrical energy can be produced from a wide variety of energy sources and can be transformed into almost any other form of energy. Furthermore, electricity can be used to distribute energy quickly and conveniently from place to place. Electricity is used for many purposes, including heating and cooling buildings, cooking food, providing lighting, and running machines and appliances. Aside from lightning, which is very hard to control, there are few natural sources of electricity. So where does the electricity we use daily come from?

Just as a flow of electric charge creates a magnetic field, a magnetic field can cause electric charges to move. If a magnet and a coil of wire move relative to one another, electric charge will flow through the wire. An ***electric generator*** is a device in which some form of energy is used to cause a magnet and a coil of wire to move relative to one another and thus produce an electric current. In other words, it converts some other form of energy into electrical energy. Figure 14.19 shows a simple hand generator. When you turn the handle, a coil of wire spins in the field created by the magnets, causing an electric current to flow through the coil.

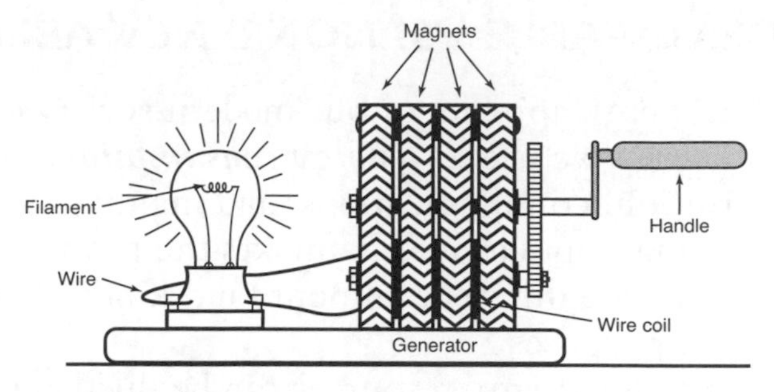

Figure 14.19
A hand generator.

If, instead of your hand, you attached the handle to a waterwheel, the energy of flowing water could be converted to electrical energy. Electrical energy produced by running water is called ***hydroelectric energy***. If you attached the handle to a windmill, the energy of moving air, or ***wind energy***, could be converted to electrical energy. Most of our electricity is generated with the heat energy released when fossil fuels are burned. In a power plant, engines convert heat energy to mechanical energy, which is then used to spin a coil of wire in a magnetic field, which converts it to electrical energy.

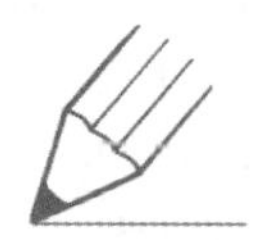

TRY THIS

1. The table below lists some advantages and disadvantages of four sources of energy. One energy source has its advantage and disadvantage listed in the wrong columns.

Source of Energy	Advantage	Disadvantage
Fossil fuels	Inexpensive	Air pollution
Nuclear fuels	No air pollution	Dangerous nuclear waste material
Sunlight	Solar cells are expensive	Renewable energy source
Wind	Renewable energy source	Cannot be used in all locations

   Which energy source has its advantage and disadvantage listed in the wrong columns?
   1. fossil fuels
   2. nuclear fuels
   3. sunlight
   4. wind

2. Which characteristic do the energy sources listed below have in common?

Wind	Solar	Nuclear	Geothermal

   1. All are used to produce electrical energy.
   2. All produce substances that harm the environment.
   3. All are renewable.
   4. All are nonrenewable.

RENEWABLE VS. NONRENEWABLE ENERGY CONSUMPTION

The energy appetite of our modern society continues to grow. Year by year, the amount of energy we use, or ***energy consumption***, gets larger. Technological advances in transportation, communications, and manufacturing create a huge demand for energy. Earth's growing population only makes the problem worse. Furthermore, different parts of the world have different amounts and kinds of energy resources to use and use them for different purposes.

Earth's energy resources can be divided into two types: renewable and nonrenewable. ***Renewable resources***, such as energy from the Sun and the wind and water energy derived from it, are replenished by natural processes. ***Nonrenewable resources***, such as coal, petroleum, and natural gas, are present in fixed amounts. Once used, they cannot be replenished. The key issue concerning energy consumption is ***sustainability***, or the ability to keep consuming energy at our present rate. Energy consumed from renewable resources can be sustained as long as natural processes keep replenishing them. Energy consumed from nonrenewable resources can last only until those resources are used up.

Unfortunately, most of the energy we consume comes from nonrenewable fossil fuels. As our consumption rises, so does the rate at which these resources are used up! Therefore, we need to increase our use of renewable resources and decrease our use of nonrenewable energy resources. However, we face problems in taking advantage of renewable resources. Solar, wind, and water energy are not equally distributed over the globe and are highly variable. Imagine being able to watch television only when the Sun shines or when the wind blows fast enough! Imagine trying to heat your house in the winter with only sunlight. How do you get heat captured by a solar collector near the equator to a city in New York? Or the energy of a river to a city in a desert? In most cases, we do these things by converting the energy to electricity, transmitting it over power lines, and then converting it back into some desired form of energy. However, we consume far more energy than such systems can currently supply. Yes, we need to develop more power plants that convert renewable energy sources into electricity. However, society will not have the ability to meet all its energy needs with renewable sources for many years. Therefore, one of the best *immediate* strategies is to conserve energy.

ENERGY CONSERVATION

Conservation means limiting our use of energy resources so that they will last longer. A limited amount of petroleum exists on Earth. Estimates range from a 30- to 100-year supply of oil at present levels of consumption. If we could cut our consumption in half by using better-designed devices that consume less fuel, we could double the effective lifetime of our oil reserves.

THE THREE Rs: REDUCE, REUSE, RECYCLE

One way to conserve a resource is to ***reduce***, or use less of it. When you insulate your attic so that less fuel is needed to heat your home, you reduce the amount of fuel your family uses. When you ride a bicycle to your friend's house instead of getting Mom to drive you, you reduce the amount of fuel your family uses.

A second way to conserve resources is to ***reuse*** them so that they do not have to be replaced. Plastics are made from petroleum. Every time you use a plastic bag to carry groceries, you are consuming petroleum. If the same bag is used over and over again before being thrown away, less petroleum would have to be used to make bags. Other reusable products include glass and plastic containers, old appliances that still work, clothing that is not worn out, and wood boxes.

A third way to conserve resources is to ***recycle***, or extract useful materials from garbage or waste. For example, almost four times as much energy is needed to extract aluminum from an ore as is needed to extract it from a used aluminum can.

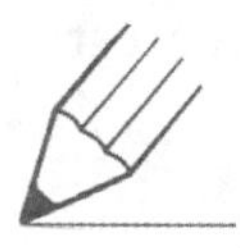

TRY THIS

The label to the right was found on a room air conditioner.

1. According to the label, what is the estimated cost of running this air conditioner for 2000 hours when electricity costs 8 cents per kilowatt-hour?
2. A family living in New York City pays 12¢ per kilowatt-hour. They usually leave for work at 8 A.M. and return home at 6 P.M. If they turn off the air conditioner while away at work for 50 days during the summer, how much money will they save?

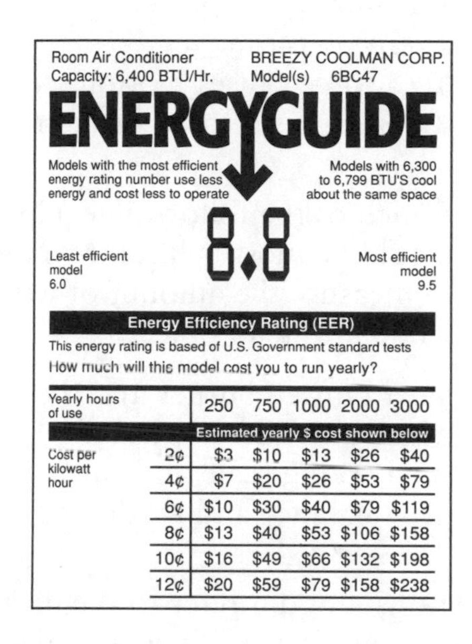

Yearly hours of use		250	750	1000	2000	3000
		Estimated yearly $ cost shown below				
Cost per kilowatt hour	2¢	$3	$10	$13	$26	$40
	4¢	$7	$20	$26	$53	$79
	6¢	$10	$30	$40	$79	$119
	8¢	$13	$40	$53	$106	$158
	10¢	$16	$49	$66	$132	$198
	12¢	$20	$59	$79	$158	$238

Practice Review Questions

1. Energy is best described as
   1. the ability to do work
   2. variations in density
   3. a force applied to an object
   4. a change in position

2. Work is done when
   1. a force is applied
   2. an object moves
   3. a force causes motion
   4. energy is created or destroyed

3. Which of the following equipment would be most helpful if you wanted to measure the amount of work done while lifting an object?
   1. a Bunsen burner and thermometer
   2. a ruler and stopwatch
   3. a spring scale and ruler
   4. a balance and graduated cylinder

Questions 4 and 5 refer to the diagram below showing a roller coaster that starts from rest at point A and moves along a track.

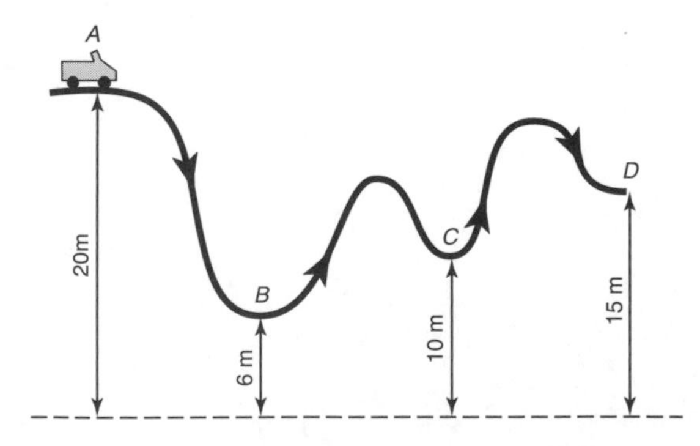

4. At which point does the roller coaster have the greatest potential energy?
   1. *A*
   2. *B*
   3. *C*
   4. *D*

5. The roller coaster has the greatest kinetic energy at point
   1. *A*
   2. *B*
   3. *C*
   4. *D*

6.* Spring 1 and Spring 2 were the same. Then, Spring 1 was pushed together a little and clamped in place. Spring 2 was pushed together a lot and clamped.

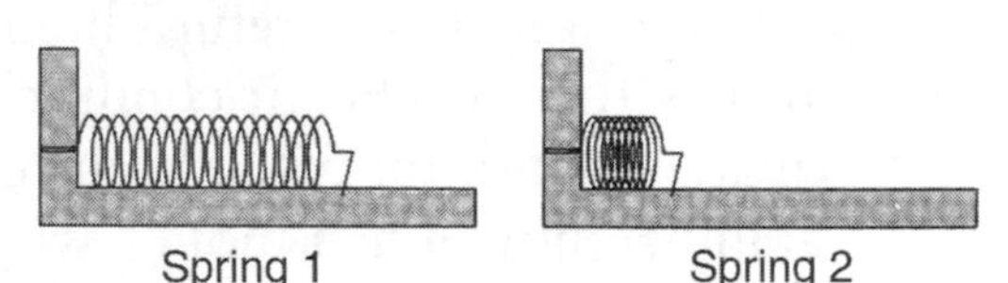

Which spring has more potential energy?
   1. spring 1
   2. spring 2
   3. Both springs have the same energy.
   4. You cannot tell unless you know what the springs are made of.

* Reproduced from TIMSS 2003 Science Released Items for Grade 8, TIMSS International Study Center, Boston College, MA

GO ON

7. The diagram below shows a device in which energy changes forms.

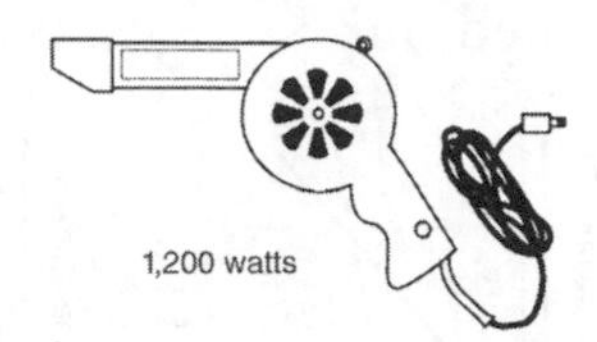

Which changes in energy occur in this device?

1. electrical → heat, mechanical, and sound
2. heat and sound → electrical and mechanical
3. heat and sound → electrical and mechanical
4. electrical → chemical and light

8. All Fahrenheit thermometers have

1. 100° between freezing and boiling points of water
2. 180° between freezing and boiling points of water
3. alcohol
4. mercury

9. The graph below shows temperature changes in 1 liter of snow as heat energy is added at a constant rate.

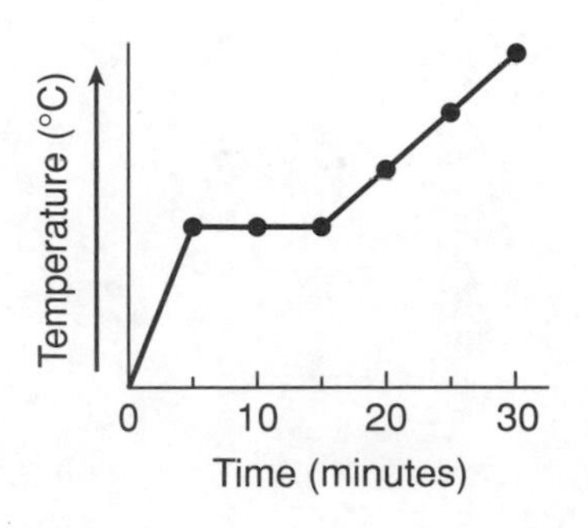

During which time interval was the latent heat of fusion added to the snow?

1. 0–5 minutes
2. 5–15 minutes
3. 15–25 minutes
4. 25–30 minutes

Questions 10 and 11 refer to the diagram below, which shows an experiment in which two insulated cups of water were connected by a metal heat transfer bar. The thermometers show the temperatures of the water in the cups at the beginning of the experiment.

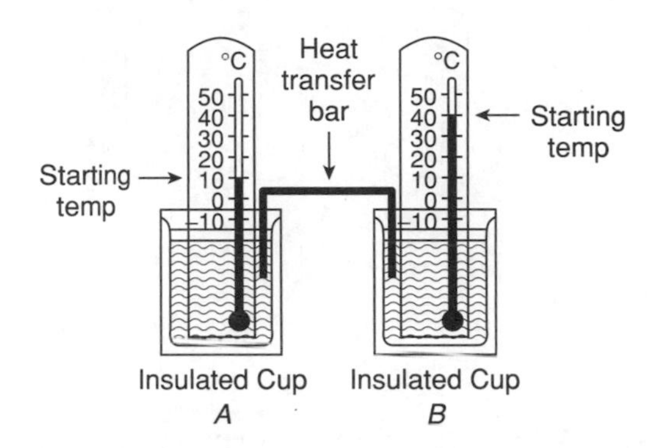

10. The greatest amount of heat energy transferred between cup *A* and cup *B* is transferred by the process of

1. radiation of heat from the water in both cups to the air
2. absorption of heat by the thermometers
3. conduction of heat through the aluminum bar
4. convection of heat within the water in both cups

GO ON ➡

11. Over the next 20 minutes, which changes are likely to occur in the water temperature in cups *A* and *B*?

1. It will decrease in cup *A* and increase in cup *B*.
2. It will increase in cup *A* and decrease in cup *B*.
3. It will increase in both cups.
4. It will decrease in both cups.

Questions 12 and 13 refer to the following, which shows a light source located at an equal distance from two air-filled cans. One can is shiny, and the other is black.

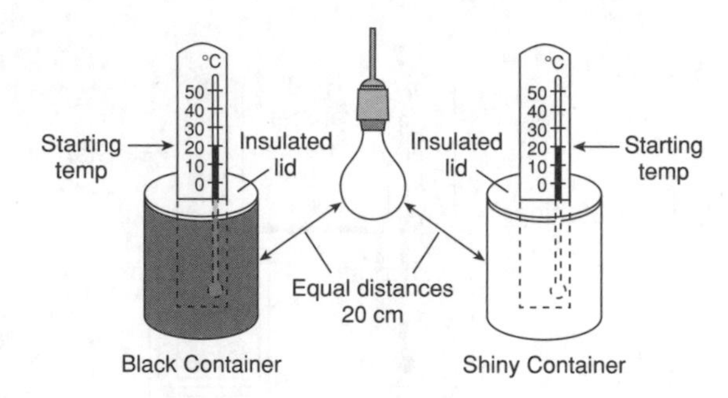

12. When the light source is on, compared to the amount of light energy striking the black container, the amount striking the shiny container is

1. less
2. the same
3. greater
4. not measurable

13. Which graph best represents the change in temperature during the first 10 minutes after the light source is turned on?

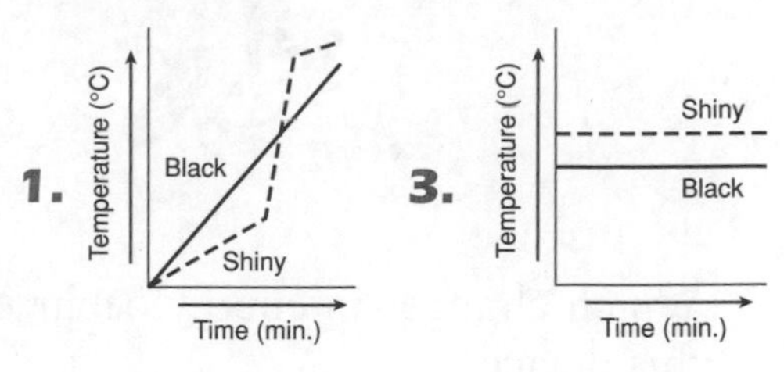

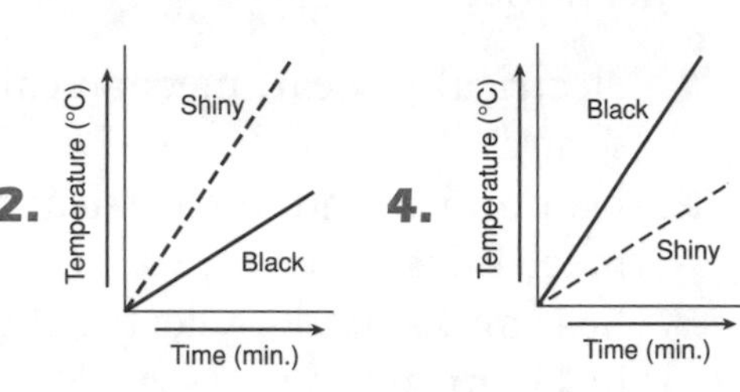

14. Sound is a form of energy produced by

1. vibrating objects
2. dry cells
3. transverse waves
4. light

GO ON ➡

Base your answers to questions 15 and 16 on the data table below, which shows the speed of sound in different materials

Speed of Sound

Material	Speed (meters per second)
Air	331
Aluminum	5000
Carbon dioxide gas	259
Water	1486
Wood (oak)	3850

15. How much faster does sound travel through aluminum than through wood?

1. 1,050 meters per second
2. 1,150 meters per second
3. 2,364 meters per second
4. 8,850 meters per second

16. Which conclusion about the speed of sound in different materials is best made from the data table?

1. Sound travels fastest through gases.
2. Sound travels fastest through solids.
3. Sound travels slowest through solids.
4. Sound travels slowest through liquids.

17. Rubber is used to cover the grips of steel pliers used by electricians because rubber differs from steel in

1. melting point
2. hardness
3. electrical conductivity
4. phase of matter at room temperature

18.* The diagrams below show a flashlight and three ways to put batteries in it.

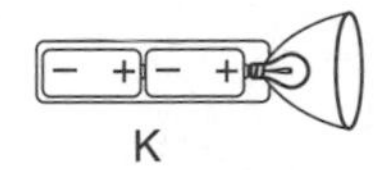

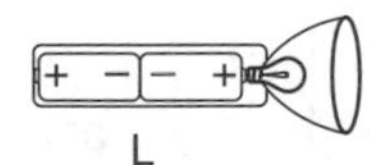

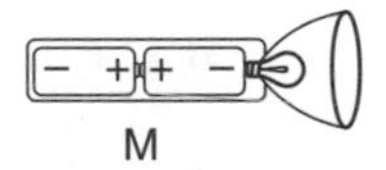

In order to make the flashlight work, which way must the batteries be placed?

1. only as in *K*
2. only as in *L*
3. only as in *M*
4. None of these ways would work.

19.* The pictures below show a lightbulb connected to a battery. Which bulb will light?

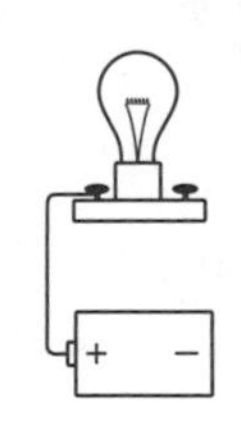

1.

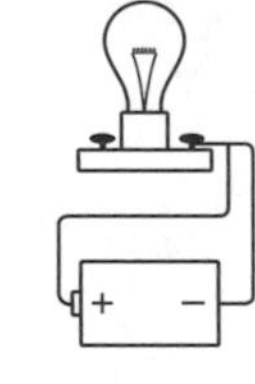

2.

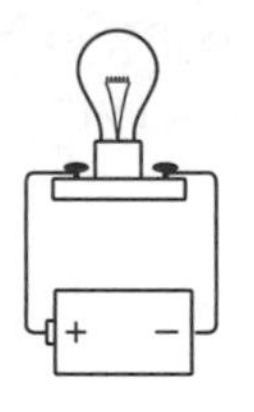

3.

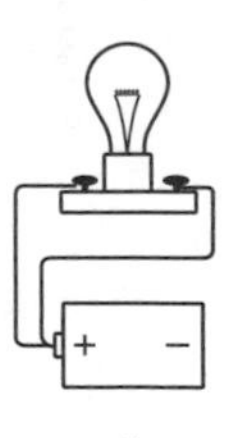

4.

GO ON ➡

20.* Which of the following shows a situation in which the magnets will repel one another?

Figure 1	S N	N S
Figure 2	S N	S N
Figure 3	N S	N S
Figure 4	N S	S N

1. Figures 1 and 3
2. Figures 2 and 3
3. Figures 1 and 4
4. Figures 1, 2, 3, and 4

21.* Which group of energy sources are ALL renewable?

1. coal, oil, and natural gas
2. solar, oil, and geothermal
3. wind, solar, and tidal
4. natural gas, solar, and tidal

22. Riding a bus to school instead of riding in a car is an example of

1. inefficient use of energy
2. recycling energy
3. reusing energy
4. conserving energy

23. The diagram below shows three cards, *A*, *B*, and *C*, and a lit flashlight on a table. Each card has a hole in it.

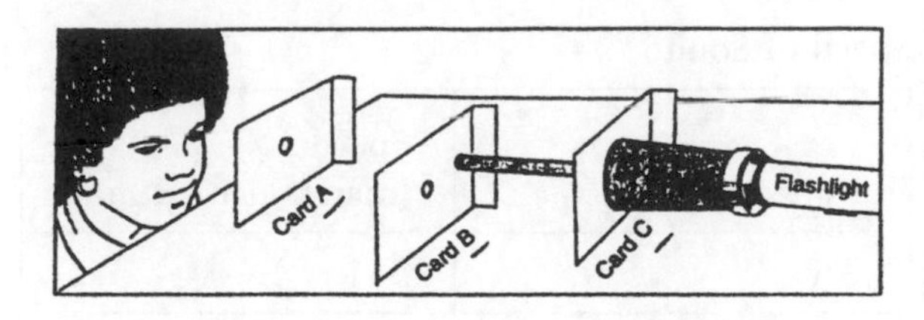

Which change would enable the student to see the direct light from the flashlight?

1. replacing the flashlight with a brighter flashlight
2. removing card *A* from the table
3. decreasing the size of the hole in card *A*
4. moving card *B* so the holes in all three cards line up

24. The diagram below shows the path of visible light as it travels from air to water to air through a glass container of water.

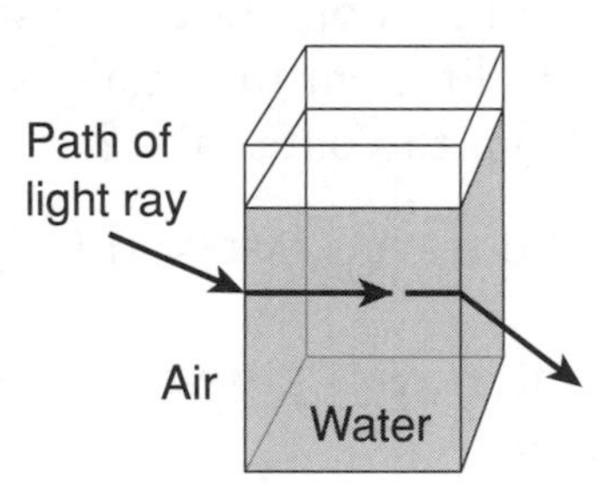

The light did not travel in a straight line because of

1. convection
2. scattering
3. refraction
4. absorption

25. The diagram below shows the frequency and wavelength of various types of electromagnetic energy.

Frequency (hertz) →
10^6 10^8 10^{10} 10^{12} 10^{14} 10^{16} 10^{18} 10^{20} 10^{22}
Visible light
Radio TV/FM Radar Heat Infrared Ultraviolet X ray Gamma ray
10^2 1m 10^{-2} 10^{-4} 10^{-6} 10^{-8} 10^{-10} 10^{-12} 10^{-14}
← **Wavelength** (meters)

Compared to ultraviolet waves, X rays have a

1. shorter wavelength and higher frequency
2. shorter wavelength and lower frequency
3. longer wavelength and higher frequency
4. longer wavelength and lower frequency

CONSTRUCTED-RESPONSE QUESTIONS

26.* The picture below shows a paint brush that is lying on a shelf in front of a mirror. Draw a picture of the paint brush as you would see it in a mirror. Use the pattern of lines on the shelf to help you.

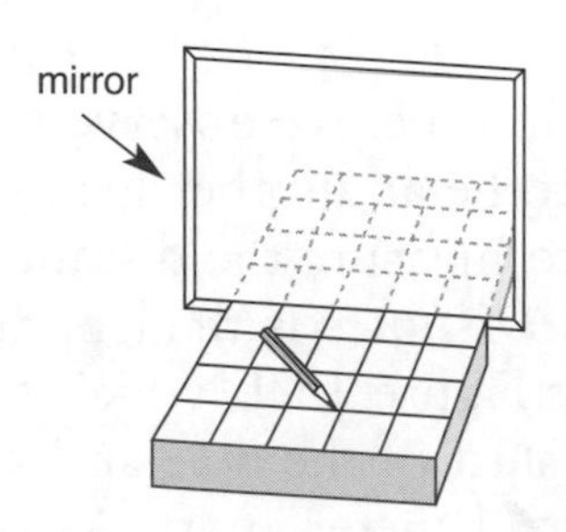

27.* The diagram below shows a compass needle with its North and South poles labeled (N and S).

It is placed next to a strong magnet as shown in the diagram below.

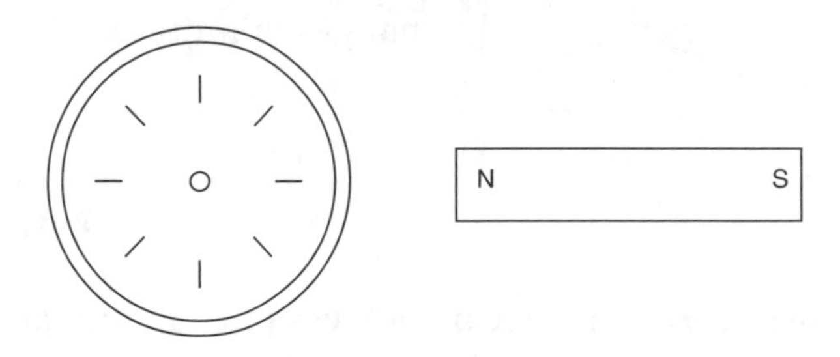

a. Draw the compass needle in the circle in the diagram above. Label the North (N) and South (S) poles of the needle.
b. Explain your answer using your knowledge of magnets.

28.* Mary was looking out her window on a stormy night. She saw lightning and then heard thunder a few seconds later. Explain why she saw lightning before she heard thunder.

29.* Identify one renewable energy source and describe one way that people make use of it.

Energy source: ____________________

Use: ________________________

Chapter 15

Motion, Forces, and Machines

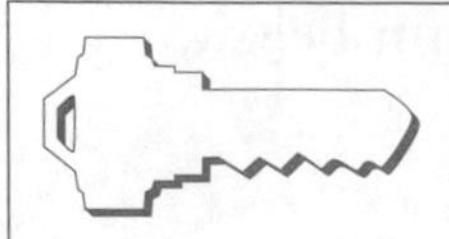

Key Idea: Energy and matter interact through forces that result in changes in motion.

MOTION

Motion is the act or process of changing position or place. When something has changed position with respect to its surroundings, we say it has *moved*. There are two separate ideas here. One is that of change; when something has moved it has *changed* position. The other idea is that of a ***frame of reference***, or what we compare an object's position with in order to know if it has moved.

As discussed in Chapter 10, the idea of absolute motion or rest is misleading. There are several possible reasons why an object may *look* like it is moving to an observer. One possibility is that the observer is standing still and the *object* is moving. Another is that the object is standing still and the *observer* is moving. Yet another is that *both* the observer and the object are moving, but one is moving faster, slower, or in a different direction than the other. Therefore, the motion of an object is always judged relative to some other object or point. The frame of reference used to measure motion depends on the situation. For a person driving a car, the road is a good frame of reference for judging motion. For a flight attendant, it is the airplane's cabin. For an astronomer studying objects in the solar system, it is the Sun. Choosing the right frame of reference is the key to solving many problems in physics.

MEASURING MOTION

How do we know when an object has moved? At one point in time, we observe it to be at one location, and at another point in time, we observe it to be at another location. One way to describe changes in the position of an object is to measure the distance it traveled, or how much ground it covered, during its motion. ***Distance*** is the length between any two points along the path of an object. Look at Figure 15.1, which shows the positions of runners in a race at two different times: the start of the race and at 2 seconds after the start of the race. The lines on the track are 1 meter apart. As you can see, both runners changed position, or moved. Runner 1 traveled a distance of 2 meters, and runner 2 traveled a distance of 8 meters.

SPEED

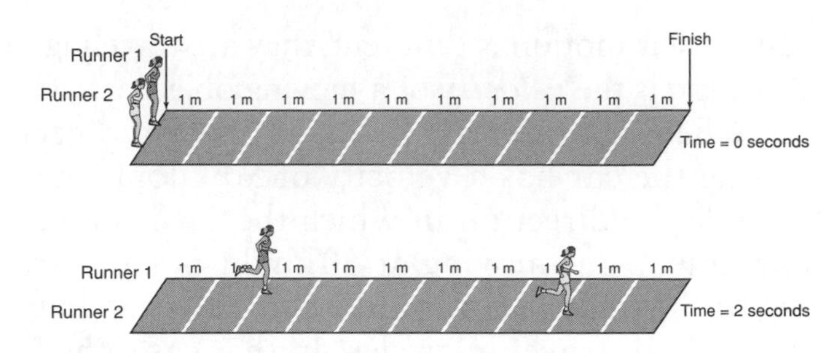

Figure 15.1
Runners on a track.

Motion is measured in terms of how much the position of an object changes in a given amount of time. ***Speed*** is the *distance* an object covers in a given amount of time. It refers to how fast an object is moving. Fast-moving objects have a high speed; slow-moving objects have a low speed; and an object that is not moving at all has zero speed. Speed is expressed by the equation

$$\text{speed} = \frac{\text{distance}}{\text{time}} \quad \textit{or}$$

$$s = \frac{d}{t}$$

For example, runner 1 covered a distance of 2 meters in 2 seconds. Therefore, the speed of runner 1 is 2 meters divided by 2 seconds, or 1 meter per second (1 m/s). Now find the speed of runner 2. Runner 2 covered a distance of 8 meters in 2 seconds.

$$\text{speed} = \frac{\text{distance}}{\text{time}} = \frac{8\ \text{meters}}{2\ \text{seconds}} = 4\ \text{m/s}$$

During the same 2 seconds, runner 2 covered a greater distance than runner 1. Runner 2 is moving at four times the speed of runner 1. In other words, runner 2 is moving faster than runner 1.

VELOCITY

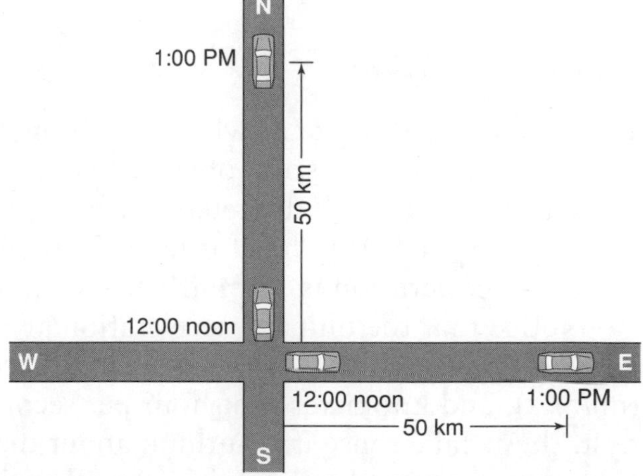

Figure 15.2
Cars traveling at the same speed in different directions.

A more complete description of an object's motion includes both speed and direction. Suppose, for example, that two cars start out at the same intersection and then travel along different roads for 1 hour. See Figure 15.2. One car travels 50 kilometers north along a road, and the other car travels east along a different road. In one hour, both cars covered a distance of 50 kilometers, so their speeds are the same—50 kilometers per hour (km/hr). Even though their speeds are the same, their motion is different; they are traveling in different directions. Speed in a particular direction is the ***velocity*** of a moving object.

When determining the velocity of an object, one must keep track of direction. Saying the car has a velocity of 50 kilometers per hour is not enough. One must include the direction in which the car is moving to describe its velocity fully. For example, a car moving at 50 km/hr north is traveling at the same speed as a car moving at 50 km/hr east, but the two are traveling at different velocities. In fact, a car can be made to travel in circles. In that case, the speed might not change at all, but the velocity (which includes direction) would be constantly changing. This is the key difference between speed and velocity. Speed has no direction; velocity is speed with a direction. Any change in an object's speed, its direction of motion, or both is a change in velocity.

Describing the direction of an object's velocity is easy. It is the direction in which the object is moving! A yo-yo moving downward has a downward velocity. If you throw a ball to your left, it has a leftward velocity. If a car is traveling south at 10 mph, it has velocity of 10 mph, south.

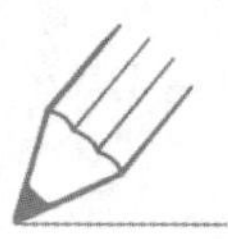

TRY THIS

1. A jet flew from Kennedy Airport in New York City to Orlando, Florida, in 3 hours. Kennedy Airport and Orlando, Florida, are about 1,500 kilometers apart. What was the velocity of the jet?
2. On the way back to Kennedy Airport, the jet ran into strong headwinds, so the flight took 5 hours. What was the velocity of the jet on the return trip?

ACCELERATION

Acceleration is the rate at which an object changes its velocity. Acceleration has to do with *changing* how fast an object is moving. For example, suppose a car is traveling at a velocity of 50 km/hr (kilometers per hour) north. To pass a truck, the driver increases the car's velocity to 60 km/hr. If it takes 2 seconds to change the car's velocity by 10 km/hr, the car's acceleration is 5 km/hr per second (10 km/hr divided by 2 s), or 5 km/hr/s.

Notice that the units of acceleration are units of velocity per time. You will often see acceleration units such as meters/second/second (m/s/s), miles per hour per second (mi/hr/s), and kilometers per hour per second (km/hr/s.) While these sound a little odd at first, they make sense if you think about the definition of acceleration—a change in velocity over time. So the units of acceleration are velocity units divided by time units, hence (m/s)/s or (mi/hr)/s.

Just because an object is moving fast does not mean that it is accelerating. A car traveling at a steady velocity of 60 km/hr has no acceleration. If an object's velocity is not changing, it is not accelerating. However, if the car increases its velocity to 80 km/hr, it accelerates. Anytime an object's velocity is changing, that object is said to be accelerating; it has acceleration.

To find the acceleration of any object, you must know how much its velocity changed and how much time the change took to occur. The change in the velocity of an object is the difference between its initial velocity (v_i) and its final velocity (v_f). The time for change to occur is the time difference between when the object had these two velocities.

The average acceleration of any object is expressed by the equation

Table 15.1 **Acceleration of a Car**

Time (s)	Velocity (kph)
0	0
1	10
2	20
3	30
4	40
5	50

$$\textbf{average acceleration } (a) = \frac{\textbf{final velocity} - \textbf{initial velocity}}{\textbf{time}} = \frac{v_f - v_i}{t}$$

The data chart in Table 15.1 shows how the velocity of a stopped car changed with respect to time as it sped up to get onto a highway.

We will calculate the acceleration of the car. The car's initial velocity was 0 km/hr, and its final velocity was 50 km/hr. The time difference between the two velocities was 5 seconds. By substituting these values into the equation, we get

$$a = \frac{v_f - v_i}{t} = \frac{50\ \text{kph} - 0\ \text{kph}}{5\text{s}} = 10\ \text{km/hr/s}$$

This means that every second, the car was going 10 km/hr faster. This is called a *constant acceleration* because the velocity is changing by the same amount each second. Be careful! An object with a *constant acceleration* should not be confused with an object with a *constant velocity*! The car's acceleration may have been a constant 10 km/hr/s, but its velocity changed from 0 km/hr to 50 km/hr. Furthermore, whether an object is changing its velocity by a constant amount or a varying amount each second does not matter. If the velocity is changing, the object is accelerating.

So far, we have considered only objects whose velocity is increasing. If an object's velocity is decreasing, it is still changing, so the object is accelerating. However, its final velocity is less than its initial velocity. Therefore, an object that is decreasing in velocity, or slowing down, has a negative acceleration (sometimes called deceleration).

GRAPHING MOTION

Another way to express motion is by using a position versus time graph. We will begin with a car moving at a constant velocity to the right of 10 km/hr. If the position of the object over time were plotted on a graph, it would look like the graph shown in Figure 15.3*a*. Note that a constant velocity results in a straight line sloping upward to the right.

Now we will look at a car moving with a constant, rightward velocity that is slower, say, 5 km/hr. See Figure 15.3*b*. Again, the constant velocity results in a straight line sloping upward to the right, but this time the slope of the line is not as steep. The steeper the line is in a position versus time graph, the higher the velocity.

Next, we will consider a car moving to the right with a changing velocity, that is, the car is speeding up or accelerating. Since accelerating objects are constantly changing their velocity, the distance traveled each second is not a constant value. See Figure 15.4*a*. Note that an accelerating object results in a line whose slope changes, so it *curves* upward to the right.

Finally, we will consider an object that is accelerating more slowly. See Figure 15.4*b*. Note that the line does not curve upward as sharply. The more sharply the line curves in a position versus time graph, the greater the object's acceleration.

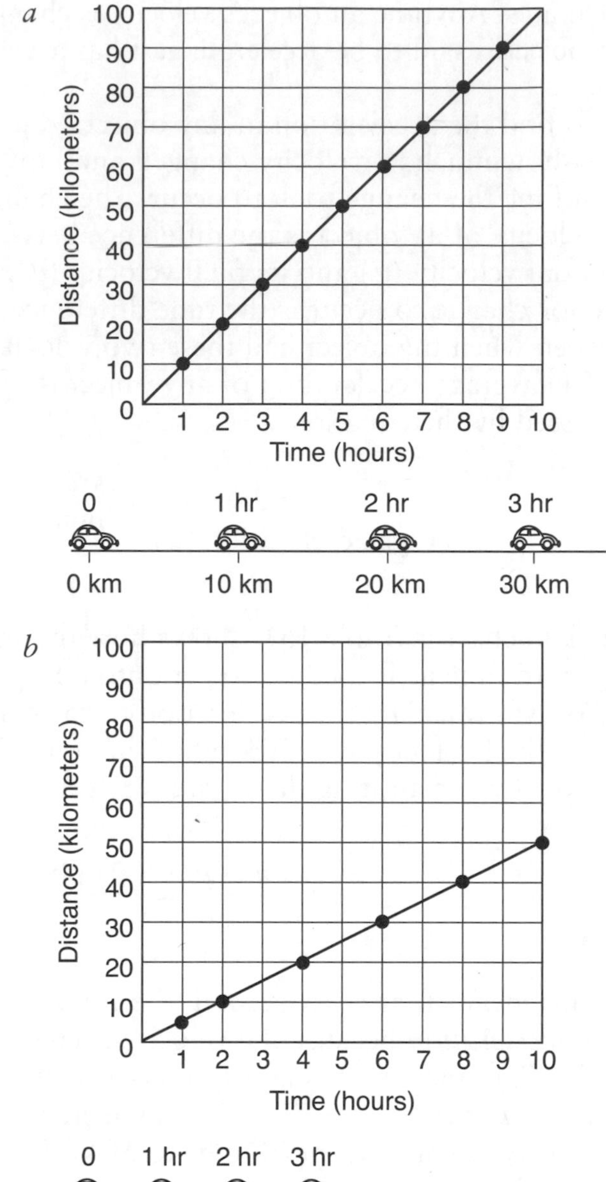

Figure 15.3
(*a*) Graph of constant velocity of 10 km/hr.
(*b*) Graph of constant velocity of 5 km/hr.

TRY THIS
Look at the graph in Figure 15.4*b*. How would you describe the object's motion?

FORCES

In the seventeenth century, the English scientist Isaac Newton proposed some ideas that explain why objects move (or do not move) as they do. These ideas have become known as Newton's three laws of motion. Newton explained motion as the result of forces exerted onto matter when energy and matter interact. When forces act on matter, they produce predictable changes.

Understanding the laws that govern motion allows us to predict these changes. Newton's first law defines what a force is. His second law defines how force can be measured. His third law explains how forces interact.

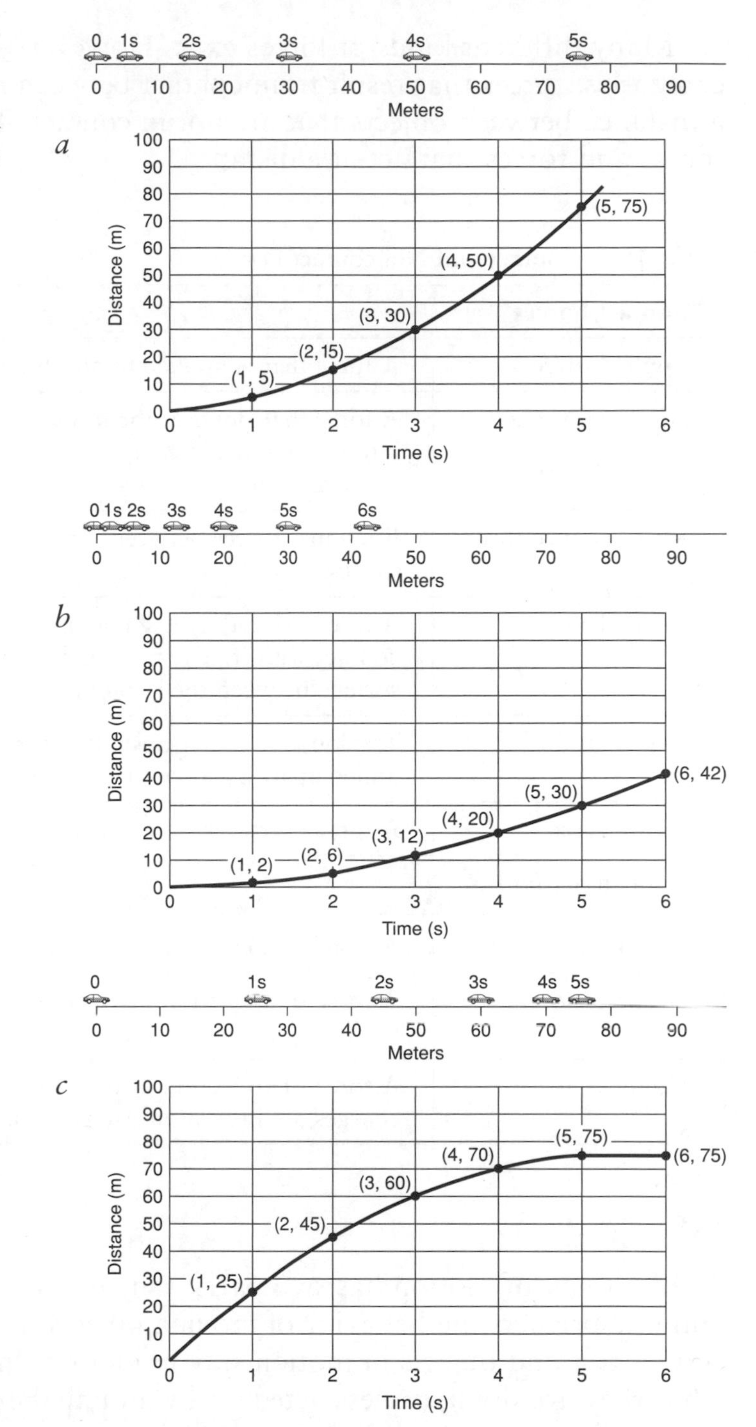

Figure 15.4
(*a*) Graph of constant, rightward acceleration of 4 km/hr/s. (*b*) Graph of constant, rightward acceleration of 2 km/hr/s. (*c*) Graph of a decelerating object.

NEWTON'S FIRST LAW OF MOTION

Newton's first law is often stated as

> **An object at rest tends to stay at rest and an object in motion tends to stay in motion with the same speed and in the same direction *unless acted upon by an unbalanced force.***

DEFINING FORCE

If we turn this statement around, it becomes a definition of force. A *force* is any influence that will cause an object at rest to move and an object in motion to change its speed or direction of motion, or both. Since velocity includes both a speed and a direction, another way to say this is "a force is any influence that will bring about a change in the velocity of an object."

Many different kinds of forces exist. However, they can be grouped into two broad categories: forces that result from contact between two objects and forces that can act at a distance between objects that are not in contact. Table 15.2 lists some common contact forces and forces that act at a distance.

Table 15.2 **Contact and Noncontact Forces**

Contact Forces	
Applied force	A force that is applied to an object by a person or another object.
Frictional force	A force that opposes the motion of two surfaces that are in contact due to intermolecular forces of attraction and irregularities in the surfaces that resist motion.
Air resistance force	Frictional forces between the air and the surface of an object moving through air.
Normal force	A force exerted by a stable object in response to another object that is in contact with it. The normal force is always directed perpendicularly to the surface between the two objects.
Tensional force	Tension is a force transmitted through a rope, string, or wire when it is pulled apart by forces acting on either end.
Spring force	A force exerted by a compressed or stretched spring.
Noncontact Forces	
Gravitational force	A force of mutual attraction that exists between all bodies of matter.
Electrical force	A force of attraction between unlike electric charges or a force of repulsion between like electric charges.
Magnetic force	A force of attraction or repulsion that comes into existence when electric charges are in motion. Unlike poles attract, and like poles repel.

INERTIA

The first law of motion has two parts. One predicts the behavior of nonmoving objects, and one predicts the behavior of moving objects. If forces are balanced, objects at rest stay at rest and objects in motion stay in motion. In other words, objects keep on doing what they are doing (unless acted on by an unbalanced force). An arrow at rest on a table will continue in this same state of rest. An arrow flying through the air at 10 m/s to the east will tend to continue in this same state of motion (flying through the air at 10 m/s east). The state of motion of an object remains unchanged as long as the object is *not* acted upon by an unbalanced force. All objects resist changes in their state of motion—they tend to “keep doing what they are doing.”

The tendency of matter to resist changes in its state of motion is called ***inertia***. What exactly, though, *is* an object’s state of motion? It is defined by the object’s velocity, a speed, and a direction. The state of motion of an object at rest is zero velocity. The state of motion of an object in motion is its speed and direction of motion. Thus, inertia could also be defined as the tendency of an object to resist changes in velocity. Since a change in velocity is an acceleration, yet another way to describe inertia is the tendency of an object to resist acceleration.

If you have ever been in a car that braked to a sudden stop, you have experienced inertia. Friction between the road and the car's locked wheels creates an unbalanced force to decrease the car's velocity. However, there is no unbalanced force to change your velocity. So, you jerk forward in your seat and will continue moving forward at the same speed and in the same direction unless acted upon by an unbalanced force—such as your seat belt. That is right; *seat belts* provide the unbalanced force that brings you from a state of motion to a state of rest. Think about what Newton's laws predict would happen to you if *no* seat belt were used.

TRY THIS

A curved groove is placed onto a level table as shown in the diagram. A ball is pushed in the groove at *P* so that it leaves at *Q*.

These diagrams show the level table and the groove from above. Which shows how the ball will move when it leaves the groove at Q?

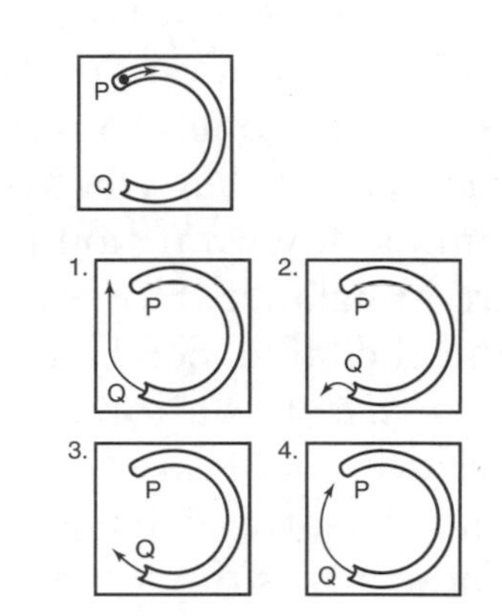

INERTIA AND MASS

The mass of a body at rest determines its inertia. The greater an object's mass, the greater its resistance to being set into motion. A soccer ball has much less mass than a solid lead ball of the same size. If you kick them both with the same force, the soccer ball will go flying but the lead ball will barely move. In fact, physicists use the inertia of an object to measure its mass.

BALANCED AND UNBALANCED FORCES

What exactly does the phrase "unbalanced force" in Newton's first law mean? We will consider a book at rest on top of a table. Two forces are acting on the book. Earth's gravity exerts a downward force on the book, and the push of the table (usually referred to as the *normal force*) exerts an equally strong upward force onto the book. Since the two forces are equal in size and in opposite directions, they balance each other. See Figure 15.5*a*. Since no unbalanced forces are acting on the book, it does not accelerate. Since it is at rest, it remains at rest. In fact, for every object at rest in the real world, there are *always* forces present (gravity, if nothing else). If the object is at rest, it is because more than one force is present, and the sum of those forces is zero, not because *no* forces are acting on it.

Can an object be in motion if the forces on it are balanced? Absolutely! Yet we rarely, if ever, see this happening in everyday life because of all the unseen forces at work around us. For example, suppose we look at a book sliding to the left across a table. At some point in time, it was probably given a push that set it into motion from a resting position. Whatever happened in the past, though, is not our concern right now. We want to focus on what is happening right now—the book is sliding to the left across a table and no one is pushing it. According to Newton's first law, it is in motion and should remain in motion. Yet the book eventually slows down and comes to rest. There must be an unbalanced force, but what is it? The force of gravity pulling downward on the book and the force of the table pushing upward on the book are in balance. However, there is an unseen force—the force of *friction* between the book and the table. See Figure 15.5*b*.

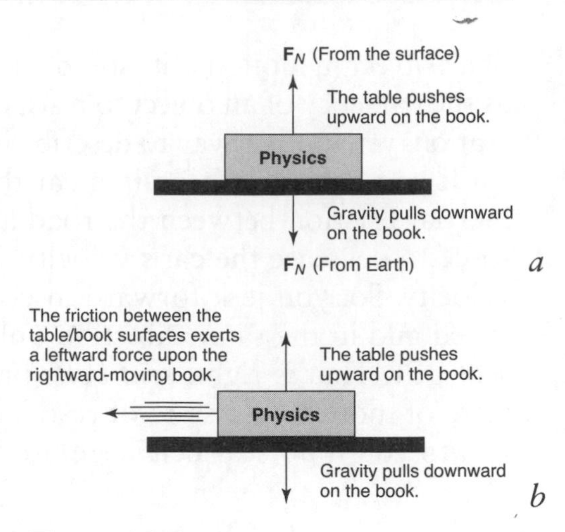

Figure 15.5
(*a*) Forces on a book at rest.
(*b*) Forces on a sliding book.

Friction is a force that opposes the motion of two surfaces that are in contact. Friction forces are the result of irregularities in surfaces that are in contact. Even very smooth surfaces have tiny bumps and jagged edges when viewed with a microscope. These act as tiny obstructions that get in the way of motion. When the surfaces try to move, the obstructions catch on one another. They also form the points of contact between the surfaces where their atoms cling together. As one surface slides past the other, the atoms snap back (which motion we feel as heat) or are torn loose from one or the other surface (which is why rubbing wears down a surface). There is even friction in liquids and gases because the body moving through them must push aside some of the liquid or gas. The direction of a frictional force is always in a direction that opposes motion. If you push a box to the left, friction acts to the right.

You can feel the effect of friction by trying to push a heavy box along different surfaces. Pushing a box along on a carpet is harder than on a smooth, waxed floor because the carpet has a rougher surface. The rougher the surface, the bigger the bumps and edges in the surface, and the greater the force of friction opposing motion.

Making surfaces smoother can reduce frictional forces. Using lubricants, such as oil or wax, can also reduce friction. ***Lubricants*** reduce friction by separating the two contacting surfaces with a layer of softer material. Instead of rubbing against each other, the surfaces rub against the lubricant. By reducing friction, lubricants reduce the heat produced when surfaces rub against one another and keep the surfaces from wearing down.

THE BIG MISCONCEPTION

The most common misconception concerning Newton's first law is that to keep an object moving, a constant force must be applied to it. The reason for this misconception can be traced to the friction that acts on all objects in the real world. Newton's law says loudly and clearly that a force is not needed to keep an object in motion. When you give a book a push and it slides along a table, it does not come to rest because of the absence of a force. It comes to rest because of the presence of a force—the force of friction. Friction opposes the motion of the book and brings it to a rest position. If there were no friction, the book would continue moving in the same direction at the same speed—*forever*!

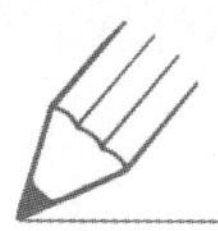

TRY THIS

The situations described below all involve several forces acting on an object.

A. A car is traveling at 20 meters per second, and the driver steps on the brake.
B. A chair is dragged across a floor at constant velocity.
C. A plane has made a turn at constant speed.
D. An object is dropped vertically toward the ground.

1. In which of the situations are there no unbalanced forces?
   1. A 2. B 3. C 4. D
2. As the mass of an object decreases, its inertia will
   1. increase 2. decrease 3. remain the same
3. If an object is moving north, the direction of the frictional force is
   1. north 2. south 3. east 4. west

NEWTON'S SECOND LAW OF MOTION

Newton's first law defines force and predicts the behavior of objects for which all forces are in balance. Newton's second law predicts the behavior of objects for which all forces are *not* in balance and gives us a way to measure force. Newton's second law states:

> **The acceleration of an object is directly proportional to the unbalanced force acting on the object and inversely proportional to the mass of the object.**

This relationship can be stated mathematically as

$$\text{acceleration} = \frac{\text{force}}{\text{mass}} \quad \textit{or} \quad a = \frac{F}{m}$$

This means that the greater the force, the greater the object's change in motion, or acceleration. It also means that the greater the mass of the object, the less a given force will change its motion.

MEASURING FORCE

Up until now, we have talked about forces in terms of how hard you kicked something or a greater force was exerted onto something. Newton's second law, however, gives us a way to define a unit for measuring force. If we rearrange the equation, we find that force can be equated to the product of mass and acceleration.

$$\text{force} = \text{mass} \times \text{acceleration} \quad or \quad F = ma$$

This equation helps us define a unit for measuring force. This unit is called the newton (N) (after you know who). One ***newton*** is the amount of force needed to give a 1-kilogram mass an acceleration of 1 (meter/second) per second (1 m/s/s.) So

$$1 \text{ newton (N)} = 1 \text{ kg} \times \text{m/s/s}$$

For example, suppose a force causes a 2-kilogram box to accelerate from rest to a velocity of 10 meters per second in one second. You can use the equation above to find the number of newtons of force that were applied to the object.

$$\text{force} = \text{mass} \times \text{acceleration}$$
$$\text{force} = 2 \text{ kg} \times 10 \text{ m/s/s} = 20 \text{ N}$$

THE EFFECTS OF UNBALANCED FORCES ON MOTION

1. The greater the force, the greater the acceleration. Newton's second law also allows us to predict the effects of unbalanced forces on an object. First, it tells us that if force is increased, acceleration will increase; and if force is decreased, acceleration will decrease. This means that the larger the force, the more acceleration it produces. Twice the force produces twice the acceleration; three times the force produces three times the acceleration, and so on. In general, a change in the amount of unbalanced force produces the same amount of change in acceleration.

We see examples of this every day. For instance, a first grader does not have as much muscular strength as an eighth grader and therefore cannot swing a baseball bat as hard. If they both hit a ball with the same mass, the ball hit by the eighth grader will accelerate more and travel farther.

2. An object is accelerated in the same direction as the applied force. Another thing this law tells us is that an object is accelerated in the direction of the force acting on it. If the force is positive, acceleration is positive; if the force is negative, acceleration is negative. By convention, we call the direction in which an object is moving positive and the opposite direction negative. If the force is applied in the direction in which the object is moving, it will increase the object's speed. If applied in the opposite direction, it will decrease the speed of the object. If the force is applied at right angles, it will change the object's direction of motion. At any other angle, the result will be a combination of change in speed and change in direction. Some examples of situations in which an applied force changes the speed of an object, its direction of motion, or both are given in Table 15.3.

Table 15.3 Examples of Acceleration Due to Unbalanced Forces

Acceleration	Balanced Forces	Unbalanced Forces
Change in speed	A car is traveling due north at a constant speed of 50 km/hr.	The car's driver steps on the gas pedal, and the car speeds up to 60 km/hr due north.
	A student is pedaling a bicycle on a sidewalk at a constant 10 km/hr in a straight line.	The bicycle enters a puddle. Although the student keeps pedaling at the same rate, friction with the water causes the speed of the bicycle to decrease to 5 km/hr.
	An apple hangs from a tree branch.	The apple falls to the ground.
Change in direction of motion	A car is traveling due north at a constant speed of 50 km/hr.	The driver maintains a speed of 50 km/hr while following a curve in the road.
	A bird is flying due east at a constant speed.	The bird continues flying due east at a constant speed but encounters a crosswind blowing due south. The bird's path curves to the southeast.
	An airplane is in level flight at constant speed at an altitude of 3,000 meters.	The pilot maintains airspeed while climbing to an altitude of 5,000 meters.
Change in speed and direction of motion	A football player is running at a constant speed along the sideline toward the goal line.	The football player is tackled from the side and brought crashing to the ground out of bounds.
	A pitched baseball flies toward home plate at constant 110 km/hr.	A batter hits the pitched baseball.
	A car is traveling along a straight stretch of highway at 70 km/hr.	The driver of the car steps on the brakes, but the left brakes are worn more than the right brakes. The car slows down but pulls to the left.

3. The greater the mass, the smaller the acceleration produced by a given force. This effect is fairly obvious to most people. It is due to an object's inertia. The greater an object's mass, the more it resists being accelerated. That is, the less the object accelerates when a given force is applied to it. Imagine a race car driver flooring the gas pedal at the start of a big race. An unbalanced force makes the car accelerate forward rapidly. Now imagine the same driver in the same car but with a bus chained to its rear bumper. What do you think will happen when the driver floors the gas pedal? You guessed it, the same unbalanced force acts on the car, but this time it will accelerate forward much more slowly—if at all!!

THE BIG MISCONCEPTION REVISITED

Remember the common misconception people have that a force is needed to keep an object moving? We will look at one more example that involves Newton's second law. Did you ever notice that you have to pedal hard to get a bike moving, but once it is moving, you do not have to pedal as hard to *keep* it moving? That is because when you first start pedaling, you have to accelerate the combined mass of your body and the bike from rest to some desired velocity. Once it is moving, you have to exert enough force only to counterbalance friction. Then the force on the bike is balanced, and it rolls along at a constant speed. If there were no friction, once the bike had accelerated to the desired speed, you could just sit back and enjoy the ride. No further force would be needed to keep the bike moving. Unfortunately, the only place we get close to such a frictionless condition is in space.

Astronauts traveling to the Moon do not have to fire their rockets throughout their entire trip. They fire them to accelerate the rocket to speed and then shut them off. The rocket's inertia then keeps it moving at the same speed and in the same direction. When the astronauts get near the Moon, they turn the ship around and fire their rockets in the *opposite* direction. This applies a force in a direction opposite to that of the rocket's motion, causing a decrease in speed until the rocket comes to rest in relation to the Moon.

TRY THIS

A uniform rod is pivoted at the center. It is acted on by two forces in the same plane. Each force has the same size, equal to 10 N (newtons). In which case is there a turning effect?

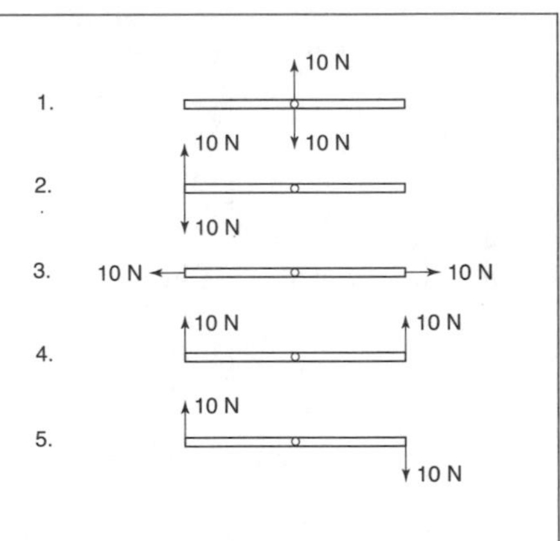

NEWTON'S THIRD LAW OF MOTION

When you fall down onto a trampoline, you apply a downward force onto the trampoline. The trampoline stretches and then applies an upward force onto you that bounces you back up. If you and a friend are on roller blades and you push against each other, you will accelerate in one direction and your friend will accelerate in the other direction. The harder you push against each other, the faster you move apart. Newton made similar observations and summed them up in his third law of motion:

For every action, there is an equal and opposite reaction.

Forces always act in pairs. There is never only a single force in any situation. Pushing on a desk with your finger is an action. When you do this, what do you feel? You feel the desk pushing back against your finger. The desk pushing back against your finger is a reaction. Now push harder. What do you feel? You should feel more pressure on your finger as the desk pushes back harder in reaction. The action force is equal in strength to the reaction force but opposite in direction.

There are many examples of action-reaction forces in everyday life. See Figure 15.6. When you swim, you use your hands and feet to push water backward. This causes the water to accelerate backward. The water reacts by pushing you forward. The harder you push the water backward, the harder the water pushes you forward, and the faster you move. When you walk, your foot pushes backward against the ground. The ground reacts by pushing forward on your foot. The reaction force of the ground causes you to move forward. When you drive, the tires of the car push backward against the pavement, and the pavement reacts by pushing forward against the car. These forces depend on friction. A person or a car trying to move on a slick, icy surface may not be able to exert enough backward-action force onto the surface to produce a reaction force large enough to accelerate them forward. The wheels of a car stuck on ice may spin, but the car does not move forward.

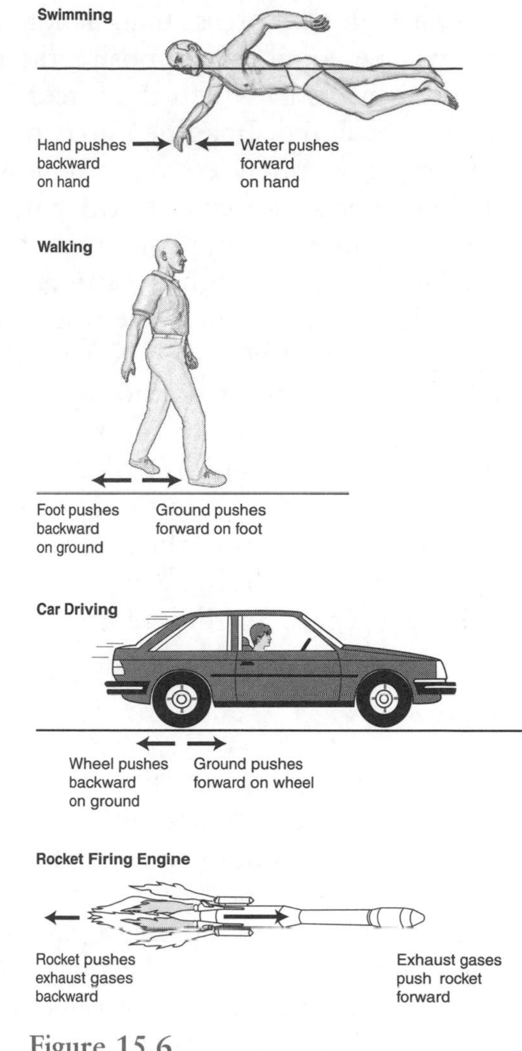

Figure 15.6
Action–reaction forces.

Action–reaction forces are what propel a rocket through space where there is no outside matter for the rocket to push against to get it moving. A rocket engine produces gases that expand. The rigid walls of the rocket engine's combustion chamber push against the expanding gases, causing them to move backward and out of the rocket. The gases, in turn, push against the rocket, causing it to move forward. Forces, no matter how large or how small, always occur in pairs, each of which is equal and opposite to the other.

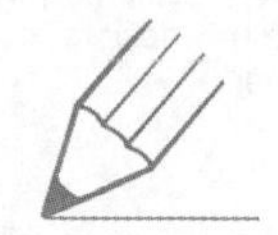

TRY THIS

Identify the pair of action–reaction forces in each of the following examples.

1. You dive off the edge of a raft. You move forward, and the raft moves backward.
2. A cannon is fired. The cannonball moves forward, and the cannon moves backward.

Although the forces in an action–reaction pair may be equal and opposite, the masses of the two objects to which the forces are applied may not be equal. Imagine a massive bowling ball striking a much less massive bowling pin. The ball exerts a force onto the pin, and the pin exerts an equal and opposite force onto the ball. However, although the forces acting on the two objects may be the same, their mass is not the same. Remember from Newton's second law that the greater the mass, the smaller the acceleration produced by a given force. Therefore, the action force of the bowling ball produces a big change in the velocity of the bowling pin. However, the reaction force of the bowling pin barely changes the velocity of the bowling ball. See Figure 15.7.

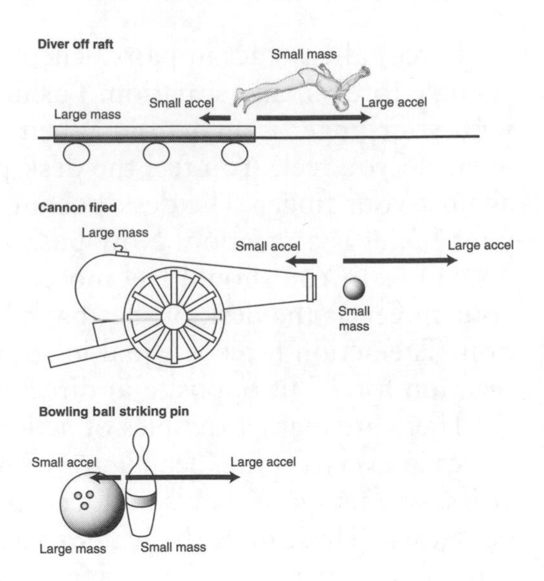

Figure 15.7
Action–reaction in objects with different masses.

BUOYANCY

An interesting action–reaction phenomenon is that of buoyancy. When an object is submerged in a fluid, it pushes the fluid aside—an action force. The fluid then pushes back on the object—a reaction force. The force with which the object pushes the fluid aside is equal to its weight—the force of gravity acting on the mass of the object. The force with which the fluid pushes back on the object equals the weight of the fluid displaced—the force of gravity acting on the mass of fluid displaced. However, the force exerted by the fluid increases with depth, so the force on the bottom of the submerged object is greater than the force on the top of the submerged object. Therefore, an unbalanced force acts upward on the object, called the ***buoyant force***. If the weight of the object is *greater* than the weight of the displaced fluid, the downward force of gravity is greater than the upward buoyant force and the object sinks. If the weight of the object is *less* than the weight of the displaced fluid, the upward buoyant force is greater than the downward force of gravity and the object floats. See Figure 15.8.

NEWTON'S LAW OF UNIVERSAL GRAVITATION

Newton's laws of motion make it possible to predict how an object will move if the forces acting on it are known. Newton thought about the forces that would be needed to keep a satellite moving in orbit around another object. In thinking about the Moon, Newton realized that the Moon would circle Earth only if some force pulled the Moon toward Earth's center. Otherwise, the Moon would

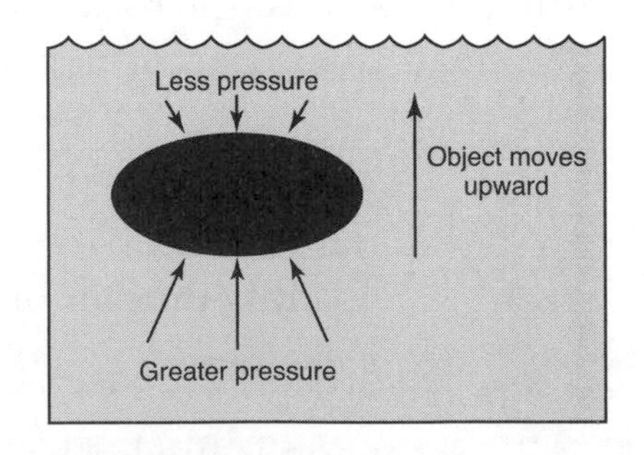

Figure 15.8
Buoyant force.

continue moving in a straight line off into space. Newton's genius was to realize that the force that kept the Moon in orbit around Earth was the same force that causes objects close to Earth to fall to the ground (like apples from a tree)—gravity. Newton defined gravity as a force of attraction that exists between *all* matter. He said that gravitational force depends on how much mass the objects have and how far apart they are. This is called Newton's law of universal gravitation. The greater the mass of the objects, the greater the force of attraction. The greater the distance between the objects, the smaller the force of attraction. See Figure 15.9.

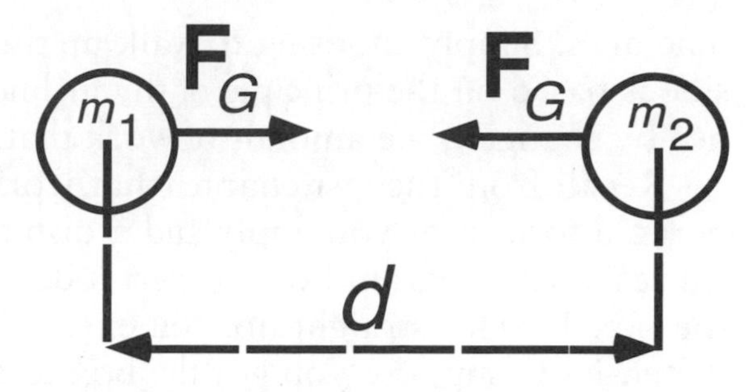

Figure 15.9
Newton's law of universal gravitation.

Newton's law of universal gravitation can be stated in a simple mathematical expression:

$$F_{\text{gravity}} \propto^{*} \frac{\text{mass}_1 \times \text{mass}_2}{\text{distance}^2} \quad \textit{or} \quad F_{\text{gravity}} \propto \frac{m_1 m_2}{d^2}$$

MACHINES

A ***machine*** is a device that allows work to be done and offers an advantage to the user. The advantages of using a machine to do work include changing the amount or direction of an applied force or changing the distance or speed of the force required to do work. An applied force is called an ***effort***. The force produced by the machine is called the ***output force***. For instance, getting the lid off a paint can using only your fingers requires a large effort. However, if you use a screwdriver to pry off the lid, you can do it with less effort. You push down on the handle of the screwdriver with a small effort, and the tip pushes up on the lid with a large output force. In this situation, the screwdriver is a kind of simple machine. However, the trade-off is that you must move the handle of the screwdriver through a greater distance to accomplish the work. The complex machines used in industry are generally made of a combination of ***simple machines***, such as the lever, pulley, wheel and axle, and inclined plane.

INCLINED PLANE

An inclined plane is simply a flat, slanted surface. A ramp is an inclined plane. We are so used to using inclined planes in everyday life that we hardly think about them as

* The symbol "∝" means "proportional to." In order to make this an equation, a special number, called a constant, must be added. You have used constants if you have ever converted from one type of unit to another. For instance, in converting from inches to centimeters, you use the constant 2.54 cm/in. Inches are proportional to centimeters, but inches × 2.54 cm/in *equals* centimeters. The gravitational constant is $6.67 \times 10^{-11}\ \text{N} \cdot \text{m}^2/\text{kg}^2$, but for general questions about gravity, you do not need to know this number.

machines. Simply choosing to walk up the gently sloping side of a hill rather than the steep side is based on the principle of the inclined plane. An inclined plane makes work easier not by changing the amount of work that must be done but by changing the *way* it is done.

Recall from the last chapter that work equals force times distance. Work has two parts, a force that you apply and a distance over which you apply the force. Gravity causes a 50-kilogram box to exert a downward force of about 1,000 newtons. In lifting the box 1 meter straight up, you exert a force of 1,000 newtons for a distance of 1 meter. Now suppose you got the box to rise 1 meter by pushing it up a 10-meter-long inclined plane. See Figure 15.10. You would do the same amount of work, but the distance over which it was done would be much greater.

What does this do to the force needed to do the work? Look at the equations and picture below:

Lifting Straight Up	Using an Inclined Plane
work = force × distance	work = force × distance
work = 1,000 newtons × 1 meter	work = 200 newtons × 5 meters
work = 1,000 newtons · meter	work = 1,000 newtons · meter

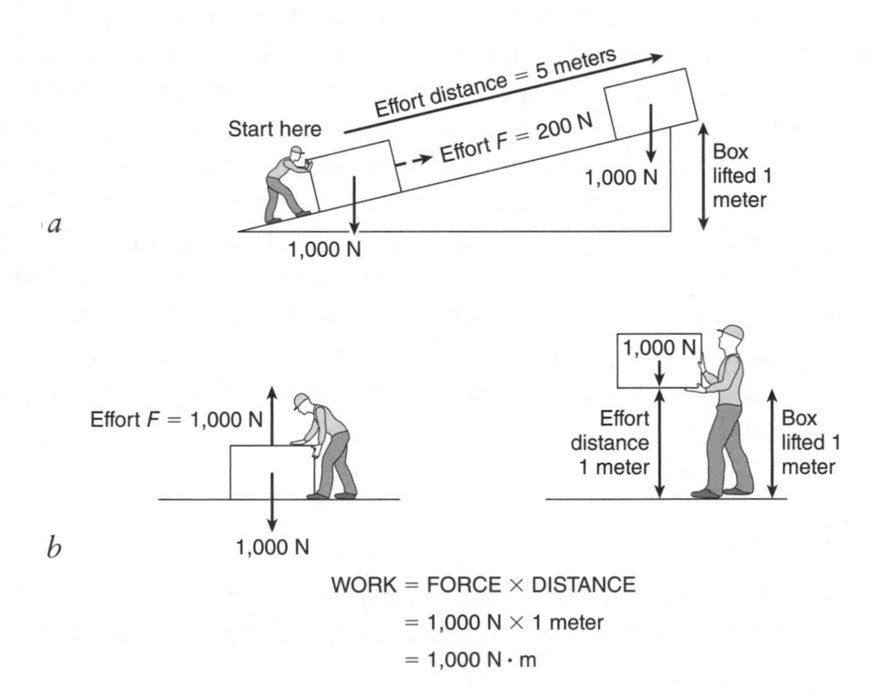

Figure 15.10
Comparison of work done raising a box to a height of one meter by lifting it straight up and by pushing it up an inclined plane.

Lifting the box straight up requires an effort of 1,000 newtons. Moving it up an inclined plane requires an effort of only 100 newtons. Less force is needed to move the box up the inclined plane. The work you do in either case is the same and equals the effort multiplied by the distance through which it was maintained. However, the amount of effort needed decreases as distance increases. An inclined plane reduces the effort needed to lift an object by increasing the distance through which the object moves.

Two common devices that are actually inclined planes are the screw and the wedge. See Figure 15.11. A screw is actually an inclined plane wrapped around a central bar. The gentler the slope of the inclined plane, the closer together the threads of the screw and the easier it is to turn. A wedge is simply two inclined planes back-to-back. When you use a wedge, you must move the inclined plane to do work. A wedge helps your effort get under heavy objects or inside them.

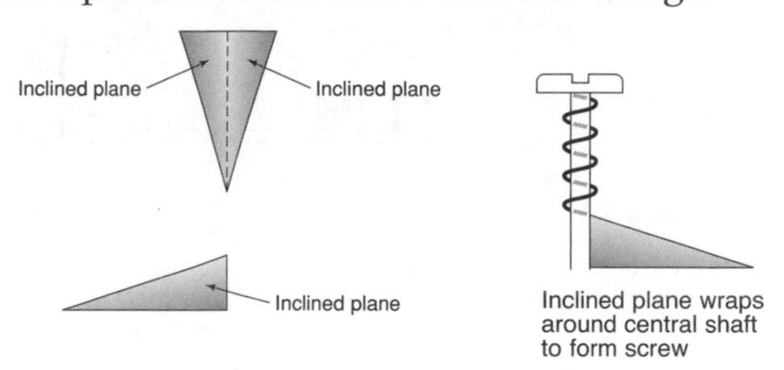

Figure 15.11
Screw and wedge as inclined planes.

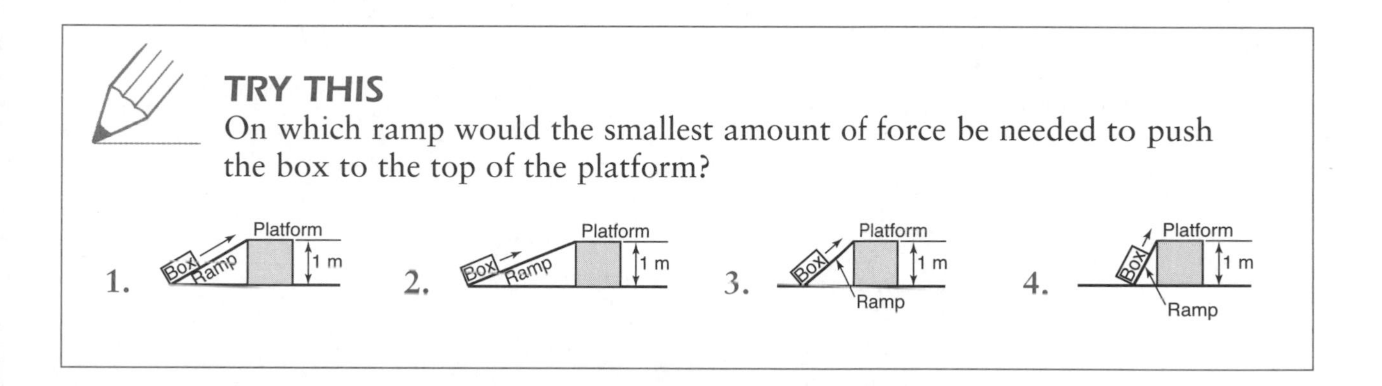

TRY THIS

On which ramp would the smallest amount of force be needed to push the box to the top of the platform?

LEVERS

A *lever* is a rigid object that is free to pivot around some fixed point. The pivot point of a lever is called the ***fulcrum***. A seesaw is a lever. The wooden plank is the rigid object, and the bar it rests on is the fulcrum.

When an effort is applied to one side of a lever, the lever swings around the fulcrum to produce an output force at another point. A lever can have the effort, output force, and fulcrum at different points along the lever. Figure 15.12 shows the positions of the effort, output force, and fulcrum in the three basic classes of levers.

A mathematical relationship exists between the amounts of the effort and output forces and their distances from the fulcrum. See Figure 15.13.

effort force × effort distance = output force × output distance

This relationship holds true for all classes of levers.

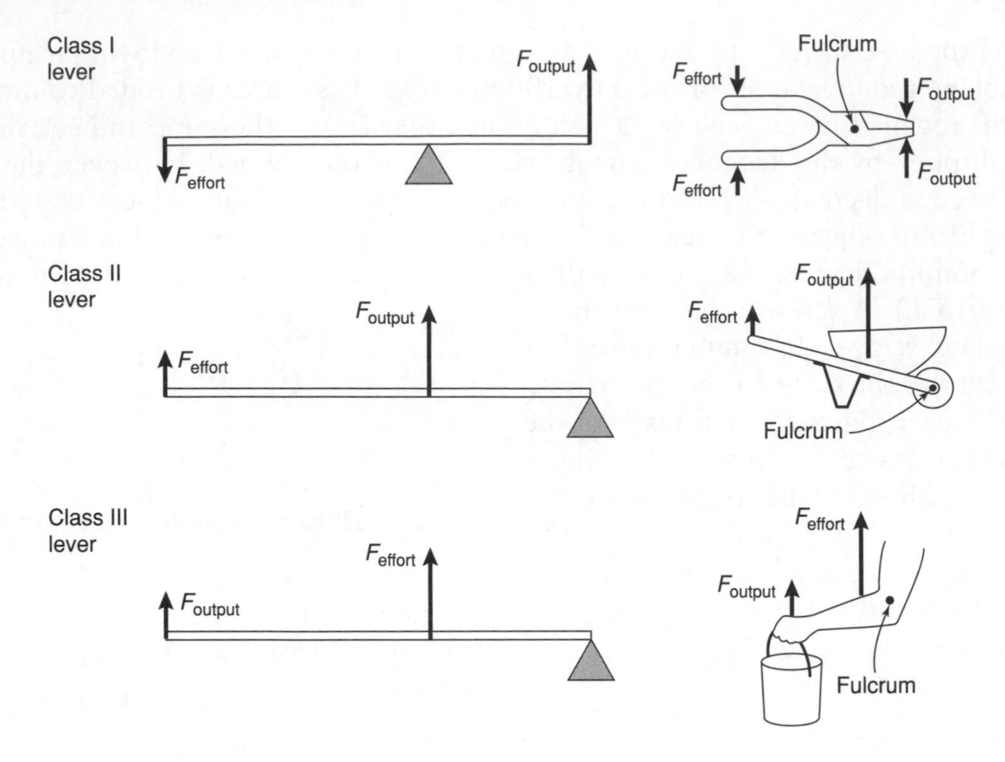

Figure 15.12
The three classes of levers.

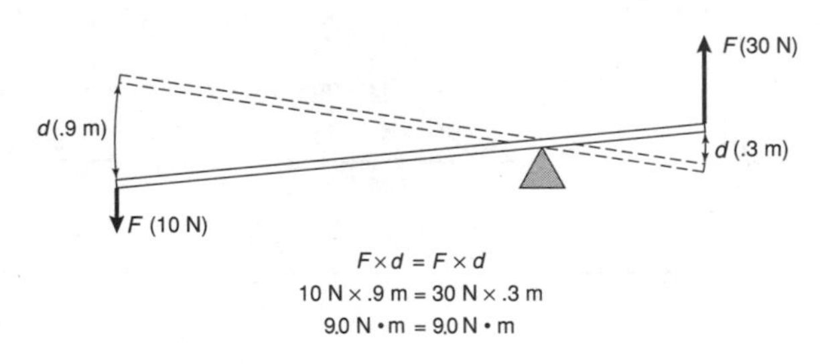

Figure 15.13
Effort and resistance arm diagram.

CLASS I LEVERS

In class I levers, the fulcrum is always placed between the effort and the output force. See Figure 15.14. Class I levers help us do work by increasing the output force produced by an effort and changing the direction of the force. Seesaws, balances, crowbars, and claw hammers are all class I levers.

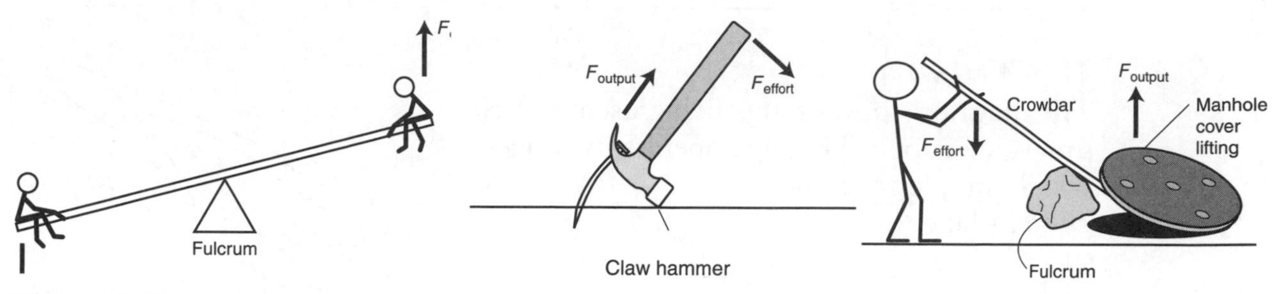

Figure 15.14
Class I levers.

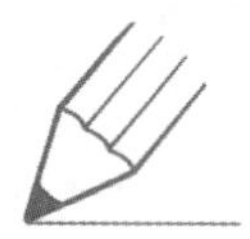

TRY THIS

1. Draw a pair of scissors. Label the fulcrum.
2. In one or more complete sentences, explain why scissors used for cutting paper have short handles and long blades, while those used to cut metal have long handles and short blades.

CLASS II LEVERS

In class II levers, the fulcrum is placed at the very end of the lever, and the effort is exerted at the other end. The output force lies between them. Class II levers do not change the direction of the force, they simply increase the output force produced by an effort. Wheelbarrows, nutcrackers, and bottle openers are all class II levers. See Figure 15.15.

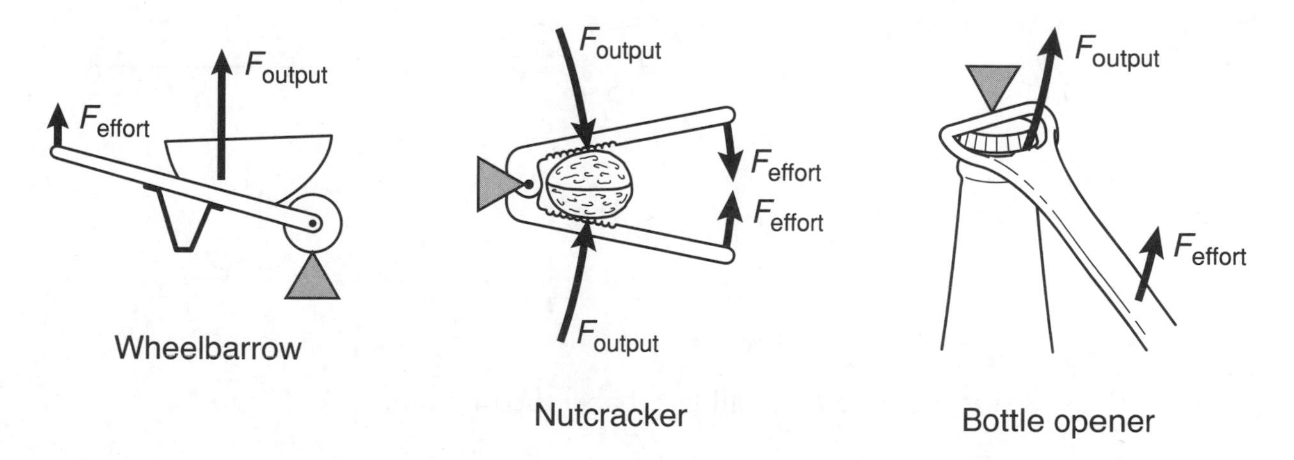

Figure 15.15
Class II levers.

TRY THIS

The diagram shows tongs being used to pick up an ice cube. The tongs operate most like which simple machine?

1. inclined plane
2. gear
3. pulley
4. lever

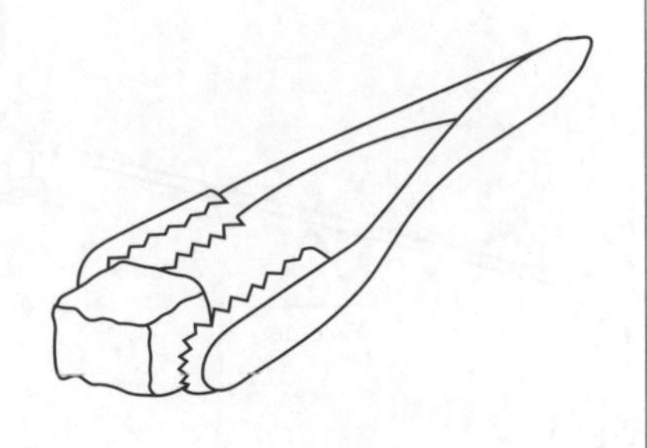

CLASS III LEVERS

In class III levers, the fulcrum is again at one end of the lever. However, this time the *output* force is at the other end of the lever, and the effort lies somewhere in between. This type of lever actually *reduces* the output force. Why would anyone want to use a class III lever? Look carefully at Figure 15.16. Do you see that the distance moved by the output force is greater than the distance moved by the effort? With a class III lever, you need to move the effort only a short distance to make the output force move a greater distance. Fishing rods, hammers, tweezers, and your forearms are all class III levers.

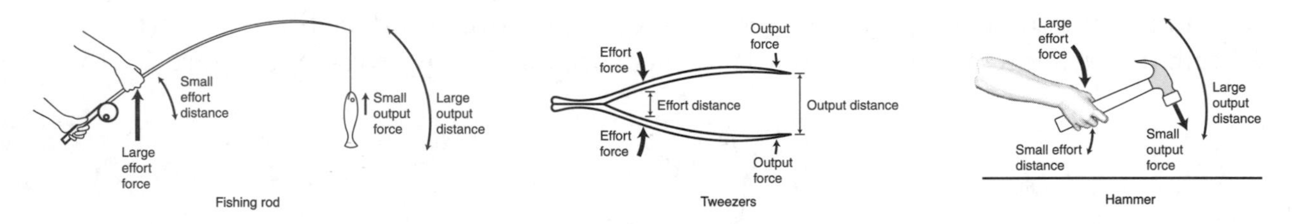

Figure 15.16
Class III levers.

TRY THIS

Draw a diagram of a baseball bat. Label the fulcrum.

WHEEL AND AXLE

A ***wheel and axle*** is a type of rotating lever. The center of the wheel and axle is the fulcrum of the rotating lever. The wheel is the outer part of the lever, and the axle is the inner part near the center. Many machines use the wheel and axle to increase output force. An effort applied to the wheel causes the axle to turn with a greater force than the wheel. Many machines, such as the winch shown in Figure 15.17*a*, use a wheel and axle in this way to increase force. On the other hand, turning the axle a short distance causes the wheel to move a greater distance, that is, at a greater speed than the axle. Clothes dryers use a wheel and axle in this way to spin the dryer drum rapidly. See Figure 15.17*b*. Screwdrivers, steering wheels, wrenches, and faucets are all examples of a wheel and axle at work.

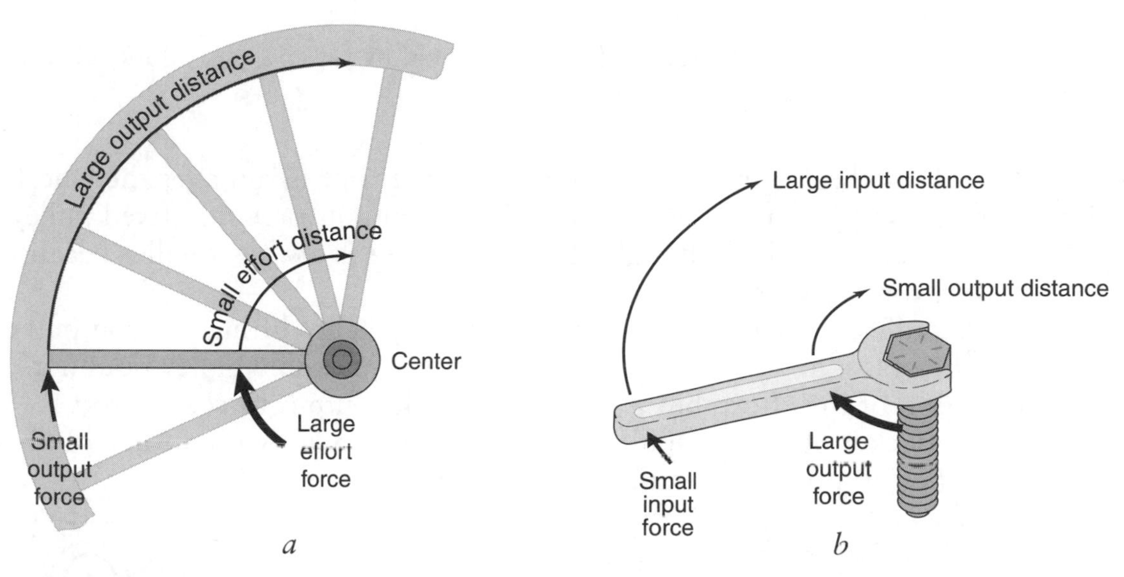

Figure 15.17
Wheel and axle. (*a*) A large effort at the axle over a small distance produces a large output distance with small output force. (*b*) A small effort over a large effort distance produces a small output distance and a large output force.

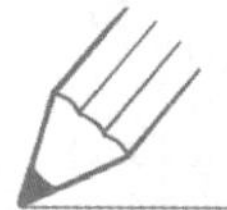

TRY THIS

1. Draw a diagram of a doorknob. Label the wheel and the axle.
2. How would you redesign a doorknob so that less force would be needed to turn it and open a door?

PULLEY

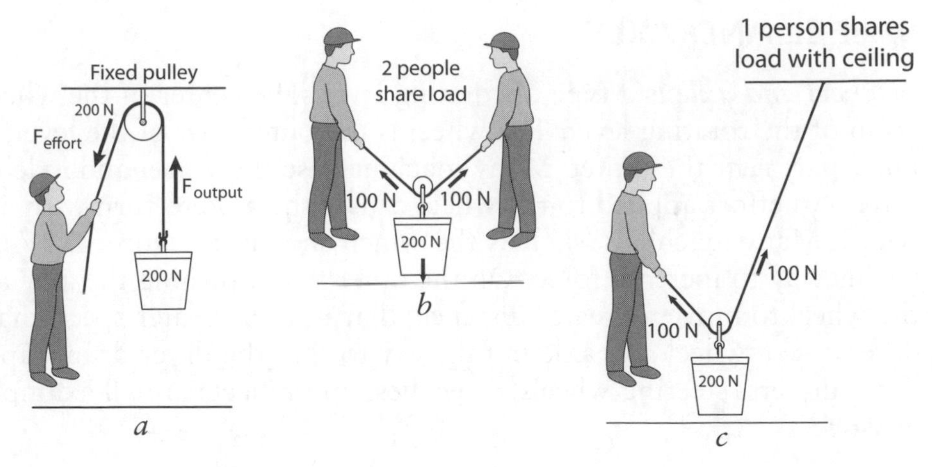

Figure 15.18
Pulley (*a*) fixed, (*b*) movable held by two people, (*c*) movable held by one person and the ceiling.

A ***pulley*** is a grooved wheel that turns by the action of a rope in the groove. There are several kinds of pulleys. Fixed pulleys change the direction of the effort force. For most people, pulling something downward is easier than lifting something upward. Roofers often use fixed pulleys to lift heavy shingles to the roof of a house. The wheel is fixed to a support, and the rope is run over the wheel to the point where an output force is needed. Ideally, the output force of a fixed pulley equals the input force. In real life though, the output force is always smaller because some of the effort force is used to overcome friction.

In a movable pulley, the wheel is fixed to the object you are lifting. As you pull on the rope, both the object and the pulley move together. A movable pulley does not change the direction of the effort, but it does increase the output force. Look at Figure 15.18*a*. A bucket of water is pulled downward by gravity with a force of about 200 newtons (about 50 pounds). A student must exert an upward force of 200 newtons to lift the bucket. Now look at Figure 15.18*b*. By lifting the bucket together, each of the two students has to exert only a force of 100 newtons. Now look at Figure 15.18*c*. The ceiling takes the place of the second student, and the first student has to exert only a 100-newton force to lift the bucket. This is how a movable pulley increases the output force. However, to lift the bucket 1 meter, you must pull a 2-meter length of rope.

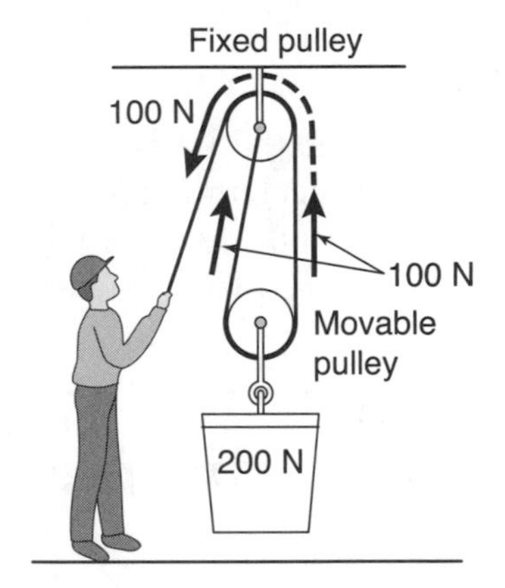

Figure 15.19
Fixed/movable pulley system.

A system of two pulleys, one fixed and one movable, changes the direction of the effort force *and* increases the output force. See Figure 15.19. The rope runs over the upper, fixed pulley. Then it runs down and around a lower pulley that is free to move and back up to the fixed pulley where it is attached. If you look carefully at the picture, you will see that two sections of rope support the bucket. One is fixed to the bottom of the upper pulley (and thus to the ceiling). The other leads over the upper pulley and out to the person exerting the effort. Since the 200-newton load is shared with the ceiling, the person has to exert only a 100-newton force to lift it.

A block and tackle is a system that uses many pairs of fixed and movable pulleys to increase the output force. As more pulley sets are added, more sections of rope support the load. Therefore, the effort needed to lift a load is greatly reduced. Can you estimate how much effort would be needed to exert the 400 newtons needed to lift the load shown in Figure 15.20?

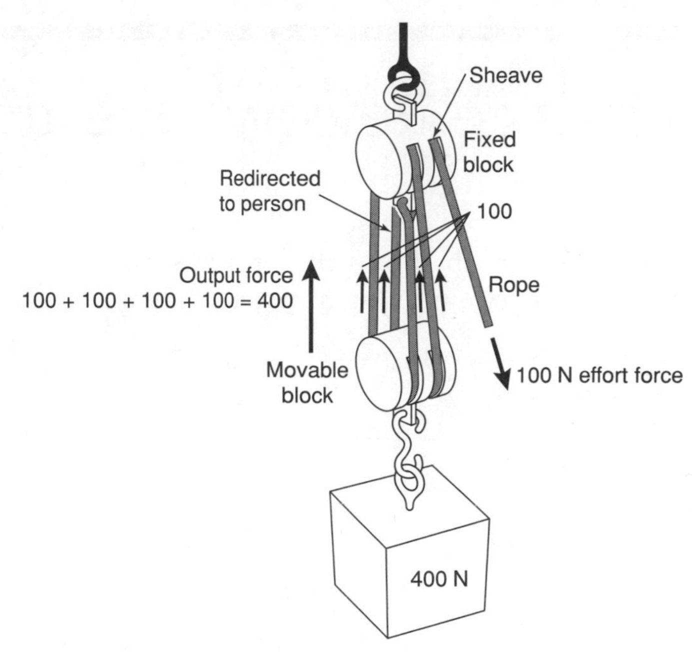

Figure 15.20
Block and tackle.

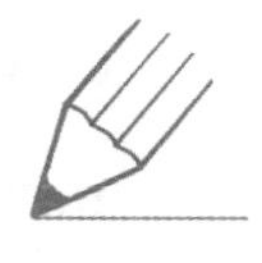

TRY THIS

Describe several ways in which you could use a pulley system to help you build a house.

MECHANICAL ADVANTAGE

Most machines are used because they increase output force. The ***mechanical advantage (MA)*** of any machine is the ratio between the output force and the effort force.

$$\textbf{mechanical advantage} = \frac{\textbf{output force}}{\textbf{effort force}} \quad or \quad MA = \frac{F_{output}}{F_{effort}}$$

A mechanical advantage greater than 1 means the output force is greater than the effort force. A mechanical advantage less than 1 means that the output force is smaller. The greater the mechanical advantage, the more the machine multiplies an effort force.

Practice Review Questions

1.* The graph shows the progress made by a beetle moving along a straight line.

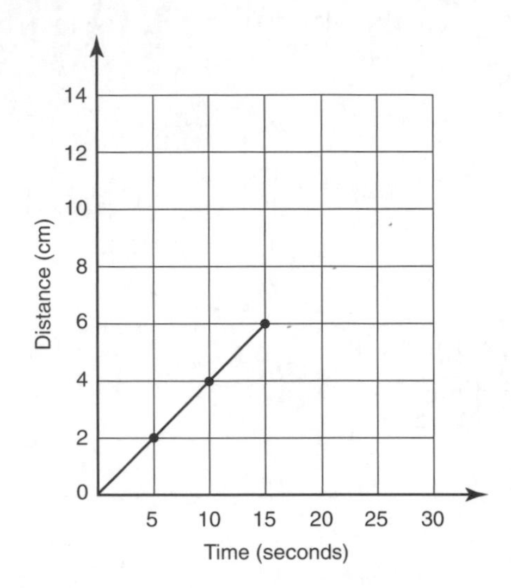

If the beetle keeps moving at the same speed, how long will it take to travel 10 cm?

1. 4 seconds
2. 6 seconds
3. 20 seconds
4. 25 seconds

2. Japan's famous "Bullet Train" travels the 550 kilometers from Tokyo to Osaka in 2.5 hours. Its average speed is

1. 220 m/s
2. 220 km/s
3. 220 km
4. 220 km/hr

* TIMSS 2003 Released Item

3. The average speed of an airplane was 600 kilometers per hour. How long did it take the airplane to travel 150 kilometers?

1. 0.25 hours
2. 0.5 hours
3. 4 hours
4. 5 hours

4. The velocity of a jet would best be expressed as

1. 100 km
2. 100 km/hr
3. 100 km/hr west
4. 100 km/hr/s

5. Which of the following is an example of acceleration?

1. the space shuttle waiting on its launch pad
2. a speed skater going around a turn at a constant speed
3. a cyclist pedaling at a constant speed in a straight lane
4. a car traveling due west at a constant speed

6. Unbalanced forces acting on a motorcycle can cause it to do any of the following *except*

1. speed up
2. come to a complete stop
3. move at a constant speed in a straight line
4. slow down

GO ON

7. Which of the following graphs shows an object speeding up?

1. 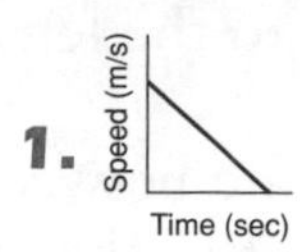

3. Speed (m/s) / Time (sec)

2. Speed (m/s) / Time (sec)

4. Speed (m/s) / Time (sec)

8.§ To keep a heavy box sliding across a carpeted floor at constant speed, a person must continually exert a force on the box. This force is used primarily to overcome which of the following forces?

1. air resistance
2. the weight of the box
3. the frictional force exerted by the floor on the box
4. the gravitational force exerted by the Earth on the box

9. A student pushes against a stalled car with 50 N of force and the car does not move. In this situation, the car

1. exerts 0 N of force
2. exerts less than 50 N of force
3. exerts 50 N of force
4. exerts more than 50 N of force

§ Source: National Center for Education Statistics, National Assessment of Educational Progress (NAEP), 2000, The Nation's Report Card: Science 2000

Questions 10 and 11 refer to the following diagram, which shows a spring scale attached to a wooden block being pulled up a wooden board.

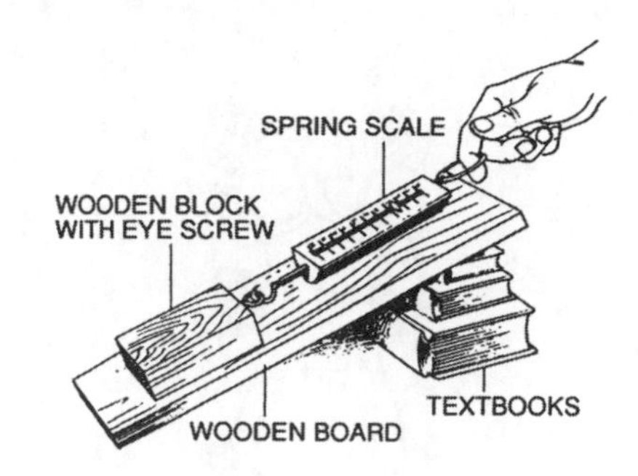

10. What is being measured by the spring scale in the diagram?

1. force
2. angle of incline
3. volume
4. temperature

11. If round pencils were placed under the wooden block as it was pulled up the wood board, what would happen to the reading on the spring scale?

1. It would remain the same.
2. It would drop to zero.
3. It would increase.
4. It would decrease.

12. Balanced forces are acting on all of the following except

1. a bus stopped at a red light
2. a student jogging at a steady pace on a straight road
3. a jet plane landing
4. a book lying on a table

GO ON →

13. Two boys wearing in-line skates are standing on a smooth surface with the palms of their hands touching and their arms bent as shown below.

If Boy *X* pushes by straightening his arms out while Boy *Y* holds his arms in the original position, what is the motion of the two boys?

1. Boy *X* does not move and Boy *Y* moves backward.
2. Boy *Y* does not move and Boy *X* moves backward.
3. Boy *X* and boy *Y* both move backward.
4. The motion depends on how hard boy *X* pushes.

14. Two students having a tug-of-war pull against each other with unequal forces as shown below

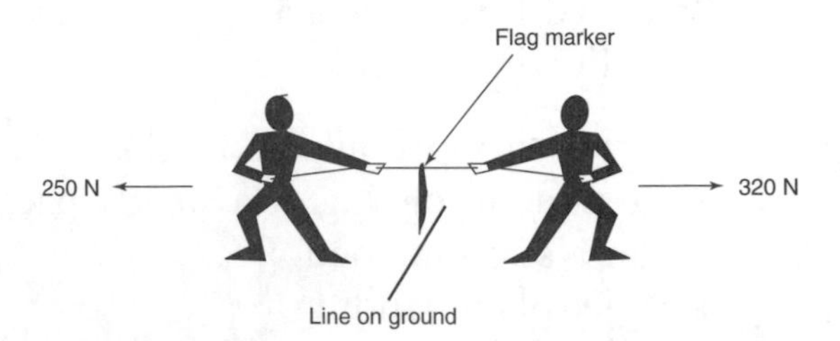

If the students' unequal opposing forces are applied to the rope at the same time, what will happen?

1. The flag will move to the right of the line.
2. The flag will move to the left of the line.
3. The flag will alternate between moving left and right of the line.
4. The flag will remain stationary over the line.

15.* The diagram on the left shows a ball on the end of a string being whirled in a circle. The diagram on the right shows the whirled ball as viewed from above.

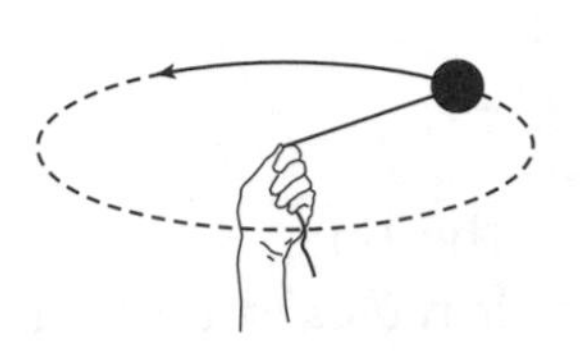

(View from above)

After several whirls, the string is released when the ball is at *Q*. Which of these diagrams shows the direction in which the ball will fly the instant the string is released?

1. 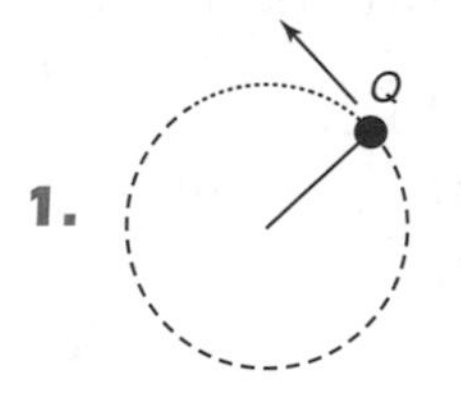

3. 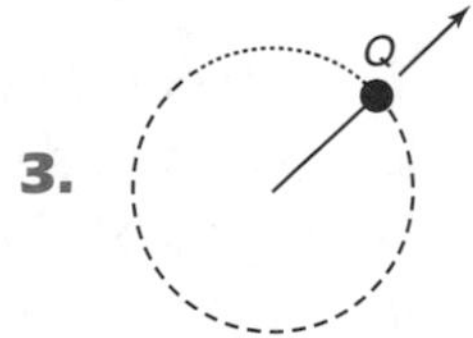

2. 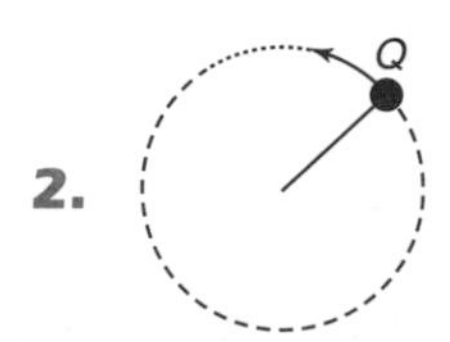

4. 

GO ON ➡

16. Which diagram shows a lever in use?

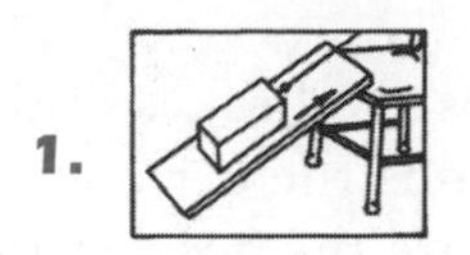
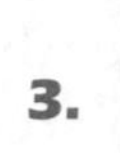

17. The picture below shows a man loading a box into a truck.

What are two simple machines being used by the man to load the box into the truck?

1. inclined plane and wedge
2. lever and pulley
3. wheel-and-axle and pulley
4. lever and inclined plane

18. Which pulley system would require the least force to raise a 200-gram mass a distance of 10 cm?

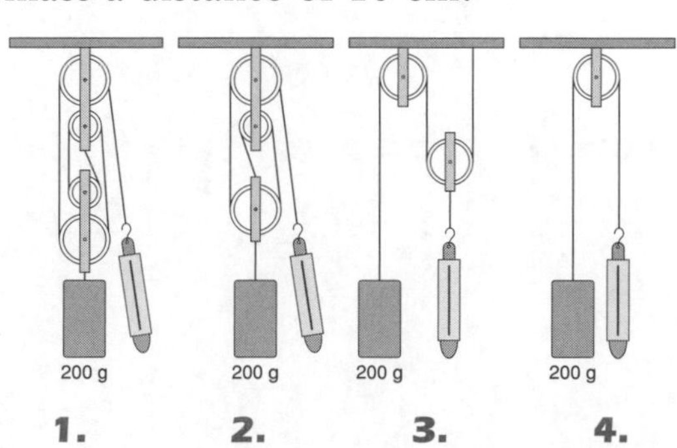

19. Gravity pulls a refrigerator downward with a force of about 1,000 N. A man exerts a 200-N force on a lever to lift the refrigerator. What is the mechanical advantage of the lever?

1. 0.2
2. 0.5
3. 5.0
4. 20.0

CONSTRUCTED-RESPONSE QUESTIONS

20.* The table below shows the results of an experiment to investigate how the length of a spring changes as different masses are hung from it.

Mass (grams)	Length of Spring (cm)
0	5
10	7
20	9
30	11
40	12
50	13
60	13

a. Describe how the length of the spring changed as different masses were hung from it.

b. What mass would be needed to stretch the spring to a length of 10 cm?

c. Predict how many centimeters the spring will stretch if a total mass of 70 grams was hung from it.

GO ON ➡

21. The diagram below shows a pulley system with a box exerting a downward force of 400 N (newtons) due to gravity.

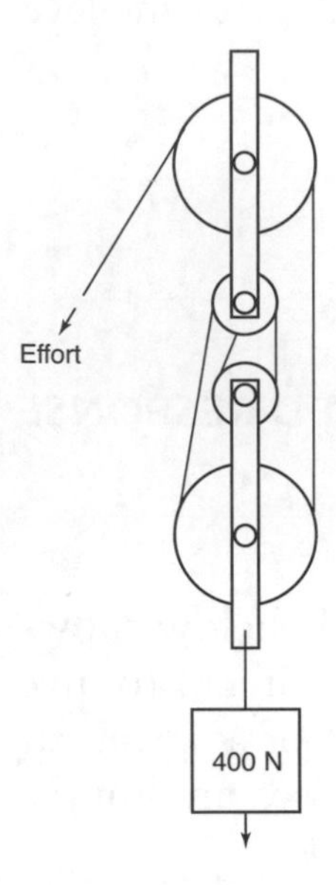

a. How much effort would be required to lift the box with this pulley system?

b. What is the mechanical advantage of this pulley system?

22. A car is traveling at a constant speed of 40 km/hr.

a. The driver hits the brakes and within 5 seconds the car comes to a full stop. State the acceleration of the car with the correct units.

b. The driver then steps on the gas and within 8 seconds is traveling at 40 km/hr again. State the acceleration of the car with the correct units.

23. The graphs below represent the motions of four different objects.

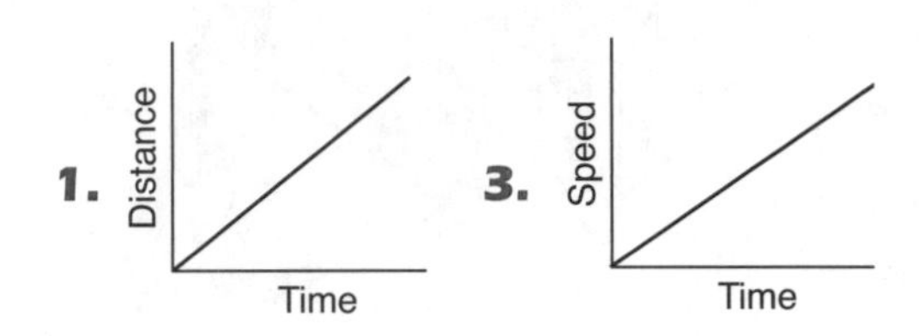

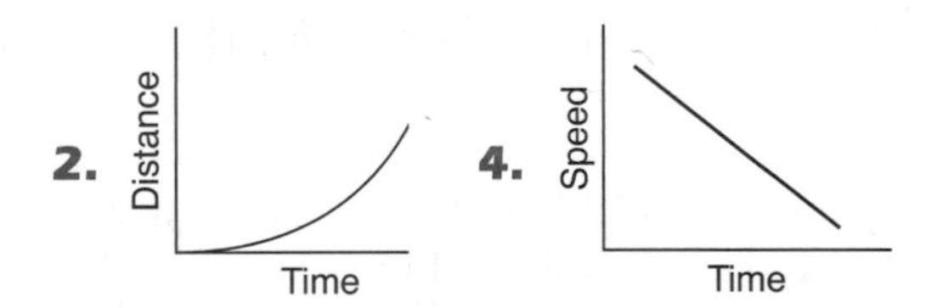

Briefly describe the motion of each of the four objects.

Glossary and Index

A

abrasion: [170, 172–173] the breakdown of rocks from rubbing against one another

absolute zero: [255] when a particle has stopped moving, it has reached its coldest possible temperature

abstraction: [2] considering something apart from a particular object or event

acceleration: [282] the rate at which an object changes velocity

adaptations: [111] the genetic changes that occur that make a species more likely to survive in a given environment

adrenaline: [64–65] a hormone produced by the adrenal glands that helps the body deal with emergencies

agents of erosion: [171] substances such as water, air, or ice set in motion by gravity that move sediment from place to place

air: [162] the mixture of gases, water droplets, ice, dust, and other particles that make up Earth's atmosphere; dry air is about 78% nitrogen, 21% oxygen

air currents: [193] the up and down motions of air

air masses: [200] large regions of air with fairly uniform characteristics

air pressure: [191–192] the amount of force exerted by the air's gas molecules hitting a given surface area

air temperature: [189] the degree of hotness or coldness at a place in the atmosphere measured by a thermometer using such scales as Celsius, Fahrenheit, or Kelvin; a measure of the amount of heat energy present in the atmosphere at a place

alkali metals: [238] a group of elements that are soft, silvery solids, easily melted, and very reactive

altitude: [151] the angle between the horizon and an object seen in the sky with the observer at the vertex

angiosperms: [101] flowering plants; seed plants having seeds enclosed in an ovary or fruit

angle of incidence: [266–267] the angle that a light ray striking a surface makes with a line perpendicular to that surface

angle of reflection: [266–267] the angle that a light ray reflected from a surface makes with a line perpendicular to that surface

anther: [98, 103] the part of a flower's stamen that bears the pollen; the plant's male sex organ

antibodies: [76] chemicals that weaken or destroy germs that enter the body

anticyclone: [203–204] a weather system centered on a high pressure region from which air moves outward and in a clockwise direction

anus: [67] the muscular ring at the end of the rectum through which feces exit the body

artery: [68] a blood vessel that carries blood away from the heart

asexual reproduction: [94] reproduction that involves only one parent and no special sex cells to begin development of a new individual

assimilation: [42] the process of converting nonliving substances into the living cells that make up the body of an organism

asteroids: [142] solid bodies having no atmosphere that orbit the Sun
asthenosphere: [179] a 100-kilometer-thick layer of Earth's upper mantle in which rock yields to pressure like a fluid
atmosphere: [161–162] the thin shell of gases, water, ice, dust, and other particles bound to Earth by gravity
atom: [219, 233] the smallest particle of an element
atrium: [69] an upper chamber of the heart

B

balance: [216] an instrument that compares the mass of two different objects by using the see-saw principle
balanced diet: [82] eating a variety of foods each day so that the body gets all of the needed nutrients
banding: [167] nearly parallel bands of different minerals in a metamorphic rock; the separation of minerals into light and dark layers during the metamorphosis of a rock
bar graphs: [11, 13] used when observations or measurements of a variable are not continuous but fall into separate categories
basalt: [164, 165] a dark, fine-grained igneous rock that makes up most of the bedrock under the oceans
beach: [174] a strip of loose materials along a coastline that forms as waves wash up against the land
Berzelius, Jons Jakob: [233] Swedish chemist who suggested chemical symbols
best-fit line: [10] a line that is drawn on a graph so that an equal number of data points fall to either side of the line
bile: [67] a substance produced by the liver that aids in digestion of fats and oils
bioclastic rocks: [166] sedimentary rocks made when materials precipitate from a solution or are left behind when water evaporates, or formed from the shells of marine life
biological environment: [120] the living things surrounding an organism
biologists: [41] scientists who study life
blizzard: [206] heavy snow and strong winds produce blinding snow, near zero visibility, deep drifts, and life-threatening wind chills
blood: [68] a complex fluid mixture including cells, dissolved food and gases, cellular wastes, proteins, enzymes, and hormones
blood vessel: [68] any of the tubes through which the blood circulates; arteries, veins, and capillaries are blood vessels
boiling: [222] the change from a liquid to a gas phase that occurs when a liquid is heated to the point where *all* of its particles are moving fast enough to become a gas
boiling point: [222] the temperature at which boiling occurs at normal pressure
bone marrow: [58] soft tissue inside bones that is the site of red and white blood cell production
bones: [58] distinct pieces making up the skeleton that consist of living cells nestled between tough, nonliving material composed of minerals; provide support for vertebrate animals
Boyle, Robert: [233] an English scientist who studied the volume of gases
brain: [62–63] large organ made of interconnected nerve cells enclosed in the skull; it is the organ of consciousness and stimulates muscles in response to sensory stimulation
bronchi: [70] two large, main branches of the windpipe
Brown, Robert: [44] a scientist who identified the cell nucleus

C

calorie: [80, 256] the amount of heat needed to change the temperature of 1 gram of water by 1 degree Celsius; unit of measurement of energy content in food

capillaries: [68] tiny blood vessels that interconnect arteries and veins
carbohydrates: [81] compounds consisting of carbon, hydrogen, and oxygen; a source of energy
carnivores: [85] organisms that feed on animals
carrier: [106] an organism that shows a dominant trait but carries the recessive gene
cartilage: [58] a tough, elastic tissue made of living cells and nonliving material; covers the ends of many bones; it is not as hard as bones and is more flexible
catalyst: [231] a substance that can increase the rate of reaction without being used up in the reaction
celestial meridian: [147] the imaginary line that passes through the north and south points on the horizon and through the zenith
celestial objects: [135–136] objects that can be seen in the sky and are beyond Earth's atmosphere
celestial sphere: [138] the model of a large spinning ball surrounding Earth containing celestial objects
cell division: [44, 93, 96] the process by which one cell divides to form two or more cells
cell membrane: [45] a sac-like lining that surrounds the cytoplasm
cell theory: [44] the explanation of a number of observations related to cells: all living things are made up of cells, all cells perform life functions, all cells arise from other living cells
cell wall: [45] a rigid, nonliving structure that protects the plant cell
cells: [44, 53] the basic units of all living things
Celsius, Anders: [255] Swedish scientist who invented the centigrade (or Celsius) thermometer
cementation: [166] the binding together of sediment by substances that crystallize or fill in spaces between loose particles of sediment
centrioles: [45] cylindrical structures located near the nucleus that are involved in cell division in animal cells
cerebellum: [62] the part of the brain that controls and coordinates muscular activity; located below the back part of the cerebrum
cerebrum: [62] the part of the brain that controls thinking, learning, memory, feeling, and voluntary muscle movements
chemical bond: [219] the attractive forces that hold atoms together
chemical change: [228–231] a change that results in the formation of a new substance with a permanent change of properties; also known as a *chemical reaction* in which one or more substances called *reactants* break down or combine chemically to form new substances with permanently changed properties called *products*
chemical digestion: [66] the breaking down of food into different substances through the action of digestive enzymes
chemical properties: [223] a characteristic or quality of an element or compound that can be observed or measured during a change in its chemical composition (e.g., reacts with acid, burns in air)
chemical rocks: [166] sedimentary rocks formed from substances that precipitate from a solution or are left behind when water evaporates
chemical weathering: [170] the breaking down of rocks by changing their chemical composition
chemistry: [215] the science dealing with composition, structure, and properties of matter
chlorophyll: [45, 82] a green pigment contained in plant cells; light absorbed by chlorophyll provides the energy needed to convert carbon dioxide and

water into carbohydrates by photosynthesis
chloroplasts: [45, 82] specialized structures in the cells of green plants that contain chlorophyll
chromosomes: [45, 93, 96] rodlike structures made of DNA that are found in the cell nucleus; store genetic material
circulatory system: [68–69] the organ system responsible for moving food, water, and oxygen to the cells and waste products away from cells; consists of the heart and blood vessels
class: [47] a group of related organisms ranking below a phylum and above an order
classification: [21, 46–49] placing data into groups; the sorting of living things into groups based on similarities and differences
clastic rocks: [165–166] sedimentary rocks made of cemented-together fragments
cleavage: [164] the tendency of a mineral to split along smooth surfaces
climate: [207–210] a term used to describe the general weather conditions in a given area over a long period of time; the most important factor affecting weathering
climax community: [124] a stable community
closed circuit: [261] a continuous pathway through which electric charge can flow
cloud: [169, 196] a visible mass of condensed water droplets that forms when air is cooled below its dew point
cold front: [201] the boundary between the leading edge of a cold air mass and the warm air mass against which it is advancing
collision theory of reaction rates: [229] the idea that collisions between atoms and molecules affect the rate at which they react
color: [163, 266] the sensation produced when light waves strike the retina of the eye; the most noticeable mineral property
coma: [142] cloud of gases surrounding a comet
comet: [142] a celestial object thought to consist of a solid nucleus of meteoroid particles embedded in ice
commensalism: [130] a relationship between two species in which one species benefits and the other is unaffected
communities: [122] all the populations living together in an ecosystem
compaction: [166] the squeezing together of sediments due to the sheer weight of layers on top of them
compass: [264] a magnet in the form of a needle suspended so that it can swing freely above a scale labeled with directions
compound: [219] a form of matter in which two or more elements are joined together in a fixed ratio
concave lens: [268] a lens that causes light to bend such that it spreads apart or diverges
concentrated solution: [226] has a relatively large amount of solute compared to the amount of solvent
condensation: [169] the process of a gas changing to a liquid
condense: [192] to change from a gas or vapor to a liquid
conduction: [257] the transfer of heat by means of collisions between molecules
conductors: [261] substances through which an electric charge flows easily
conservation: [272] limiting the use of energy resources so that they will last longer
constants: [3] factors that do not change
consumer: [84–85, 126] an organism that cannot make its own food; an organism that ingests other organisms or food particles

control: [4] a standard of comparison for testing a hypothesis; a parallel experiment in which no variables have been purposefully changed
controlled experiment: [25] a test of a hypothesis planned so that only one variable at a time purposefully changed
convection: [257] the transfer of heat by movement of a volume of liquid or gas from one region to another due to density differences
convection cell: [193] a circulation pattern of motion in which warm air rises, expands outward, cools, and sinks back to the surface
convergent boundaries: [181] places where tectonic plates move toward each other and collide
convex lens: [268] a lens that causes light to bend so that it comes together or converges; a lens that curves outward at the center
coordinate system: [144] a way of locating points by labeling them with numbers called *coordinates*
coordinates: [144] two numbers that define the position of a point with respect to a system of intersecting lines
Copernicus, Nicolaus: [138] an astronomer who devised the heliocentric model of the universe
core: [178] the central, innermost, and densest layer of Earth, made mostly of nickel and iron
Coriolus effect: [195] objects moving freely with respect to Earth's surface appear to curve toward the right in the northern hemisphere and toward the left in the southern hemisphere due to Earth's rotation
cross-pollination: [98] the transfer of pollen from the anther of one flower to the stigma of another flower
crust: [179] the solid, outermost layer of Earth
current electricity: [259, 261] refers to the flow of electric charge through matter
currents: [173] movements of water
cyclone: [203] a weather system centered on a low pressure region in which air moves inward and in a counter clockwise direction
cytoplasm: [45] the living, jellylike substance of a cell in which the cell nucleus is embedded

D

Dalton, John: [234] an English chemist; proposed the modern atomic theory
data: [21] scientists' observations; may be measurements, descriptions, or other facts
data pairs: [9] representation of coordinate points on a graph
day: [148] the cycle of daylight and darkness that is produced by Earth's rotation
decomposers: [85, 126] organisms that feed on dead plants and animals
deductive reasoning: [9] reasoning from general theories to account for specific experimental results
deflation: [173] occurs when wind picks up and carries away loose particles of rock
delta: [174] a large, flat, fan-shaped pile of sediment deposited at the mouth of the stream
density: [217–218] a measure of how closely the matter of a given substance is packed into a given volume
dependent variable: [3] variable that changes in response to manipulations of the independent variable
deposition: [174] the process by which sediment is dropped into a new place when an agent of erosion slows down and can no longer carry it
dew point: [193] the temperature at which the water vapor in a given part of the air is in equilibrium
diaphragm: [70] the muscle that gets air into and out of the lungs

digestion: [42, 66] the breaking down of nutrients into simpler substances that cells can use
digestive system: [66–67] the organ system that mechanically and chemically breaks down food
dilute solution: [226] a relatively small amount of solute compared to the amount of solvent
disease: [76] any harmful change that interferes with the normal structure, function, or appearance of the body or any of its parts
distance: [280] the length between any two points
divergent boundaries: [181] places where tectonic plates are moving apart
diversity of life: [46] the great variety of different organisms, each of which performs life functions in different ways
DNA: [93] deoxyribonucleic acid; molecules shaped like a spiral ladder that contain genetic information
domain: [262] a group of atoms locked together by aligned magnetic forces
dominant: [105] in genetics, one of a pair of genes that control a trait that masks that of its recessive counterpart
dunes: [175] mounds of sand deposited by wind
dwarf planet: [142] a solar system object that has enough gravity to form a round or nearly round shape, but has not swept up everything near its path

E

Earth: [141–142] a terrestrial planet, almost a perfect sphere
ecological succession: [123–124] the process by which one community is replaced by another over time
ecology: [119] the scientific study of the relationship between living things and their environment
ecosystem: [120–123] the interrelationships between living and nonliving things in a particular place
egg cells: [74] female sex cells containing one-half the normal number of chromosomes
electric cells: [261] devices that use chemical reactions to separate positive and negative charges
electric charge: [259] a fundamental property of matter
electric current: [261] a flow of electric charge from one place to another
electric field: [260] the region of space in which a force due to electric charges exists
electric generator: [271] a device in which some form of energy is used to cause a magnet and a coil of wire to move relative to one another and thus produce an electric current
electrical energy: [259] the ability to do work by virtue of the forces of the attraction and repulsion between electric charges
electromagnet: [264] a piece of soft iron that becomes a temporary strong magnet when electricity flows through a wire coiled around it
element: [219] a substance composed of one kind of atom; one of the basic building blocks of all matter
embryo: [98] an organism in the early stages of development
endocrine system: [64–65] a system of glands that release chemical messengers into bloodstream
energy: [80, 246] what something has if it can do work; in food, energy exists in the chemical bonds of molecules
energy consumption: [272] the amount of energy we use
energy pyramids: [126] a model showing how energy moves through a food chain
engineering design: [33] using scientific knowledge to solve practical problems
enzymes: [66] substances that chemically break apart food molecules into simpler molecules

epicenter: [177] the point on Earth's surface directly above an earthquake's focus; where an earthquake is first felt

epidermis cells: [54] outer cells that cover and protect the leaf

epithelial cells: [54] flat, thin animal cells that cover and protect inner and outer surfaces of the body

equator: [144] an east-west reference line for latitude, midway between the North Pole and South Pole

equinox: [151] occurs twice a year (March 21, September 22) when Earth's axis is tilted neither toward nor away from the Sun and night and day are about equal in length

erosion: [171] the natural process that moves sediments from place to place

esophagus: [66] a long muscular tube connecting the mouth to the stomach

estrogen: [65] a hormone produced in the ovaries that regulates the normal growth and development of sex glands, regulates reproduction, and controls the development of sex characteristics

evaporate: [169, 192] the process of a liquid changing to a gas

evolution: [49, 113] the idea that life forms have developed from earlier, different life forms

excretion: [42] the process by which waste products of life functions are removed from an organism

excretory system: [72–73] organ system responsible for the disposal of waste products

experiment: [21] a situation set up by a researcher to test an idea

external fertilization: [97] eggs are fertilized outside the body of the female parent

extrapolation: [6] a form of inductive reasoning used to estimate or extend beyond known information

extrusive igneous rock: [164] solidified after pouring onto the surface, cooling rapidly, forming small crystals

F

Fahrenheit, Gabriel Daniel: [255] German physicist who invented the Fahrenheit thermometer

family: [48] a group of the most similar members of an order

fats: [81] nutrient molecules that contain carbon, hydrogen, and oxygen; often also contain phosphorus or nitrogen; a concentrated source of energy; a storage facility for energy

feces: [67] the solid waste material resulting from the digestive process; contains undigested and undigestible food

feedback mechanism: [86] the method organisms use to control their internal environment in response to changes in the external environment

fertilization: [96] the merging of the nuclei of two gametes

field: [198] a region of space in which a measurable quantity is present at every point

field studies: [21] the collection of data under conditions that cannot be perfectly controlled

focus: [176–177] the point where crustal rock breaks from internal pressure, resulting in an earthquake

foliation: [167] parallel layers of crystals formed when rocks recrystalize under great pressure

food chain: [126] a model of feeding relationships in an ecosystem

force: [246] any influence that changes the speed of a body of matter, its direction of motion, or both

fossil fuels: [270] fuels formed from the remains of once-living things, such as coal, petroleum, or natural gas

fossils: [113, 166] the remains of once-living things

Foucault, Jean: [146] a French physicist who proved the rotation of Earth
Foucault Pendulum: [146] a device used to demonstrate proof that Earth rotates
freezing: [222] the change from a liquid phase to a solid phase
freezing point: [222] the temperature at which a substance changes from liquid to solid
frost action: [170] the breaking down of rocks due to forces exerted when water that seeps into cracks and crevices freezes and expands

G

galaxies: [139] clusters of billions of stars
gametes: [96] special sex cells; an egg or sperm cell
gas exchange: [70] the process by which oxygen is supplied to the bloodstream and carbon dioxide is removed
gas: [220] a phase of matter having no fixed shape or volume
gastrin: [65] a hormone produced by the stomach lining that stimulates secretion of gastric juices
generation: [102] offspring that are at the same level of descent from a common ancestor
genotype: [104] the genetic makeup of an organism
genus: [48] a group of the most similar members of a family
geocentric model: [150] a model of the universe that positioned Earth at the center
geothermal energy: [270] the heat energy within Earth
germs: [76] microscopic organisms that cause infectious diseases
gestation: [100] the development of young in the uterus from fertilization to birth
glaciers: [173, 175] large masses of ice that form where snowfall exceeds melting for extended periods of time
glands: [64] organs that produce fluids the body needs
graduated cylinder: [217] a device used to measure the volume of a liquid
granite: [164–165] a coarse-grained igneous rock that forms most of the bedrock under the continents
graph: [8–12] a diagram representing mathematical relationships between data
gravity: [143] the force of attraction that exists between all matter
greenhouse effect: [209] the process by which carbon dioxide in the air lets sunlight pass through to Earth's surface, warming it but blocking heat rays from escaping
groundwater: [169] water from precipitation that seeps downward into, and fills, empty spaces between particles of soil or rock
groups: [235] vertical columns on the Periodic Table of the Elements
growth: [43] the process by which living things increase in size and in the amount of material they contain
gymnosperms: [102] nonflowering plants; plants whose seeds are exposed, not enclosed in an ovary or fruit

H

habitat: [121] the place where a plant or animal lives
halogens: [238] nonmetals that react with metals to form compounds that are called salts
hardness: [163] a mineral's resistance to being scratched
heart: [68] the muscular organ located in the chest cavity that forces blood through the body
heartbeat: [68] the regular rhythm of the heart contracting and relaxing
heat energy: [252–253] the total kinetic energy of all particles that make up a sample of matter

heliocentric model: [138] a model of the universe that positions the Sun at the center

hemisphere: [147] one-half a sphere

herbivores: [85] organisms that feed on plants

heredity: [103] the passing on of traits from parents to children

homeostasis: [43, 86] the livable condition that living things maintain regardless of their surroundings; the maintenance of proper internal conditions

Hooke, Robert: [44] the scientist who, using an early microscope, observed structures he called cells

horizon: [147, 171] the circle formed by the intersection of the sky and the ground or sea; recognizable layers in soil caused by weathering and plant growth

hormones: [64] fluids produced by endocrine glands

humidity: [192] the amount of water vapor in the air

hurricanes: [205] huge cyclonic storms that form over oceans near the Equator during summer months

hybrids: [105] having two different genes for the same trait

hydroelectric energy: [271] electrical energy produced by running water

hydrosphere: [162] the layer of water covering Earth's surface

hypothesis: [23–25] a tentative explanation that accounts for a set of observations and that can be tested by further investigation

I

ice storm: [206] a storm in which rain freezes when it hits the ground, creating a coating of ice on roads and walkways

igneous rocks: [164] rocks formed by the cooling and crystallization of minerals from a molten state

immunity: [76] resistance to infection by a particular germ

impulse: [62] an electrochemical change in nerve cells that is transmitted from neuron to neuron

independent variable: [3–4] the variable that is purposefully changed or manipulated in an investigation

inductive reasoning: [5] reasoning from specific observations and experiments to more general hypotheses and theories

infectious diseases: [76] diseases caused by microscopic organisms that are passed from one person to another

inference: [22] a conclusion that logically follows from a set of observations

ingestion: [42, 66] the taking in of food from the environment

instruments: [20] devices that give more information about things than can be observed with our senses alone

insulators: [261] substances through which electric charge can flow only with great difficulty

insulin: [64] hormone produced by the pancreas that regulates the storage of sugar

internal fertilization: [97–98] the fertilization of eggs inside the body of the female parent

interneurons: [62] carry impulses from sensory neurons to motor neurons

interpolation: [6] a form of deductive reasoning used to estimate a value between two known values

intrusive igneous rock: [164] rock that solidified below the surface, cooling slowly to form large crystals

investigations: [21] systematic approaches to collecting observations

involuntary actions: [62] responses of the body that are not under conscious control

involuntary muscles: [60] muscles not under conscious control, they work automatically

isoline: [198] a line connecting points of equal field value (e.g., *isotherms* connect points of equal temperature;

isobars connect points of equal barometric, or air, pressure)

iterative process: [33] a repetitive process

J

joints: [58] places where bones are connected to one another

journals: [26] periodicals containing research articles on a particular subject

Jupiter: [141–142] a Jovian planet; the largest planet in the solar system

K

Kelvin temperature scale: [255] a temperature scale in which zero is assigned to absolute zero (the coldest possible temperature); the scale increases in increments that are the same size as the Celsius scale

kidneys: [72] excretory organs that remove liquid wastes from the blood

kinetic energy: [248] the energy that a body has because of its motion

kinetic theory of matter: [220–221] an explanation of the phases of matter in terms of their motion

kingdom: [47] large groups into which all living things are divided; most biologists use a five-kingdom system to classify living things (i.e., Animalia, Plantae, Fungi, Protista, Monera)

L

large intestine: [66–67] an organ that reabsorbs water from digested food mass and eliminates digestive waste

larva: [99] the wormlike form of insects that undergo metamorphosis, from the time they emerge from the egg until they form a pupa

latent heat: [256] heat gained or lost during a phase change

latent heat of fusion: [256] the quantity of heat given off or absorbed when water changes between the solid and liquid phases

latent heat of vaporization: [256] the quantity of heat given off or absorbed when water changes between the liquid and gas phases

latitude: [144] angular distance north or south of the Equator

latitude lines: [144] lines that run in an east-west direction around Earth, making up part of Earth's coordinate system; also called *parallels*

lava: [164] molten rock at Earth's surface

Lavoisier, Antoine Laurent: [234] a French chemist who proposed the law of conservation of matter

law, scientific: [28] something that is always observed to happen the same way

law of conservation of energy: [250] states that energy cannot be created or destroyed

law of conservation of matter: [232] during an ordinary chemical reaction, matter cannot be created or destroyed

law of electric charges: [259] like charges repel, unlike charges attract

law of magnetic poles: [263] like poles repel, unlike poles attract

Leeuwenhoek, Anton van: [44] Dutch scientist who built early microscopes

lens: [268] part of the eye that focuses light; a curved piece of glass that bends or refracts light

life functions: [42–43] the activities that are required to stay alive; all living things carry out all of the life functions

ligaments: [58–59] strips of tough tissue that hold bones together

light: [46, 250, 264–266] a form of energy made up of electromagnetic waves; the main source of energy for all living things

Linnaeus, Carolus: [47] a scientist who developed the modern classification system

liquid: [220] phase of matter that has a definite volume, but does not have a definite or rigid shape

lithification: [166] the hardening of layers of loose sediment into rock
lithosphere: [161, 179] Earth's solid, outer layer of rock
liver: [72] an organ that removes toxic substances from the blood
locomotion: [43, 57] the process by which an organism moves
logical reasoning: [5] having a valid explanation for forming a conclusion
longitude: [144] angular distance east or west of the prime meridian
longitude lines: [144] semicircles connecting the north and south poles, making up part of Earth's coordinate system; also called *meridians*
longitudinal wave: [258] wave in which individual particles of a medium vibrate back and forth in the same direction in which the wave travels
longshore current: [173] a flow of water formed parallel to the shore when waves strike the shore at an angle and are reflected, interfering with incoming waves
lunar eclipse: [153] an event in which the Moon moves through Earth's shadow at full moon
lustre: [163] the way light reflects from the surface of a mineral

M

magma: [164] molten rock beneath Earth's surface
magnet: [262] a substance in which the domains are aligned; has the property of attracting iron or steel
magnetic field: [263] region of space around a magnet in which magnetic forces exist
magnetic forces: [262] forces that come into existence between electric charges when they are in motion
mammary glands: [74] glands in mammals that produce milk to nourish young
mantle: [178] the layer of Earth above the core and beneath the crust, made of dark, dense minerals; not completely solid
Mars: [140–141] a terrestrial planet
mass: [215] the amount of matter an object contains
mass wasting: [172, 174] the downhill movement of sediments under the direct influence of gravity
mathematical models: [34] models that use symbols such as letters or numbers to represent objects or systems and the ways in which they relate to one another
matter: [215] anything that has mass and takes up space
mechanical digestion: [66] the physical breaking down of food into smaller pieces
mechanical models: [34] models that have working parts that permit them to function or move like the real object
medulla: [62] the part of the brain stem that controls involuntary activities of the body
meiosis: [96] the formation of specialized sex cells so that each sex cell has only half the total number of chromosomes in the parent cell
melting: [222] the change from a solid phase to a liquid phase
melting point: [222] the temperature at which a substance changes from a solid to a liquid
Mendel, Gregor: [103] an Austrian monk who did scientific experiments with pea plants that became the basis for genetics
Mendeleev, Dmitri: [235] a Russian chemist who developed the periodic table of the elements
meniscus: [217] the curved surface that most liquids form in a graduated cylinder
mental models: [34] exist only in your mind, representing things that cannot be seen
Mercury: [140–141] a terrestrial planet
meridians: [145] *see:* longitude lines
metabolism: [43, 85] the totality of chemical reactions that are required for life functions

metalloids: [239] elements that have some metallic and some nonmetallic properties
metamorphosis: [99] a series of changes in physical form that occur during the development of some organisms
metamorphic rocks: [167] rocks that formed from other rocks under the influence of heat, pressure, or chemical action
meteor: [142] the streak of light given off by the vaporization of a meteoroid
meteoroid: [142] a chunk of matter orbiting the Sun that hits Earth's atmosphere
meteorite: [142] any surviving matter from a meteoroid that falls to Earth
mid-ocean ridges: [181] a line of volcanic ridges formed by magma welling up through the mid-ocean rift
mid-ocean rift: [181] a continuous series of cracks formed by divergent boundaries
mimicry: [111–112] having a body shape, color, or other trait that resembles another living organism
mineral composition: [164] the minerals that make up a rock; determines the physical and chemical properties of a rock
minerals: [162] inorganic compounds; naturally occurring, inorganic, crystalline substances with a fixed chemical composition
mitosis: [93] the division of the cell nucleus into two identical daughter nuclei
mitotic cell division: [93–94] the combination of two processes: mitosis and cytoplasmic division; results in the production of two daughter cells identical to one another and the parent cell
mixtures: [219, 224] two or more substances that are not chemically combined and are capable of being separated
models: [34] anything that represents the properties of an object or system
molecule: [234] the smallest unit of a compound; consists of two or more atoms bound together
moons: [141] satellites; solid bodies that orbit planets
moraine: [175–176] a deposit of unsorted sediments that piles up along the sides or end of a glacier
motor neurons: [62–63] carry impulses from brain and spinal cord to muscles and glands
mouth: [66] the place where digestion begins by food being crushed and chewed by teeth, jaws, and tongue
moving water: [172] water set in motion by gravity; responsible for more erosion than all other agents of erosion combined
multicellular organisms: [53] living things made of many cells
muscular system: [60–61] the system of muscles responsible for producing motion in an organism; includes voluntary and involuntary muscles
mutation: [110] an organism born with a trait that none of its ancestors have
mutualism: [129] a relationship in which both species benefit from the other's presence

N

natural selection: [114] Charles Darwin's theory about how evolution occurs; suggests that individuals with traits better adapted to the environment are more likely to survive and reproduce, thereby passing on the more adaptive traits
neap tides: [155] when there is very little difference between high and low tides; occur during the first and third quarter moons when the sun's and the moon's gravitational pulls are at right angles to each other
nebulae: [140] clouds of gas and dust

Neptune: [140–141] a Jovian planet
nervous system: [62–63] an organ system that receives and processes information from the senses and transmits messages to control and coordinate the body
neuron: [62] an individual nerve cell
niche: [121] the role of each organism in a habitat
noble gases: [238] a group of elements that are gases at room temperature, have low melting and boiling points, and almost never react with other elements
noninfectious diseases: [76] diseases that are not spread from one person to another
nonrenewable resources: [272] energy resources that are present in fixed amounts
nuclear energy: [250, 270] energy contained within the atoms that make up matter
nucleus: [45] the control center for the cell's activities
nutrients: [42, 80] substances that can be used to build and repair body parts, provide energy, or control body functions
nutrition: [42, 80] the intake and use of food by living things
nymph: [99] an immature form of an insect that doesn't have a pupal stage; resembles the adult but lacks wings

O

obsidian: [165] volcanic rock that hardened quickly, before crystals could form; volcanic glass
oblate spheroid: [144] a slightly flattened sphere
observation: [20] taking careful notice of natural phenomena in a systematic way
occluded fronts: [202] boundaries between air masses that occur when a warm air mass is trapped between two cold air masses and is lifted completely off the ground
offspring: [103] an organism's children
omnivores: [85] organisms that feed on both plants and animals
opaque substances: [267] substances that do not allow any light to pass through them
open circuit: [261] a circuit in which there is a break in the conducting pathway; prohibits electric current from flowing
optimization: [37] the evaluation of an alternate solution to a problem to find the most functional and efficient solution
orbit: [146] the path that a revolving object follows
order: [48] a group of related organisms, ranking below a class and above a family
organ system: [55] a group of organs that work together to carry out a specific set of life functions
organism: [42] an entire living thing
organs: [62] a group of tissues working together to perform a life function
ovary: [74, 98] part of the female reproductive system; a structure that contains eggs
ovulation: [74] the process by which an egg is released from the ovaries
oxygen: [46] a colorless, odorless gas that makes up 21% of Earth's atmosphere; needed by living things for respiration

P

pancreatic juice: [67] a substance produced by the pancreas that aids in digestion
parallel circuit: [261] a circuit with more than one conducting pathway through which electric charge can flow
parallels: [144] *see:* latitude lines
parasitism: [129] a relationship between two species in which one species benefits and the other is harmed
pathogens: [76] microscopic organisms that cause infectious diseases

peer review: [26] a process by which scientists work together to review each other's work
penumbra: [153] the part of a shadow in which light is only partially blocked so that it is dim but not totally dark
precipitation: [169, 196] condensed moisture that falls to the ground; occurs when condensed water droplets become too large to remain suspended, and fall back to the surface
Periodic Table of the Elements: [235–239] system devised by Dmitri Mendeleev to organize the elements by both their atomic weights and properties
periods: [235] the horizontal rows on the Periodic Table of the Elements
peristalsis: [66] a series of wavelike muscular contractions that move food along the digestive tract
perspective: [135] how things appear from a certain point of view
perspiration: [72] a liquid waste material that is excreted through the skin
phases: [220] the physical state or form in which matter exists; solid, liquid, or gas
phases of the Moon: [152] changing shapes of the illuminated portion of the Moon visible to an observer on Earth; varies cyclically as the Moon orbits Earth
phenotype: [105] the expression of a trait; the way an organism looks
phloem: [54] a complex plant tissue that transports food materials to all parts of the plant
photosynthesis: [82] the process by which green plants use energy from the Sun in the presence of chlorophyll to change carbon dioxide and water into sugars and oxygen
phylum: [47] a major group within a kingdom, whose members share at least one special characteristic
physical environment: [120] the nonliving things surrounding an organism
physical model: [34] an object built to scale that represents in detail another object, event, or process
physical properties: [223] characteristics or qualities that can be observed or measured without changing the chemical composition of the sample that distinguishes or identifies it
physical weathering: [170] natural processes that break down rock without changing its chemical composition, i.e., a process that changes only the size and shape of rocks
pistil: [98] the female reproductive organ in plants
planets: [138, 140–141] bodies that are at least partly solid, orbit the Sun, and give off mainly reflected sunlight
plasma: [68] the fluid portion of blood
plate tectonics: [179–182] a new model of Earth's structure in which the lithosphere is seen as broken into several large slabs
pollen: [98] a fine, yellowish powder made of tiny grains that contain the male sex cells in plants
pollen tubule: [98] a structure that grows from the pollen grain to reach the egg in the ovary
pollination: [98] the transfer of pollen to the sticky pad at the top of the pistil (stigma)
population: [122] a group of organisms of the same species that are found together at a given place and time
potential energy: [248] the energy that a body has because of its position or structure rather than its motion
predators: [85] animals that capture other animals for food
pressure: [221] the amount of force exerted on a given area
prey: [85] the animals that are captured by predators for food
primary waves: [177] a wave resulting from an earthquake that vibrates in the same direction as it travels

producer: [82] an organism that makes its own food from inorganic substances
products: [228] the substances that form as a result of a chemical reaction
property: [223] a characteristic or quality that distinguishes or identifies something
prostate gland: [74] produces secretions that form semen
proteins: [81] complex organic compounds containing carbon, hydrogen, oxygen, and sometimes sulfur; made of amino acids, the building blocks of cells
Proust, Joseph Louis: [234] a French chemist who developed the law of fixed proportions
pulse: [68] the surge of blood through the arteries caused by each contraction of the heart
pumice: [165] volcanic rock that contains holes formed by trapped gas bubbles
punnet square: [104] a simple model used to show the findings of Gregor Mendel's genetic experiments
pupa: [99] the stage of development between larva and adult in the metamorphosis of many insects
pupil: [269] an opening in the center of the iris through which light enters the eye
pure substance: [219] matter having the same composition and properties throughout

R

radiation: [257] energy carried by electromagnetic waves
rate of dissolving: [227] the amount of solute that a solvent dissolves per second
reactants: [228] substances that combine chemically to form a new substance
reasoned hypothesis: [24] includes an explanation of why the researcher thinks changing the independent variable will affect the dependent variable in a particular way
recessive: [105] a trait that does not show up in the presence of a dominant trait
rectum: [67] the lowest part of the large intestine; where feces are stored
recycle: [273] a conservation method in which useful materials are extracted from waste
red blood cells: [68] blood cells that contain hemoglobin and carry oxygen from the lungs to the body cells and carbon dioxide away from body cells
reduce: [272] to use less of a resource
reduction division: [96] a special kind of cell division that happens only in sex cells and results in each sex cell having only half the total number of chromosomes required
reflex actions: [62] quick, involuntary actions that do not involve the brain
refraction: [267–268] the bending of light when it passes at an angle from one transparent substance into another
regulation: [42] includes all of the processes by which an organism responds to changes within itself and in its surroundings
relative humidity: [192] the current amount of water vapor in the air compared to the amount of water vapor in the air at equilibrium
relative motion: [137] a change in the position of two objects in relation to one another
renewable resources: [272] energy resources that are replenished by natural processes
repeated trials: [4] testing an idea several times to reduce the effects of chance errors or unknown variables on the overall result
reproduction: [43, 91–97] the process by which living things produce offspring
reproductive system: [74–75] the organ system responsible for producing sex cells necessary for the production of offspring
residual soil: [171] soil that forms by the weathering of the underlying bedrock

respiration: [42, 70, 80] includes the processes by which energy stored in food is changed into a form that can be directly used by living cells; process by which nutrients are combined with oxygen to release energy
respiratory system: [70–71] the organ system responsible for gas exchange: supplying oxygen and removing carbon dioxide
retina: [269] the part of the eye that contains sensory neurons that are sensitive to light
reuse: [273] to conserve resources by using an item multiple times instead of replacing it
revolution: [146] the motion of one body around another body on a path called an orbit
rock cycle: [168] the constant changing of rocks from one type to another by natural processes
rocks: [164] naturally formed, nonliving, Earth material that holds together in a firm, solid mass
rotation: [145–146] a motion in which every part of an object spins around a central line called the axis of rotation
runoff: [169] the water from precipitation that flows downhill into rivers and streams and eventually back to the ocean

S

salivary glands: [66] glands located in the mouth, under the tongue; produces a liquid called saliva
sandbars: [175] long, narrow piles of sand deposited in open water
saprophyte: [85] plants that feed on dead plants and animals
saturated: [226] what a solution is said to be when a maximum amount of solute has dissolved in a solvent at a given temperature
Saturn: [140–141] a Jovian planet
scale: [9, 34] a series of marks made at regular distances along a line or curve to use in measuring; the intervals into which axes are divided
scavengers: [85] animals that feed on dead plants and animals
Schleiden, Matthias: [44] a scientist who suggested that all plants are made of cells
Schwann, Theodor: [44] a scientist who suggested that all animals are made of cells
scientific diagrams: [14] pictures that display information
scientific inquiry: [19] the particular ways in which scientists observe, think, experiment, and check whether their ideas and information are valid
scrotum: [74] sac outside of the body in which the testes are located
secondary waves: [177] a wave resulting from an earthquake that vibrates at right angles to its direction of travel
sedimentary rocks: [165–167] consist of fragments of other rocks that have been eroded by running water, wind, or glaciers
sediments: [171] the fragments or particles of rock produced by weathering
seed: [102] a young plant with food stored in a cover
seismic waves: [176–177] vibrations that spread out from the point rock breaks due to internal pressure; the focus
selective breeding: [110] the process of breeding plants and animals for particular genetic traits
self-pollination: [98] pollination that occurs within a single flower
seminal glands: [74] produces secretions that form semen
sense organs: [62] the organs through which the body obtains information
sensory neurons: [62] send impulses from sense organs to the spinal cord and brain
sequencing: [21] arranging data into some kind of order

series circuit: [261] a circuit in which there is only one conducting pathway along which charge can flow

sex-linked traits: [106] traits that are carried on the chromosomes that also carry sex, specifically, the X and Y chromosomes

sexual reproduction: [94, 96–98] involves two parents and the merging of special sex cells to begin development of a new individual

skeletal-muscular system: [60–61] an organ system that provides support, shape, protection, and locomotion; in humans, it comprises 206 bones and their attached muscles

skin: [72] the outer layer of tissue of the human or animal body; contains sensory neurons that respond to touch, pressure, heat, cold, or pain; assists in the excretion of liquid wastes

sleet: [196] rain that freezes into ice pellets before reaching the ground

small intestine: [66] part of the digestive system, following the stomach, where chemical digestion and nutrient absorption take place

small solar system body: [142] an object in the solar system that is neither planet, dwarf planet, nor satellite; includes asteroids, meteoroids, and comets

soil horizons: [171] distinct layers in soil that vary in thickness and composition

solar eclipse: [152] occurs when the Moon passes directly between Earth and the Sun, casting a shadow on Earth and blocking our view of the Sun

solar energy: [269–270] energy contained in the electromagnetic waves that emanate from the Sun

solar system: [140–143] a system made up of the Sun and all of the bodies that orbit it: the eight planets, their moons, and many smaller bodies such as comets and asteroids

solid: [220] the phase of matter having a definite shape and a definite volume

solubility: [224, 226] the maximum amount of a solute that will dissolve in a given amount of solvent at a given temperature

soluble: [226] able to be dissolved

solute: [226] a substance present in the smallest amount in a solution

solution: [226] a mixture that is a uniform blend of two substances, having the same composition and properties throughout

solvent: [226] the substance present in the largest amount in a solution

sound: [250, 257–259] a form of energy produced by vibrating objects

sound waves: [258] the successive layers of squeezed together and spread-out matter that travel outward from the source of continued vibrations

source region: [200] the surface over which an air mass forms

species: [48–49] the most similar organisms within a genus

sperm ducts: [74] tubes that carry sperm to the urethra

sperm cells: [74] male sex cells containing one-half the normal number of chromosomes

spinal cord: [62] a thick bundle of nerves through which the brain sends and receives messages

spring tides: [155] very high and very low tides occurring every new and full Moon when the Sun and Moon's gravitational pulls align

stamens: [98] the male reproductive organs in plants

starches: [81] complex carbohydrates

stars: [139] hot, glowing balls of gas

static electricity: [259] forces exerted on matter because of an imbalance of electric charge

station model: [197] a symbolic way to show a summary of the weather conditions at a weather station

stationary fronts: [202] the boundary between a warm air mass and a cold air mass when neither has enough force to move the other
stomach: [66] a strong, muscular organ that mixes and churns food with digestive juices
streak: [163] the color of a mineral's powder
streams: [172] bodies of water that flow downhill in relatively narrow but clearly defined channels
subduction: [181] the process by which a dense ocean plate slides under a less dense continental plate and plunges into the mantle, where it melts
substance: [219] matter having the same composition and properties throughout
sugars: [81] simple carbohydrates
summer solstice: [151] June 21, the date when the altitude of the Sun at noon stops increasing
Sun: [140] average-sized star around which Earth and other planets revolve
superposition: [166] the principle that each layer of a sedimentary rock is younger than the layer beneath it
surface waves: [177, 178] waves resulting from an earthquake that are like water waves; matter moves in circles; occur only at Earth's surface
surveys: [21] observations that describe things the way they are without drawing inferences
sustainability: [272] the ability to keep consuming energy at a present rate
sweat glands: [72] small coiled tubes that open to the outside of the skin and secrete sweat
switch: [261] a device that allows one to control the flow of electric charge through a circuit
symbolic representation: [2] the use of symbols to stand for other things and how they are related
symbols: [2] things that stand for other things
symptoms: [76] characteristic effects of a specific disease on the body
synapse: [62] the small gap between neurons
synoptic weather map: [197–199] a map that summarizes weather variables measured at many places at the same time
synthesis: [42] the processes by which large molecules are built from small molecules

T

taxonomy: [47] the branch of science that deals with classifying living things into groups
technology: [36] the application of scientific knowledge to industrial, agricultural, or business-related problems
temperature: [253] a measure of the average kinetic energy of each particle in a body
temperature: [255] a specific degree of hotness or coldness as indicated on a standard scale; a measure of the average kinetic energy of the particles in a body; governs climate, and influences life in many ways
tendons: [58] strong cordlike fibers that attach muscles to bones
testes: [74] part of the male reproductive system, produces sperm cells
testosterone: [65] a hormone produced in the testes that regulates normal growth and development of sex glands, regulates reproduction, and controls the development of sex characteristics
texture: [164] the size, shape, and arrangement of crystals or grains in a rock
theory: [28] a set of ideas that explains related observations

theory of evolution: [40–50] the idea that life began with very primitive life forms that have, over time, changed into all the different types of organisms on Earth today
thermometer: [254] an instrument that measures temperature
thunderstorms: [204–205] a violent storm of lightning and thunder often accompanied by heavy rain and sometimes hail; likely to occur wherever and whenever there is strong heating of Earth's surface
tides: [154–155] the motion of the oceans caused by the gravitational pull of the Moon and Sun
tissues: [59–61] a group of cells that are similar and act together to perform a life function
tornado: [206] a rotating column of air whirling at speeds up to 800 km/hr; usually form over land from intense thunderstorms and have a funnel-shaped downward extension of a cumulonimbus cloud
trade-offs: [37] features that must be sacrificed to get others
traits: [103] characteristics, often shared by family members
transform boundaries: [180–181] places where tectonic plates slide sideways past each other, forming faults
translucent substances: [267] substances that allow some but not all of the light that reaches them to pass through
transparent substances: [267] substances that allow almost all of the light that reaches them to pass through
transpiration: [169] a process by which plants release from their leaves water that they have taken in through their roots
transport: [42] the process by which the simple substances obtained from nutrients are absorbed by the organism and circulated throughout the organism
transported soil: [171] soil that forms from sediments deposited by agents of erosion
trench: [181] the deepest places in the oceans, reaching depths of more than 11 kilometers
trend: [10] a pattern of observations in which changes occur in a particular direction

U

umbra: [153] the part of a shadow in which all light has been blocked so that it is totally dark
unicellular: [53] living things made of a single cell
unity of life: [41] the way in which all living things are similar to one another
Uranus: [140–141] a Jovian planet
ureter: [72] the tubes that carry urine from the kidneys to the urinary bladder
urethra: [72] the tube that carries urine outside the body
urinary bladder: [72] a muscular bag where urine is stored until it is removed from the body
urinary system: [72–73] the system that removes liquid waste from the body
urine: [72] a fluid waste composed mostly of water and some salts, sugars, vitamins, and amino acids
uterus: [74] the organ of the female reproductive system in which a fertilized egg is implanted, and which holds and nourishes the young until birth

V

vagina: [74] birth canal
vaporization: [222] the change from a liquid state to a gas state
variable: [3–4] something that can change or be changed

variation: [**110**] a characteristic in an individual that is different from other individuals of the same species
vascular tissue: [**54**] conducting cells that transport materials throughout a plant
vein: [**68**] a blood vessel that carries blood toward the heart
ventricles: [**69**] lower chambers of the heart
Venus: [**140–141**] a terrestrial planet
villi: [**67**] tiny fingerlike projections along the lining of the small intestine
Virchow, Rudolph: [**44**] a scientist who suggested that cells reproduce
visual model: [**34–35**] something that looks like the real object
vitamins: [**81**] organic compounds that are micronutrients; necessary for normal growth and nourishment; often act as coenzymes
volcano: [**164**] the opening in the crust through which magma erupts, and the mountain built up by erupted material
volume: [**217**] the amount of space an object takes up
voluntary muscles: [**60**] muscles under conscious control; controlled by your will

W

warm front: [**201–202**] boundary that forms when a warm air mass pushes up and over a cold air mass
warning: [**207**] a report released by the National Weather Service stating that hazardous weather has been sighted or shown on radar
watch: [**207**] a report released by the National Weather Service stating that hazardous weather conditions are likely to develop in your area
water: [**46, 81**] the medium in which all chemical reactions involved in life functions occur; necessary for life; dissolves or breaks apart other chemicals
water cycle: [**169**] the movements and phase changes by which Earth's water cycles between the hydrosphere, atmosphere, and lithosphere
wave: [**173**] a moving ridge or swell of water
wavelength: [**165**] the distance between ripples in the electromagnetic field
weather: [**189**] the conditions in the atmosphere at a given place at a particular time
weather fronts: [**201–202**] the line on the ground marking a boundary between air masses
weather variables: [**189**] the changing characteristics of weather: air temperature, air pressure, humidity, wind speed, wind direction, clouds, and precipitation
weathering: [**170**] the breakdown of rocks into smaller particles by natural processes
weight: [**215**] the amount of the force of gravity between one object and another
white blood cells: [**68**] blood cells that protect the body against infection
wind: [**193–194**] horizontal movement of the air
wind direction: [**193**] the direction from which a wind blows
wind energy: [**271**] the energy of moving air
wind speed: [**194**] the speed at which air is moving
windpipe: [**70**] large breathing tube through which air travels to the lungs
winter solstice: [**151**] December 21, the date when the altitude of the Sun at noon stops decreasing
work: [**246–247**] the amount of change that a force brings about when it acts upon something; the amount of force exerted onto an object multiplied by the distance the object moves in the direction of the force

X

xylem: [54] complex plant tissue that conducts water and minerals upward from the roots of the plant

Y

year: [150] the time Earth takes to complete one revolution around the Sun

Young, Joseph: [247] an English scientist who suggested that stored work be called energy

Z

zenith: [147] the point in the sky directly overhead to an observer

zygote: [74, 96] the single cell that forms as a result of fertilization

Answer Key

CHAPTER 1

TRY THIS

Page 3

1. $P = M + F$
2. $D = M/V$
3. Drop in T (°C) = [increase in height (m)/1,000] × 6 (°C/m)
4. $F + 1 = R - 100$

Page 5

Independent variable: brand
Dependent variable: height of water
Constants (and variables that were deliberately held constant): composition of water, size and shape of strip of paper towel, amount of water in jar, depth towel is immersed in water, time paper towel is immersed in water.

Page 6

1. 18°C, extrapolation
2. 94 sit-ups, interpolation

Page 13

1. line graph

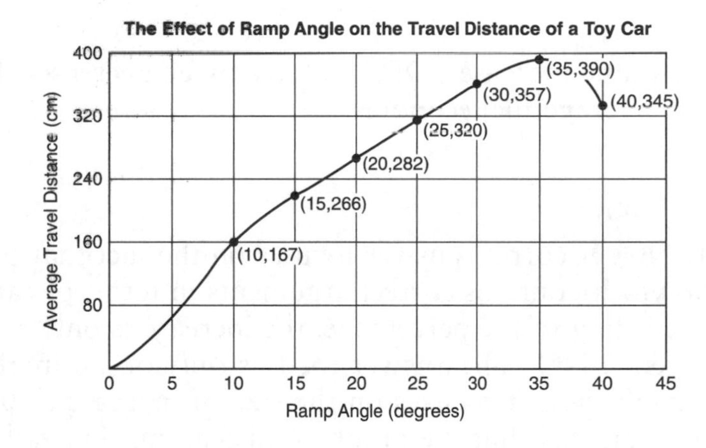

As ramp angle increases from 0 to 35 degrees, the average travel distance increases.

2. bar graph

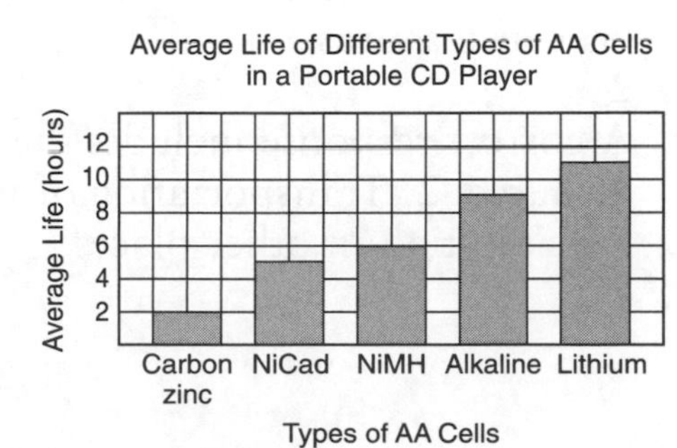

Ordinary carbon-zinc cells have the shortest average life, rechargeable NiCad and NiMH cells have intermediate life, and alkaline and lithium cells have the longest average life.

Page 14

June; leaves; roots; July; in ground; August; holes and ragged edges; August through October; they dig deeper into the ground; May through June.

PRACTICE REVIEW QUESTIONS

1. 1. bar; 2. line; 3. line
2. 2
3.

a.

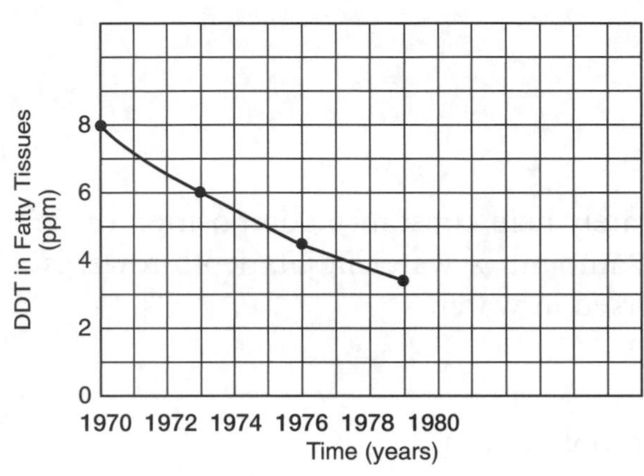

b.

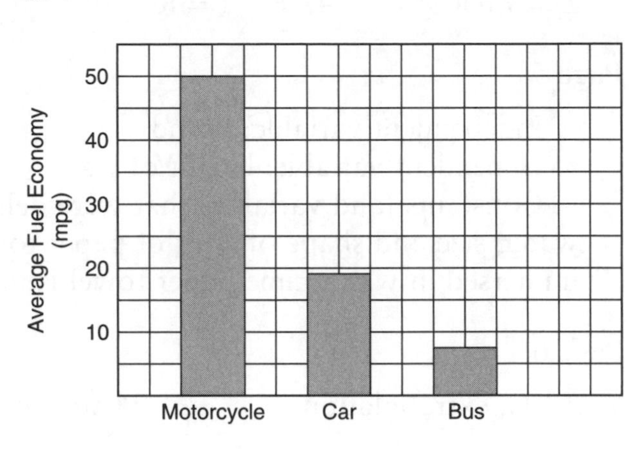

4. a. inverse—as time increases, DDT in fatty tissue decreases; b. the larger the vehicle, the lower the average fuel economy
5. 3
6. 2
7. 1. 60 km/h; 2. 9:07
8. Full Credit: NO. A correct answer focuses on the fact that only a small part of the graph is shown. It contains correct arguments in terms of ratio or percentage increase. According to the percentage, the increase is only about 2%.

   Partial Credit: NO. The answer focuses *only* on an increase given by the exact number of robberies. It focuses on the size of increase *without the use of numbers*. It indicates that the graph is misleading, but fails to point out the crucial features.
9. 2
10. A correct response includes the three following steps: 1. Evaporation of water from a source. 2. Transportation of water as vapor/clouds to another place. 3. Precipitation in other places.
11. 1

CHAPTER 2

TRY THIS

Page 19

Examples include: The age of Earth, the origin of life, the cause of disease.

Page 20

1. pitch, loudness, direction to sound, source of sound.
2. Examples include: Observing the surface of the Moon using a telescope, determining changes in air pressure in order to forecast weather.

Page 23

1. O; 2. I; 3. O; 4. I; 5. O; 6. O; 7. O; 8. I; 9. O; 10. I

Page 26

1. Controls include: Add a tray with no coloring, use same color of food coloring in each tray, use same amount of water in each tray, measure temperature at same time each day.
2. Controls include: Measure strength of magnet before being dropped, drop magnet the same way each time, use fresh paper clips for each trial (some may become slightly magnetized), use the same technique for picking up clips each time, increase the number of trials.

Page 27

Answers will vary but should include restatement of hypothesis in an "if, then" format, identification of variables that need to be controlled, recognition that design of the control is flawed, and specification of how "growth" will be measured.

PRACTICE REVIEW QUESTIONS

1. 3	4. 4	7. 2	10. 3	13. 2	16. 4
2. 4	5. 3	8. 3	11. 4	14. 3	17. 2
3. 3	6. 1	9. 4	12. 4	15. 4	

18. Light does not appear to affect the germination of corn.
19. a. If the surface of the tennis court is changed, then the height a tennis ball will bounce changes.
    b. Independent variable: tennis court surface
    Dependent variable: height tennis ball will bounce
    c. Constants: tennis ball, height from which ball is dropped.
    d. Answers may vary. Example: For each surface tested, drop a tennis ball from a height of 1 meter and record the height to which it bounces.

20. a. Independent variable: amount of salt added
    b. Dependent variable: time (until boiling starts)
    c. Constants: volume of water, heating rate, starting temperature of water, characteristics of pots

CHAPTER 3

TRY THIS

Page 36

1. Mechanical model, visual model; scale would determine size and placement of solar system objects.
2. The ratio of surface area to volume of the model and actual lake would be different; thus the model would heat up and cool off faster than the actual lake.
3. Mechanical, toy car; Visual, drawing; Mental, pain of burning your hand; Mathematical, speed = distance/time.

Page 37

Method	Trade-offs
Walking	Time, exposure to weather
Bicycling	Exposure to weather, storage at school
Car	Availability of family member to drive, pollution
Bus	Comfort, speed, more stops

PRACTICE REVIEW QUESTIONS

1. 4 2. 3 3. 2 4. 3 5. 3 6. 4 7. 2

8. a. A correct answer refers to rusting, or corrosion.
   b. Any plausible consequences are correct, for example:
      1. Increased profit (for the painting company or community): It is cheaper for the company; fewer painters are needed; they can paint more bridges
      2. Painters do not need to paint so often or work so hard: they can wait 2 years before starting again; longer vacations for workers; they can have another job in the meantime
      3. Other correct answers: fewer problems with traffic; increased unemployment or lower salary for workers—workers may be laid off until the bridge needs to be painted.
9. a. Machine *B*
   b. Because machine B uses 1 liter of gas per 1,000 liters of water and machine *A* uses 1.25 liters per 1,000 liters of water *or* because it uses less gasoline per liter of water *or* because it can pump the same amount of water using less gas.

CHAPTER 4

TRY THIS

Page 43

An automobile does not grow or reproduce; therefore it does not carry out all of the life functions.

Page 45

1. nucleus, chromosomes, cytoplasm, cell membrane, nucleolus, mitochondria, nuclear membrane
2. In plant cells—cell wall, chloroplasts, chlorophyll, large vacuole; in animal cells—centrioles, small vacuoles
3. Each plant cell is surrounded by a rigid cell wall, which provides stability to the plant cell.

Page 50

Mink and sea otter.

PRACTICE REVIEW QUESTIONS

1. 3	4. 3	7. 3	10. 4	13. 3
2. 3	5. 1	8. 1	11. 2	14. 4
3. 2	6. 2	9. 3	12. 4	15. 2

CHAPTER 5

TRY THIS

Page 55

1. A cell is a single basic unit, tissue is a group of similar cells that work together to perform a life function.
2. Tissues in the hand include: epithelial, bone, muscle, blood, and nerve.

Page 57

1. 2 2. 1
3. Tissue is a group of similar cells; an organ is a group of tissues working together to perform a life function.

Page 60

1. 2
2. Part C is a tendon. Tendons connect muscle to bone, allowing movement of the bone when the muscle contracts.

Page 64

1. The endocrine system sends messages by releasing chemicals directly into the bloodstream where they circulate and stimulate target organs. The nervous system sends messages in the form of impulses (very small electric charges) that travel along the neurons that make up nerves.
2. Refer to Figure 5.6. Acceptable answers include, but are not limited to: The hormone adrenaline is produced by adrenal glands. It controls the release of sugar from liver and the contraction of arteries, and it affects blood pressure. The hormone insulin is produced by the pancreas. It regulates the storage of sugar in the liver and speeds up the oxidation of sugar in cells.

Page 70

1. Exhaled air has more carbon dioxide and less oxygen than inhaled air.
2. Oxygen levels would decrease, and carbon dioxide levels would increase.

Page 73

1. Cooling; excretion of liquid wastes.
2. 1. as feces 2. perspiration and urine 3. exhaled air

PRACTICE REVIEW QUESTIONS

CHAPTER 6

TRY THIS

Page 81

1. To increase blood sugar levels in the blood so muscles have more food available to them to be oxidized to release energy.
2. Proteins contain amino acids essential to the synthesis of new cellular material needed for growth.

Page 82

Any diet that contains the correct number of servings of each food type shown in the MyPlate chart is acceptable.

Page 84

1. Green plants are the underlying food source for all other living things.
2. Carbon dioxide, water, light energy.

Page 85

elephant—herbivore; tiger—carnivore; lynx—carnivore; coyote—omnivore.

PRACTICE REVIEW QUESTIONS

1. 3
2. 1
3. 3
4. 4
5. 2
6. 2
7. 1
8. 4
9. 3
10. 3
11. 4
12. 1
13. 1
14. 1
15. 1
16. 2
17. 4
18. 3
19. 1
20. 1

CHAPTER 7

TRY THIS

Page 93

1. One-celled: daughter cells are two distinctly separate individuals. Multicellular: daughter cells remain part of original organism.
2. Growth, reproduction.

Page 95

1. binary fission *or* asexual reproduction
2. The genetic material of the daughter cell is identical to the genetic material of the parent cell.

Page 97

1. 3
2. cell division
3. 3

Page 100

Arrow *D*

Page 101

Typical answers might include:

Internal Fertilization	External Fertilization
Egg cells remain in reproductive system of female until fertilized by sperm inserted into the female by the male.	Both types of sex cells shed into surrounding water, and sperm swim or are carried by water currents to the egg.
Effective in water and on land.	Limited to aquatic environments.
Used by most land animals.	Used by aquatic invertebrates, most fish, and many amphibians.
Many sperm, few eggs.	Many sperm, many eggs.
Zygote protected by egg or female's body.	Zygote exposed to environment.

Page 105

1.

	T	*t*
T	*TT*	*Tt*
t	*Tt*	*tt*

2. 50%
3. 25% pure tall, 25% pure short
4. *tt*
5. *Tt, TT*

Page 106

1. 4
2. 50%

PRACTICE REVIEW QUESTIONS

1. 3
2. 1
3. 1
4. 3
5. 2
6. 3
7. 2
8. 1
9. 4
10. 2
11. 1
12. 4
13. sexual reproduction
14. 100%
15.

	D	*d*
D	*DD*	*Dd*
d	*Dd*	*dd*

25%

CHAPTER 8

TRY THIS

Page 111

Mutation due to exposure to radiation in space.

Page 112

1. 3
2. 3

Page 113

2. Fossilization

Page 115

3. Natural selection has favored this trait.

PRACTICE REVIEW QUESTIONS

1. 2	4. 2	7. 3	10. 1	13. 3
2. 3	5. 1	8. 4	11. 3	14. 1
3. 1	6. 2	9. 4	12. 3	15. 4

CHAPTER 9

TRY THIS

Page 119

Living: fish, plant, snail. Nonliving: rock, water, gravel, sunlight.

Page 120 (*top*)

Materials include: nutrients, water, oxygen.

Page 120 (*bottom*)

1. decrease 2. increase

Page 122

Lettuce plants: 11; humans: 2

Page 123

1. Ponds undergo succession.
2. Habitats include but are not limited to shallows near shore, pond bottom, deep water, among plants, water surface.
3. To avoid competition; they are better adapted to some habitats.

Page 125

1. The biosphere is that portion of Earth in which living things exist and includes *all* ecosystems. An ecosystem consists of the living and nonliving things *in a particular place* that interact with one another.
2. A population consists of individuals of the same species, a community consists of populations of many species.

Page 126

algae—zooplankton—bluegill—turtle
algae—plant-eating insects—bluegill—human

Page 127

1. Many producers are required to supply the nutrient needs of a single consumer.
2. Producers.
3. Every step along a food chain or food pyramid involves a loss of energy and a decrease in the number of organisms.

Page 128

All of the other organisms in the food web would be affected by the change.

The rattlesnake, fox, and owl populations would probably decrease because less food would be available to them. Rabbits and chipmunks would then probably increase due to less predation by rattlesnakes, foxes, and owls. Lettuce and populations producing seeds might then decrease because there would be more rabbits and chipmunks to eat them. Nut-bearing trees might increase because fewer nuts would be eaten by mice or decrease because fewer nuts would be buried by mice in places where they could germinate. However, the exact effects would be difficult to predict precisely.

Page 129

1. Water, oxygen, and nitrogen are replenished by natural cycles in nature.
2. All living things require energy to carry out life functions. The ultimate source of energy is sunlight that is converted to food by producers.

Page 130

1. *B* and *D*	4. (+, −)
2. *A* and *E*	5. (+, +)
3. *C* and *F*	6. (+, −)

PRACTICE REVIEW QUESTIONS

1. 3	4. 1	7. 1	10. 1	13. 1	15. 4
2. 2	5. 4	8. 3	11. 2	14. 1	16. 1
3. 2	6. 1	9. 1	12. 2		

CHAPTER 10

TRY THIS

Page 135

Closer: house, tree; farther: cloud, sun

Page 136

Earth rotates from west to east at 15° per hour.

Page 137

1. a tree
2. a car in the adjacent lane moving in the same direction and at the same speed
3. a car traveling toward the observer's car from the opposite direction

Page 138

According to the geocentric model, star trails would be the result of the celestial sphere on which stars are located revolving around a stationary Earth.

Page 140

In a star, matter is highly concentrated and hot enough to support fusion. In a nebula, matter is widely dispersed and too cool to support fusion. A galaxy is a cluster of billions of stars.

Page 141

1. Mars
2. Jupiter
3. Earth and Venus; Uranus and Neptune
4. Planets near the Sun are denser than those farther from the Sun.

Pagc 142

1. A meteoroid is a chunk of matter in space. A meteorite is the matter in a meteoroid that survives passage through the atmosphere to strike the ground.
2. The constant stream of particles emitted by the Sun (solar wind) drives gas molecules surrounding the comet away from the Sun.

Page 145

1. Niagara Falls—*A* 2. London—*C* 3. Bermuda—*B* 4. Kiev—*D*

Page 149

1. 3 2. 2

Page 154

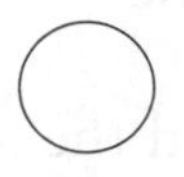

Sun Earth

Page 155

1. and 2.

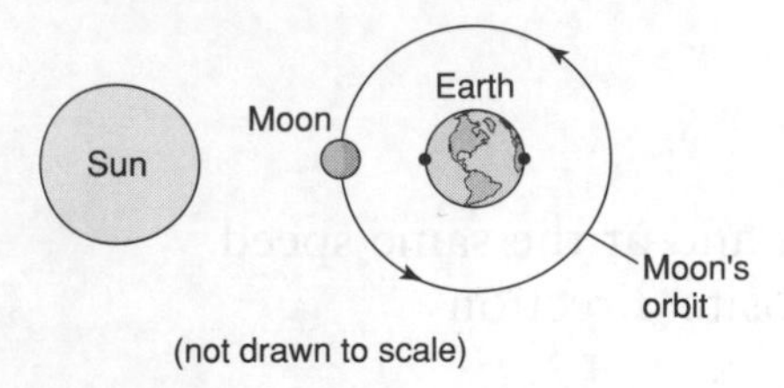

3. Approximately one complete revolution each month.

PRACTICE REVIEW QUESTIONS

1. 4	5. 4	9. 3	13. 1	17. 1	21. 3
2. 4	6. 2	10. 2	14. 1	18. 2	22. 3
3. 1	7. 3	11. 2	15. 3	19. 3	
4. 1	8. 4	12. 4	16. 3	20. 2	

23. and 24.

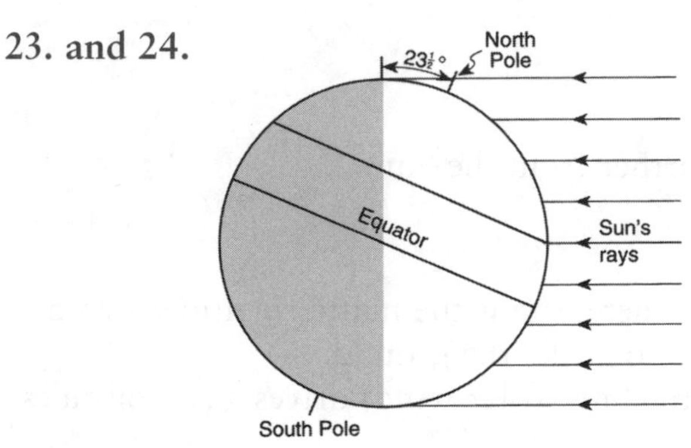

25. June

26.

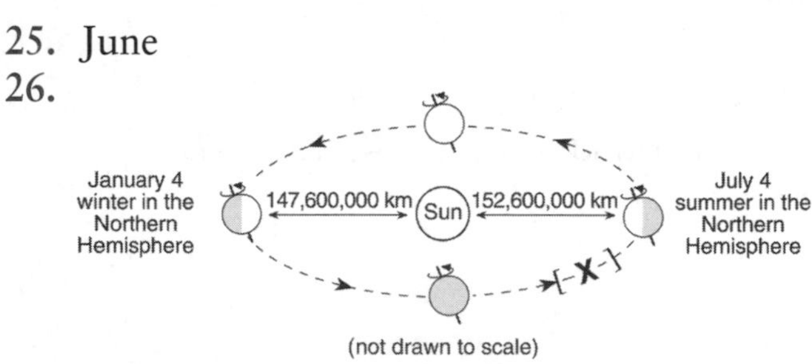

27. ellipse, or slightly elliptical
28. The Sun's apparent size decreases as Earth moves from perihelion (closest to the Sun) to aphelion (farthest from the Sun).
29. When Earth is farthest from the Sun, the Northern Hemisphere is tilted toward the Sun and New York State receives the most direct (most intense) sunlight of the year and also has the greatest number of hours of sunlight each day

CHAPTER 11

TRY THIS

Page 163

1. topaz
2. The same mineral may occur in several colors, and different minerals may have the same color.

Page 165

1. Magma is molten rock beneath Earth's surface, lava is molten rock that has escaped onto Earth's surface.
2. Igneous intrusions form underground where the surrounding rock traps heat from the magma so that the rock cools slowly, resulting in large crystals.

Page 168

sedimentary rock → heat and/or pressure → metamorphic rock → melting → magma → solidification → igneous rock
sedimentary rock → melting → magma → solidification → igneous rock

Page 169

The written piece should include references to evaporating, being transported by winds, condensing to form clouds, falling to Earth as precipitation, and runoff back to the ocean.

Page 173

1. Answers include: in solution; in suspension; by rolling, bouncing, or sliding along the streambed.
2. Abrasion causes the sediment particles to become smaller and rounder.

Page 174

When a glacier advances, ice is forming at the glacier's source faster than it is melting at the glacier's end. When a glacier retreats, its end is melting faster than the ice is flowing outward from its source.

Page 175

Both have similar triangular, fanlike shapes, and both form when sediment-laden water slows down.

Page 177

1. The focus of an earthquake is the place where rock first breaks or moves. The epicenter is the place on Earth's surface directly above the focus—the first place on the surface to be struck by seismic waves leaving the focus.
2. P-waves vibrate in line with the direction of wave motion. S-waves vibrate at right angles to the direction of wave motion. P-waves travel at a faster speed than S-waves.

Page 179

1. ocean floors—3.0; continents—2.7.
2. Density increases with depth.

Page 182

1. 1. continental United States—North American Plate
   2. India—Indian-Australian Plate
   3. Hawaii—Pacific Plate
   4. Greenland—North American Plate.
2. New crust is forming along the midocean ridges that form the north and west edges of the Nazca Plate. Crust is being destroyed in the subduction zone along its eastern edge.

PRACTICE REVIEW QUESTIONS

1. 3
2. 3
3. 2
4. 3
5. 3
6. 2
7. 1
8. 2
9. 2
10. 2
11. 3
12. 3
13. 3
14. 3
15. 4
16. 2
17. 1
18. 2
19. 4
20. 2
21. 3
22. 1

23. Continental crust is thicker and less dense than oceanic crust.
24. size, shape, density
25. Glossopteris
26. matching coastlines, matching fossils in rocks along corresponding coastlines
27. rock layer *D*

CHAPTER 12

TRY THIS

Page 191

During January, sunlight strikes the ground in New York State indirectly and therefore has little warming effect.

Page 192

The warm, moist air carried over India during the summer results in lower air pressures, while the cool, dry air carried over India during the winter results in higher air pressures.

Page 193

As the air temperature approaches the dew point temperature, the probability of rain increases.

Page 195

At 30°N and 30°S, sinking air heats due to compression, causing these regions to have a hot climate. Hot air increases evaporation, so these regions also have dry climates.

Page 196

If Earth's direction of rotation were reversed, all of the global wind belts would curve in the opposite direction due to a reversal of the Coriolis effect.

Page 197

1. The number 65 should be written to the upper left and the number 53 to the lower left of the station model.
2. Partly cloudy or 25% cloud cover.

Page 201

1. Compared with a continental polar air mass, a maritime tropical air mass would have higher temperatures, lower air pressures, and higher humidity.
2. The North Atlantic

Page 203

1. 1   2. 2

Page 206

Solar energy absorbed by evaporating water released when the water condenses.

Page 207

1. A tornado watch means that a tornado is likely to form in your area, and you should stay alert and be prepared to take shelter. A tornado warning means that a tornado has been sighted, and you should take immediate cover.
2. When a warning has been issued, you should immediately take shelter.

PRACTICE REVIEW QUESTIONS

1. 3	5. 4	9. 2	13. 3	17. 3
2. 1	6. 2	10. 1	14. 3	18. 2
3. 2	7. 4	11. 1	15. 3	19. 2
4. 2	8. 1	12. 3	16. 2	20. 1

21.

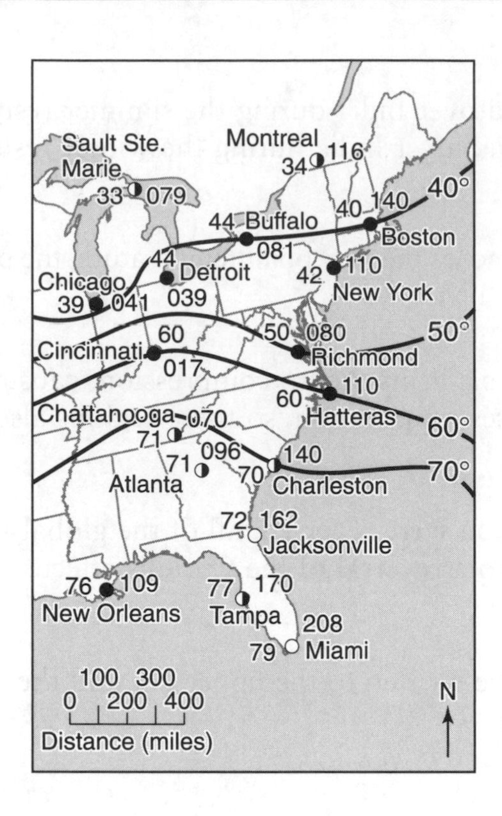

22. Temperatures generally decrease as latitude increases.
23. 1,020.8 mb
24. northwest
25. The hurricane lost its source of energy as it moved over land.
26. flooding, wind damage
27. Some preparations include: identify an evacuation route, obtain emergency supplies, assess your property to ensure that landscaping and trees do not become a wind hazard, protect all windows by installing commercial shutters, properly brace roof and secure to framing.

CHAPTER 13

TRY THIS

Page 216

1. 4
2. mass = 454 grams

Page 218

1. 3
2. 2

Page 219

1. A compound is a substance consisting of two or more elements chemically combined in a fixed ratio. A mixture is made of two or more substances that are not chemically combined and are capable of being separated by physical means.
2. Both compounds and mixtures consist of more than one element.

Page 220

Phases of Matter

Property	Solid	Liquid	Gas
Takes up space	✓	✓	✓
Has mass	✓	✓	✓
Takes the shape of its container		✓	✓
Fills container			✓

Page 221

1. Answers include: solids—wood, plastic, brick, glass, paper; liquids—water, oil, syrup, alcohol, milk; gases—methane (natural gas), oxygen.
2. As the temperature of a gas is increased, its molecules move faster and strike the walls of the container with greater force.

Page 223

1. The ice cube can absorb latent heat from the surrounding drink while melting without increasing in temperature; water increases in temperature as soon as it absorbs heat from the drink.
2. Heat added to boiling water is used to change phase and does not increase the temperature of the water until all of it has vaporized. Therefore, turning up the heat will not increase the temperature of the water (though it will increase the rate of boiling), and thus the eggs will not cook faster.

Page 227

1. 75°C 2. Any two of the following: hydrochloric acid (HCl), ammonia (NH_3), sulfur dioxide (SO_2)

Page 229

1. 2 2. 3

Page 231

In one test tube but not the other, one of the following occurs: a large amount of heat or light is given off; odors and/or colors change; bubbles of gas form; a clear solution turns cloudy and an insoluble solid settles out of the solution.

Page 233

300 grams; conservation of mass

Page 238

ammonia—nitrogen (N) and hydrogen (H)
baking soda—sodium (Na), hydrogen (H), carbon (C), and oxygen (O)
hydrogen peroxide—hydrogen (H) and oxygen (O)
natural gas—carbon (C) and hydrogen (H)
sugar—carbon (C), hydrogen (H), and oxygen (O)
table salt—sodium (Na) and chlorine (Cl)

PRACTICE REVIEW QUESTIONS

1. 3
2. 3
3. 4
4. 1
5. 1
6. 2
7. 1
8. 3
9. 3
10. 3
11. 2
12. 1
13. 3
14. 2
15. 1
16. 4
17. 4
18. 2
19. 3
20. 2
21. 2
22. 4
23. 2

24. W: iron filings
    X: cork
    Y: sand
    Z: salt

25. a.

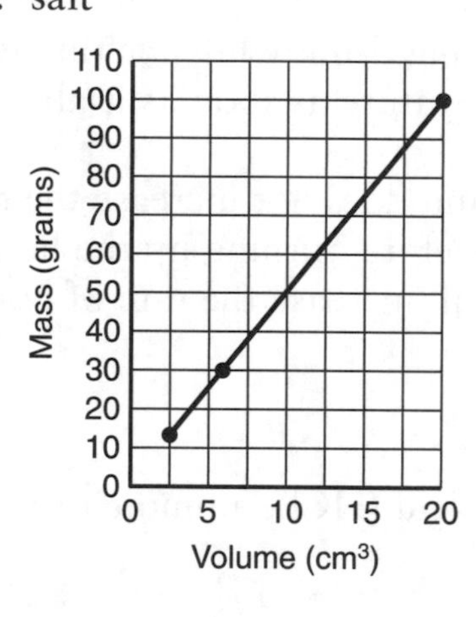

b. 50 g

26. Phases of Matter

Property			
Has a definite shape	✔		
Has a definite volume	✔	✔	
Takes the shape of its container		✔	✔

27. Compounds, such as water, have physical and chemical properties that are different from those of the elements of which they are composed.

CHAPTER 14

TRY THIS

Page 247

1. A force must be exerted onto an object, and the object must move in the direction of the applied force.
2. Since the building does not move, no work is done.

Page 249

Figures *a*, *c*, and *e* have potential energy; figures *b*, *d*, and *f* have kinetic energy.

Page 251

Device	Starting Form	Transformed To
Fan	Electrical	Mechanical, heat, sound
Guitar	Mechanical	Sound, heat
Truck	Chemical	Mechanical, heat, sound
Candle	Chemical	Light, heat

Page 252

The mechanical energy transferred to the nail when it is struck by the hammer causes its molecules to move faster and thus increases its temperature.

Page 254

1. 1 2. 4 3. 2

Page 255

1. 3 2. 85°F = about 30°C

Page 256

When sweat evaporates, it absorbs latent heat of vaporization. Heat absorbed from the skin lowers its temperature.

Page 257

1. Metal is a better conductor of heat than wood. Therefore, when you touch metal, heat is conducted away from your warm skin faster than when you touch wood.
2. Solar energy would not be able to reach Earth. Earth's energy of formation and any energy that reached Earth (e.g., kinetic energy of objects hitting Earth) would be

unable to escape. Earth would probably still be in a molten state. Earth would probably be devoid of life as we know it.

3. Air is a poor conductor of heat. Therefore if you place your finger next to a flame, little heat will reach it by conduction. However, heat from the flame would be carried upward by convection and burn your finger.

Page 259

1. 2 2. 2

Page 262

1. 3
2. Similar to Figure 14.9*b*, *c* but with one more lamp. If circuit drawn is similar to *b*, all three will go out. If circuit drawn is similar to *c*, two will remain lit.

Page 264

1. See the diagram of domains in a weak and a strong magnet in Figure 14.11.
2. Mars would have to be surrounded by a magnetic field.

Page 266

1. transverse 2. straight 3. reflected
4. microwaves, orange visible, blue visible, ultraviolet rays, gamma rays

Page 269

1. opaque 2. green 3. refracted
4. *A*—incident light ray, *B*—reflected light ray, *C*—normal, *D*—angle of incidence, *E*—angle of reflection.

Page 271

1. 3 2. 1

Page 273

1. $106
2. $40

PRACTICE REVIEW QUESTIONS

1. 1	6. 2	11. 2	16. 2	21. 3
2. 3	7. 1	12. 2	17. 3	22. 4
3. 3	8. 2	13. 4	18. 1	23. 4
4. 1	9. 2	14. 1	19. 3	24. 3
5. 2	10. 3	15. 2	20. 3	25. 1

26. 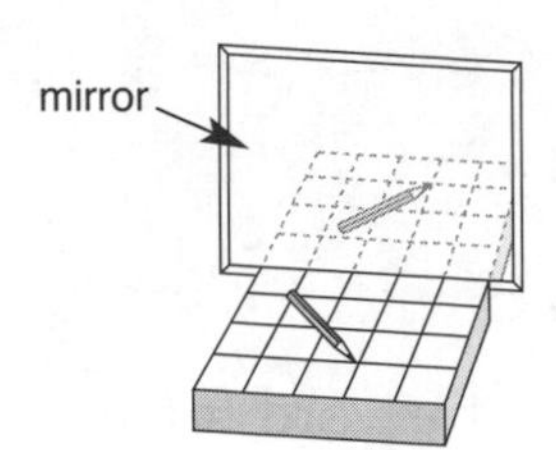

27. a. 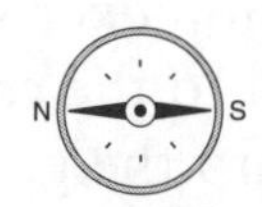

b. A correct response explains that opposite poles attract (N toward S, for example) or like poles repel.

28. Thunder is caused by lightning. The sound is made as the air around the lightning bolt rapidly heats and cools causing a shock wave known as thunder. Since light travels faster than sound, light from the lightning reaches Mary before the sound of the thunder.

29. Energy Source: Correct answers include, but are not limited to: Sun or sunlight. Use: Heat water by solar panels, converted to electricity by solar cells; wind (or windmills). Use: grinding corn, pumping water, generating electricity; water (waves, tides, water wheels, etc.). Use: To generate electricity, tidal barrage. (Note: Fossil fuels (gas, coal, oil) are not renewable and are therefore incorrect.)

CHAPTER 15

TRY THIS

Page 282

1. 500 km/hr, south   2. 300 km/hr, north

Page 284 (*top*)

For the first 5 hours, the velocity of the car was 10 km/hr to the right. For the second 5 hours, it was 0 km/hr.

Page 284 (*bottom*)

The object is decelerating.

Page 287

3

Page 289

1. 2   2. 2   3. 2

Page 292

5

Page 293

1. You exert a backward force onto the raft, and the raft exerts a forward force onto you.
2. After the gunpowder explodes, expanding gases exert a backward force against the cannon and an identical forward force onto the cannonball. The cannon exerts an equal and opposite forward force against the gases that is transmitted to the cannonball. The cannonball exerts an equal and opposite backward force onto the gases that is transmitted to the cannon. Since the cannonball has less mass than the cannon, the cannonball undergoes a larger forward acceleration, and the cannon undergoes a smaller backward acceleration.

Page 297

Ramp 2

Page 299

1. 

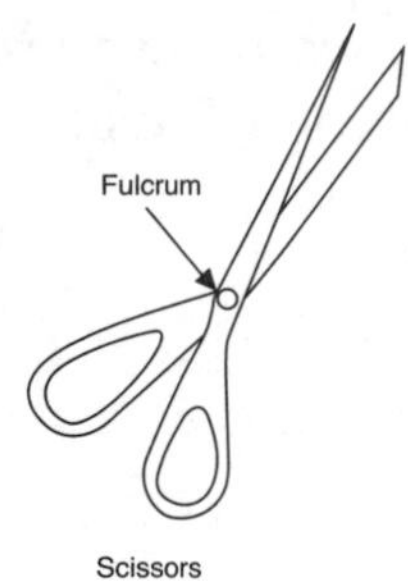

2. Scissors are double levers. The length of the handle is the effort distance, and the length of the blade is the output distance. The force exerted by the hand is the effort force, and the force exerted by the blade is the output force. Cutting metal requires a much larger output force than cutting paper. Since effort force × effort distance = output force × output distance, increasing the effort distance and decreasing the output distance will result in a larger output force for a given effort force.

Page 300 (*top*)

1. 4

Page 300 (*bottom*)

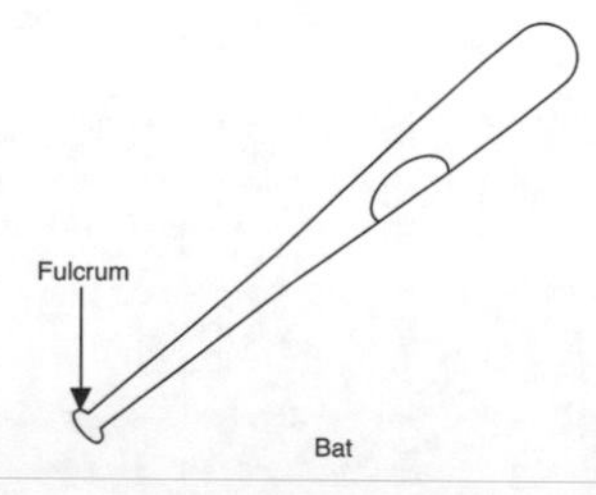

Page 301

1. 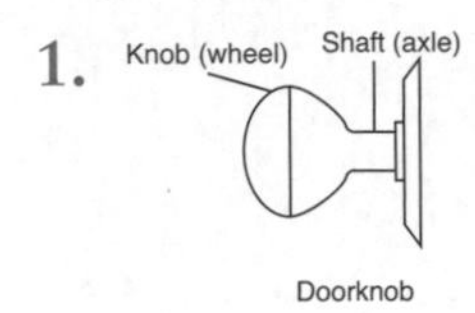

2. Increase the diameter of the knob (wheel), and decrease the diameter of the shaft (axle).

Page 303

Answers will vary, but may include examples such as
Lift heavy shingles to the roof.
Raise prefabricated walls.
Hoist bricks or mortar to workers.

PRACTICE REVIEW QUESTIONS

1. 4
2. 4
3. 1
4. 3
5. 2
6. 3
7. 3
8. 3
9. 3
10. 1
11. 4
12. 3
13. 3
14. 1
15. 1
16. 4
17. 4
18. 1
19. 3

20. a. At low mass, it increased 2 cm for every 10 g. Then it changed and only increased by 1 cm at 40 g. Then, from 50 g on, it did not increase anymore.
    b. 25g
    c. 13 cm
21. a. 100 N
    b. 4
22. a. –8 km/hr/s
    b. 5 km/hr/s
23. 1. constant speed
    2. acceleration
    3. acceleration
    4. deceleration

Model Examination and Answers

NEW YORK STATE
INTERMEDIATE-LEVEL SCIENCE TEST
PART I: SAMPLE QUESTIONS 1–45

Directions (1–45): Each question is followed by four choices numbered 1 through 4. Read each question carefully. Decide which choice is the best answer. Mark your answer in the space provided by writing the number of the answer you have chosen.

1.* A son can inherit traits

1. only from his father
2. only from his mother
3. from both his father and his mother
4. from either his father or his mother, but not from both **1.** ______

2.* A person sorted some animals into the two groups listed in the table.

Group 1	Group 2
Humans	Snakes
Dogs	Worms
Flies	Fish

Which characteristic of animals was used for the sorting?

1. legs
2. eyes
3. nervous system
4. skin **2.** ______

3.* Fossil fuels were formed from

1. volcanoes
2. remains of living things
3. gases in the atmosphere
4. water trapped inside rocks **3.** ______

4.* One of the main causes of acid rain is

1. waste from nuclear power plants
2. spills from chemical manufacturing plants
3. gases from burning fossil fuels
4. gases from aerosol spray cans **4.** ______

5.* When chlorine gas reacts with sodium metal, what type of substance is formed?

1. an element
2. a compound
3. a mixture
4. a solution **5.** ______

* Reproduced from IEA TIMSS 2003 Science Released Items for Grade 8, TIMSS International Study Center, Boston College, MA

6.* A girl has an idea that green plants need sand in the soil for healthy growth. In order to test her idea she uses two pots of plants. She sets up one pot of plants as shown below.

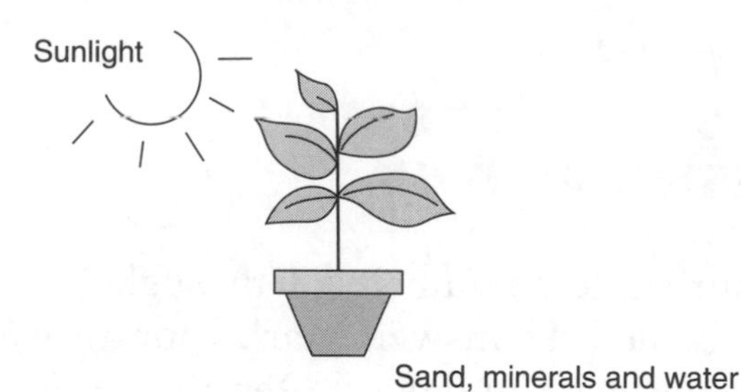

Which ONE of the following should she use for the second pot of plants?

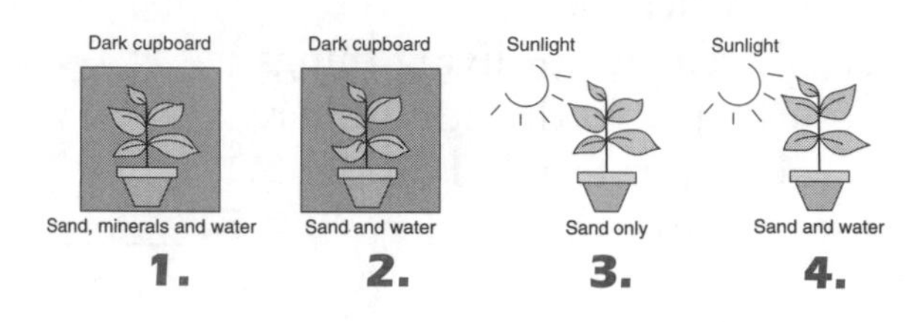

6. ______

7.* A powder made up of both white specks and black specks is likely to be

1. a solution
2. a pure compound
3. a mixture
4. an element **7.** ______

8.* Which of the following organs in fish has the same function as the human lung?

1. kidney
2. heart
3. gill
4. skin **8.** ______

* Reproduced from IEA TIMSS 2003 Science Released Items for Grade 8, TIMSS International Study Center, Boston College, MA

9.* The picture shows how a student set up some apparatus in a laboratory for an investigation. The inverted test tube was completely filled with water at the beginning of the investigation as shown in Figure 1. After several hours, the level of water in the test tube had gone down as shown in Figure 2.

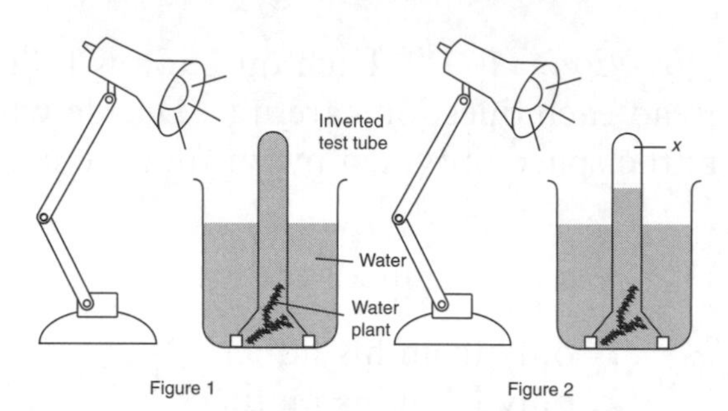

What is contained in the top part of the test tube labeled *X* in Figure 2?

1. air
2. oxygen
3. carbon dioxide
4. vacuum **9.** ______

10.* Which of the following takes place during fertilization in animals?

1. production of sperm and egg
2. joining of sperm and egg
3. division of egg
4. development of embryo **10.** ______

11.* In humans, where does the absorption of food into the blood stream mainly take place?

1. stomach
2. mouth
3. large intestines
4. small intestines **11.** ______

12.* Animals and plants are made up of a number of different chemical elements. What happens to all of these elements when animals and plants die?

1. They die with the animal or plant.
2. They evaporate into the atmosphere.
3. They are recycled back into the environment.
4. They change into different elements. **12.** ______

13. What do all living things have in common?

1. All produce most of Earth's oxygen.
2. All are made of one or more cells.
3. All can make food by photosynthesis.
4. All cause various types of human diseases. **13.** ______

14. During a race a runner's body temperature increases. The runner responds by sweating, which lowers body temperature. This process is an example of

1. nutrition
2. growth
3. homeostasis
4. classification **14.** ______

15. The scientific name of the common housefly is *Musca domestica*. This name indicates the housefly's

1. kingdom and phylum
2. genus and species
3. phylum and family
4. class and order **15.** ______

16. You are examining a small letter "e" printed on paper. Which of the following correctly shows how the "e" will appear when viewed through a microscope?

1. e

2.

3. ə

4. **16.** ______

17. The life process being carried out by the amoeba in the diagram below is

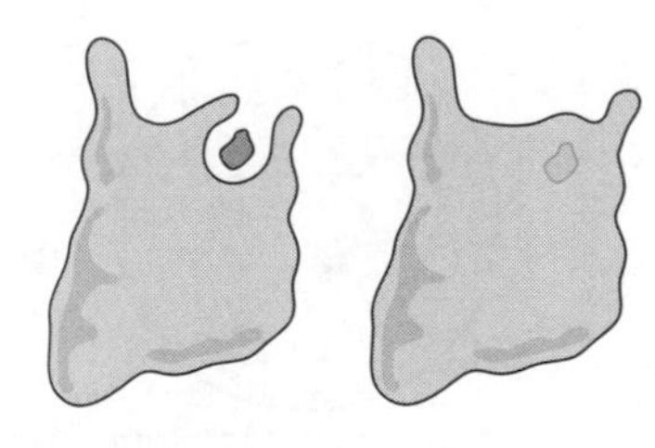

1. obtaining food
2. getting rid of wastes
3. reproducing
4. inheritance **17.** ______

* Reproduced from IEA TIMSS 2003 Science Released Items for Grade 8, TIMSS International Study Center, Boston College, MA

18. An organism was added to a test tube containing water, which was then sealed and placed in sunlight. The graph below shows how the oxygen content of the test tube changed over a period of time.

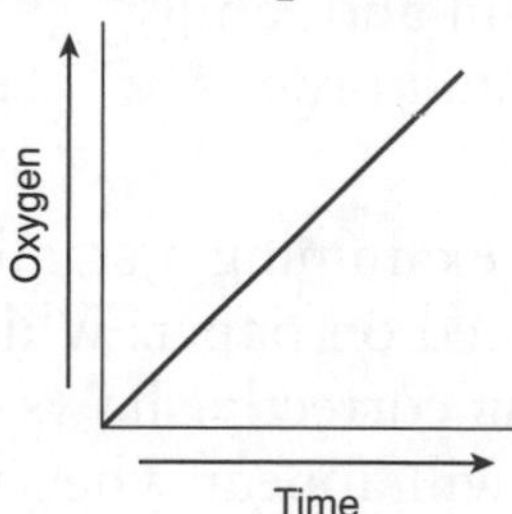

Which type of organism was probably added to the test tube?

1. amoeba
2. green algae
3. fish
4. virus **18.** ______

19. A student looks at a transparent ruler under the microscope and finds that the diameter of the field of view is 1 millimeter. The student then places a slide containing scallion skin stained for viewing under the microscope and sees the following:

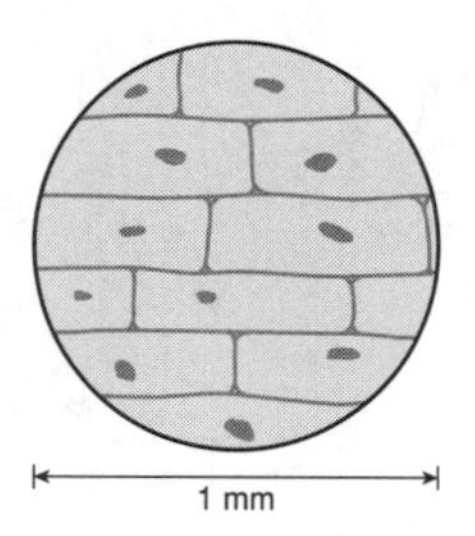

The length of one cell is most nearly

1. ½ millimeter
2. 1 millimeter
3. 2 millimeters
4. 3 millimeters **19.** ______

* Reproduced from IEA TIMSS 2003 Science Released Items for Grade 8, TIMSS International Study Center, Boston College, MA

20.* The table below shows the temperature and precipitation (rain or snow) in four different towns on the same day.

	Town *A*	Town *B*	Town *C*	Town *D*
Lowest temperature	13°C	–9°C	22°C	–12°C
Highest temperature	25°C	–1°C	30°C	–4°C
Precipitation (rain or snow)	0 cm	5 cm	2.5 cm	0 cm

Where did it snow?

1. town *A*
2. town *B*
3. town C
4. town *D* **20.** ______

21.* A butterfly sitting on a leaf laid some small eggs. The pictures below show the changes that took place in the eggs.

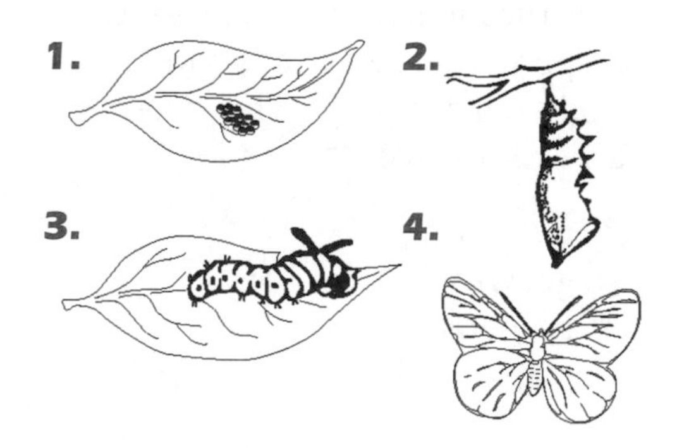

In what order did the changes take place?

1. 1, 2, 3, 4
2. 1, 3, 2, 4
3. 1, 4, 3, 2
4. 1, 4, 2, 3 **21.** ______

22.* Why would male insects be treated to prevent sperm production?

1. to increase the number of female insects.
2. to reduce the total population of insects.
3. to produce new species of insects.
4. to prevent insects from mating. **22.** ______

23.* Cats are most closely related to which of the following animals?

1. crocodiles
2. whales
3. frogs
4. penguins **23.** ______

24. The equations below represent the processes of photosynthesis and respiration.

Photosynthesis

Light energy + carbon dioxide + water $\xrightarrow{\text{(chlorophyll)}}$ food + oxygen

Respiration

Food + oxygen $\longrightarrow$ energy + carbon dioxide + water

Which statement best compares these processes?

1. Both processes release energy.
2. Neither process releases energy.
3. Both processes use light energy.
4. One process uses energy and one releases energy.

24. ______

25. The diagrams below show onion root tip cells at various stages of mitosis.

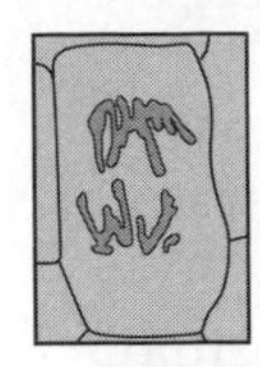

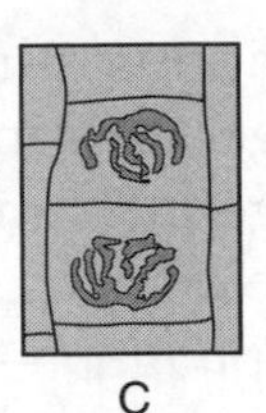
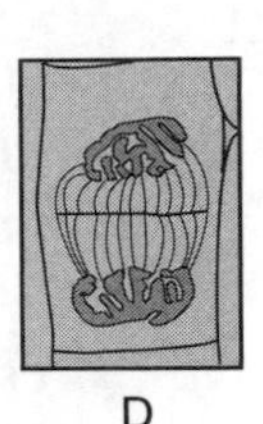

A B C D

Which of the following lists the pictures in the correct chronological order:

1. A, B, C, D
2. B, D, C, A
3. C, D, B, A
4. B, A, D, C **25.** ______

26. Which structures in the diagram below enable a person observing this cell to identify it as a plant cell?

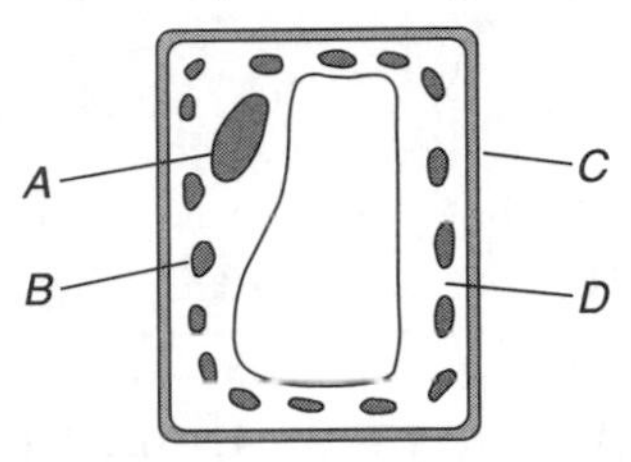

1. *A* and *B*
2. *B* and *C*
3. *A* and *C*
4. *B* and *D* **26.** ______

27. The cardiovascular system consists of

1. heart, blood vessels, and blood
2. atria, ventricles, and valves
3. lungs, body cells, and digestive organs
4. plasma, red blood cells, and white blood cells

27. ______

* Reproduced from IEA TIMSS 2003 Science Released Items for Grade 8, TIMSS International Study Center, Boston College, MA

28. Which list shows the order in which food passes through organs of the human digestive system?

1. small intestine → stomach → large intestine → esophagus
2. large intestine → small intestine → esophagus → stomach
3. esophagus → stomach → small intestine → large intestine
4. stomach → small intestine → esophagus → large intestine

28. ______

29.* The pictures show two different mountains. The mountains in picture *A* are rough and jagged. The mountains in picture *B* are smooth and rounded.

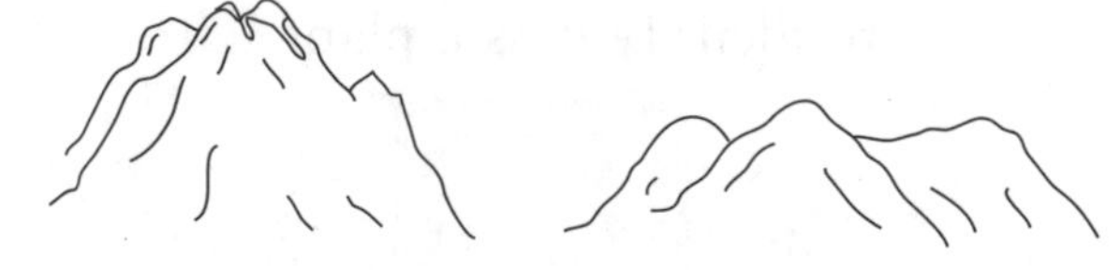

Which statement about these mountains is probably true?

1. The mountains in Picture *A* are older.
2. The mountains in Picture *B* are older.
3. The mountains are about the same age but were formed in different ways.
4. The mountains are about the same age but are in different hemispheres.

29. ______

* Reproduced from IEA TIMSS 2003 Science Released Items for Grade 8, TIMSS International Study Center, Boston College, MA

30.* The shape of the Moon appears to change regularly during each month. Which of the following best explains why the shape of the Moon appears to change?

1. The Earth turns on its axis.
2. The Moon turns on its axis.
3. The Moon orbits around the Earth.
4. Clouds cover the Moon.

30. ______

31.* The diagram below shows a map of the world with lines of latitude marked. Which of the following places marked on the map is most likely to have an average yearly temperature similar to location **X**?

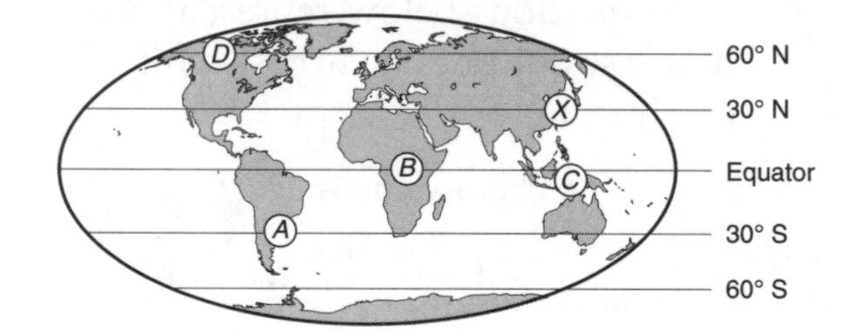

1. *A*
2. *B*
3. *C*
4. *D*

31. ______

32.* The diagrams show nine different trials Usman carried out using carts with wheels of two different sizes and different numbers of blocks of equal mass. He used the same ramp for all trials, starting the carts from different heights.

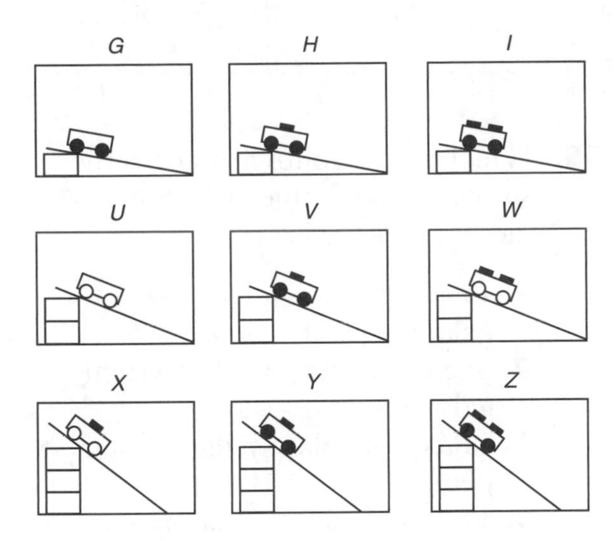

He wants to test this idea: The higher the ramp is placed, the faster the cart will travel at the bottom of the ramp. Which three trials should he compare?

1. *G*, *H*, and *I*
2. *I*, *W*, and *Z*
3. *U*, *W*, and *X*
4. *H*, *V*, and *Y*

32. ______

33.* The diagram below shows the Pacific Ring of Fire. Earthquakes and volcanic activity occur along the Ring of Fire.

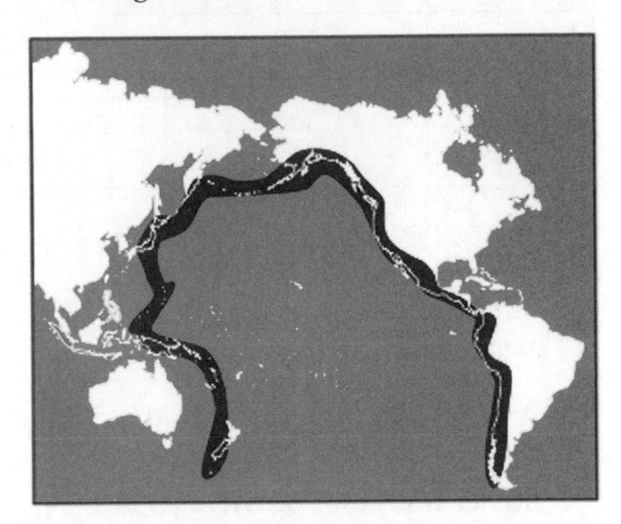

Which of the following best explains why?

1. It is located at the boundaries of tectonic plates.
2. It is located at the boundary of deep and shallow water.
3. It is located where the major ocean currents meet.
4. It is located where ocean temperature is the highest.

33. ______

* Reproduced from IEA TIMSS 2003 Science Released Items for Grade 8, TIMSS International Study Center, Boston College, MA

	Average Surface Temperature (°C)	Atmospheric Composition	Mean Distance from the Sun (millions of km)	Time to Revolve Around the Sun (number of days)
Venus	470	Mostly carbon dioxide	108	225
Mercury	300	Trace amounts of gases	58	88

34.* The table above shows some information about the planets Venus and Mercury.

Which of the following explains why the surface temperature of Venus is higher than that of Mercury?

1. There is less absorption of sunlight on Mercury because of the lack of atmospheric gases.
2. The high percentage of carbon dioxide in the atmosphere of Venus causes a greenhouse effect.
3. The longer time required for Venus to revolve around the Sun allows it to absorb more heat from the Sun.
4. The Sun's rays are less direct on Mercury because it is closer to the Sun.

34. ______

35. Which of the following correctly shows celestial objects in order of increasing size?

1. solar system → planet → galaxy → star
2. planet → star → solar system → galaxy
3. galaxy → solar system → star → planet
4. star → planet → solar system → galaxy

35. ______

36.* A small, fast-moving river is in a V-shaped valley on the slope of a mountain. If you follow the river to where it passes through a plain, what will the river most likely look like compared with how it looked on the mountain?

1. much the same
2. deeper and faster
3. slower and wider
4. straighter

36. ______

* Reproduced from IEA TIMSS 2003 Science Released Items for Grade 8, TIMSS International Study Center, Boston College, MA

37.* Ken put a thermometer in a glass filled with hot water. Why does the liquid inside the thermometer rise?

1. Gravity pushes it up.
2. Air bubbles are released.
3. Heat from the water makes it expand.
4. Air pressure above the water pulls it up.

37. ______

38.* Which of the boxes *X*, *Y*, or *Z* has the LEAST mass?

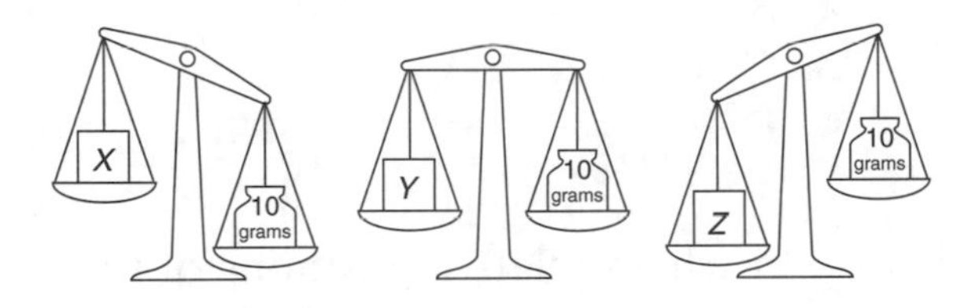

1. *X*
2. *Y*
3. *Z*
4. All three boxes have the same mass.

38. ______

39.* A ray of light strikes a mirror as shown.

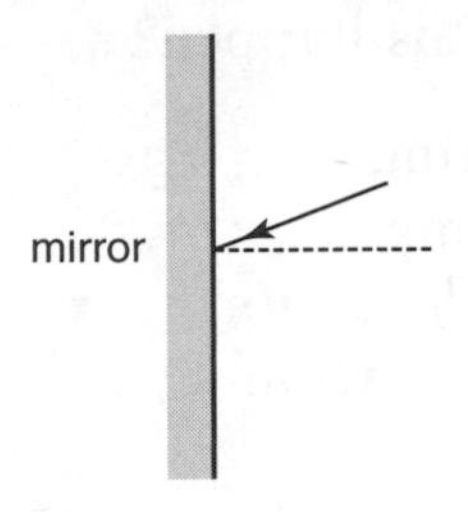

Which picture best shows the direction of the reflected light?

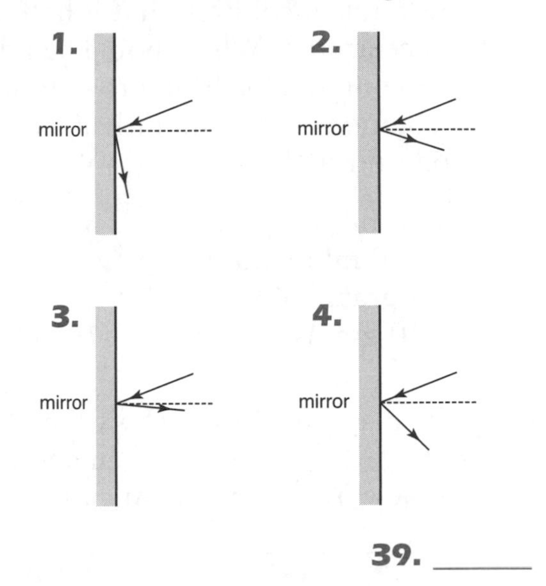

39. ______

40.* A balloon filled with helium gas is set free and starts to move upward. Which of the following best explains why the helium balloon moves upward?

1. The density of helium is less than the density of air.
2. The air resistance lifts the balloon up.
3. There is no gravity acting on helium balloons.
4. The wind blows the balloons upward.

40. ______

* Reproduced from IEA TIMSS 2003 Science Released Items for Grade 8, TIMSS International Study Center, Boston College, MA

41.* A wet towel will dry when it is left in the Sun. Which process occurs to make this happen?

1. melting
2. boiling
3. condensation
4. evaporation **41.** ______

42.* David makes a solution by dissolving 10 grams of salt in 100 ml of water. He wants a solution that is half as concentrated. What should he add to the original solution to obtain a solution that is about half as concentrated?

1. 50 ml of water
2. 100 ml of water
3. 5 grams of salt
4. 10 grams of salt **42.** ______

43. The diagrams below show the phases of the Moon as seen by an observer in New York State in August.

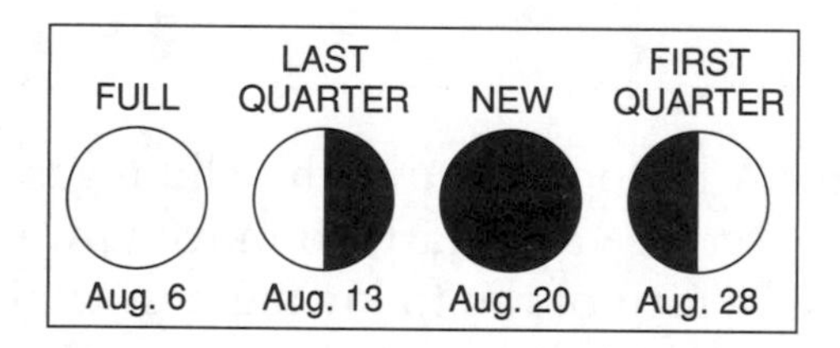

Which phase could have been observed on August 17?

43. ______

44. The diagram below shows several positions of Earth as it moves around the Sun.

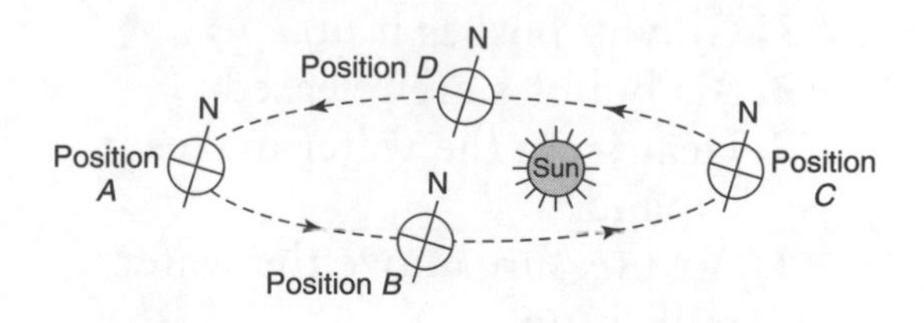

Which position shows Earth during summer in the Northern Hemisphere?

1. *A*
2. *B*
3. *C*
4. *D* **44.** ______

45.§§ In which relative positions of the Sun (S), Earth (E), and Moon (M) is a lunar eclipse most likely to happen?

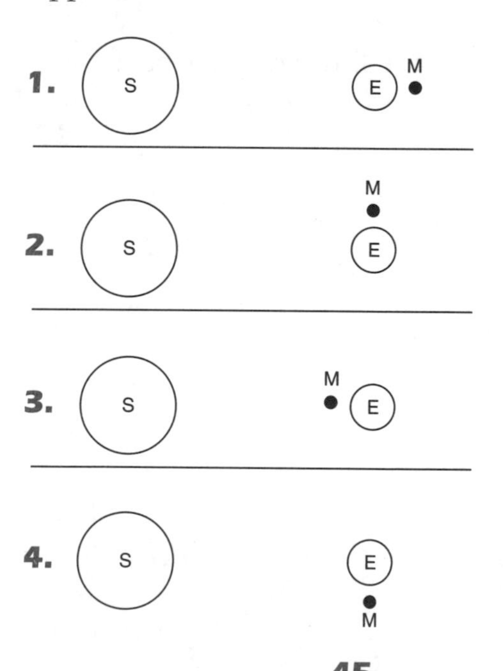

45. ______

* Reproduced from IEA TIMSS 2003 Science Released Items for Grade 8, TIMSS International Study Center, Boston College, MA

§§ Source: National Center for Education Statistics, National Assessment of Educational Progress (NAEP), 1996, The Nation's Report Card: Science 1996

NEW YORK STATE INTERMEDIATE-LEVEL SCIENCE TEST PART II: SAMPLE QUESTIONS 46–75

Directions (46–75): Record your answers in the spaces provided.

Base your answers to questions 46–49 on the pond ecosystem illustrated below. The members of this ecosystem interact with one another and with the nonliving environment.

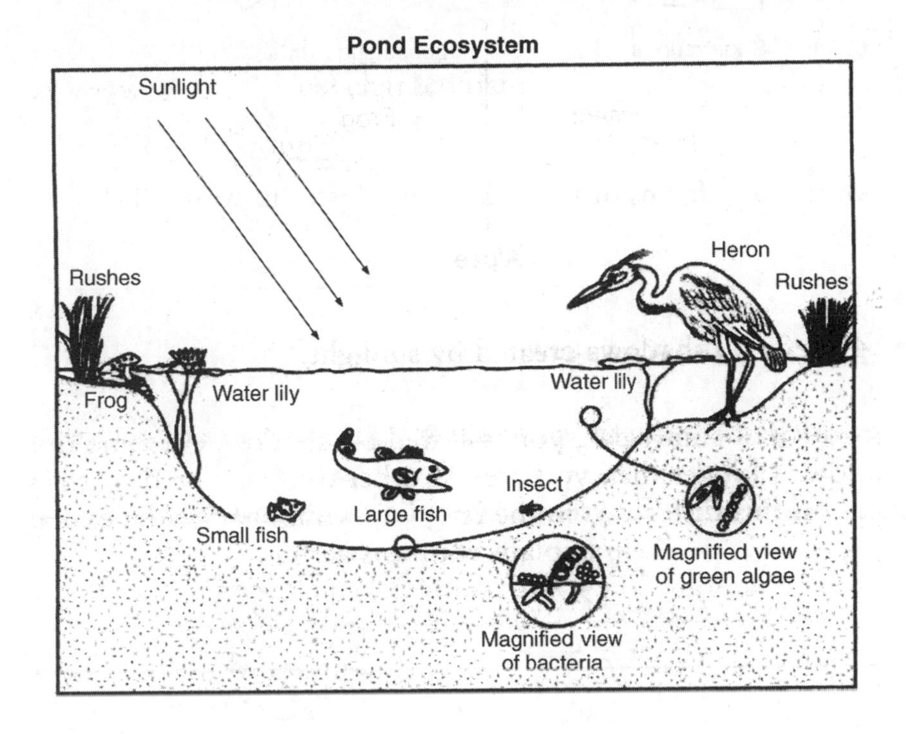

46. What is the main source of energy for this ecosystem? [1]

____

47. Identify one primary consumer in this ecosystem. [1]

____

48. Identify one secondary consumer in this ecosystem. [1]

____

49. Identify one organism in this ecosystem that produces chlorophyll. [1]

____

50.* The picture below shows two forms of sugar—solid cubes and packets of loose crystals. One cube has the same mass of sugar as one packet.

a. Which of the two forms of sugar dissolves faster in water? [1]

____

b. State one reason for your answer. [1]

____

51.* The pictures below show different stages in a frog's growth.

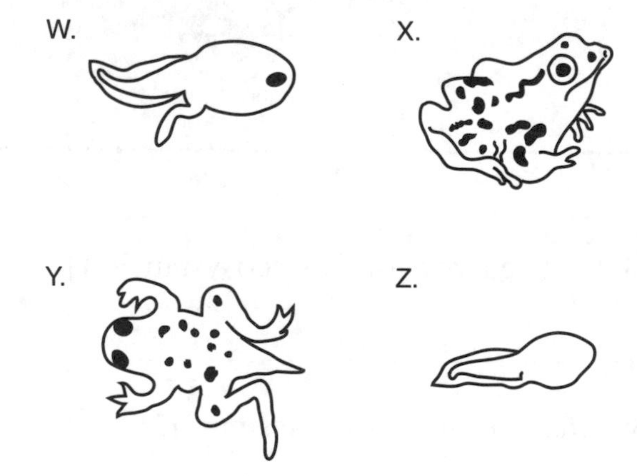

Write the letters in the boxes below to show the pictures in order of the frog's growth. [1]

Young ⟶ Old

* Reproduced from IEA TIMSS 2003 Science Released Items for Grade 8, TIMSS International Study Center, Boston College, MA

Base your answers to questions 52–54 on the diagram below, which shows an incomplete food web.

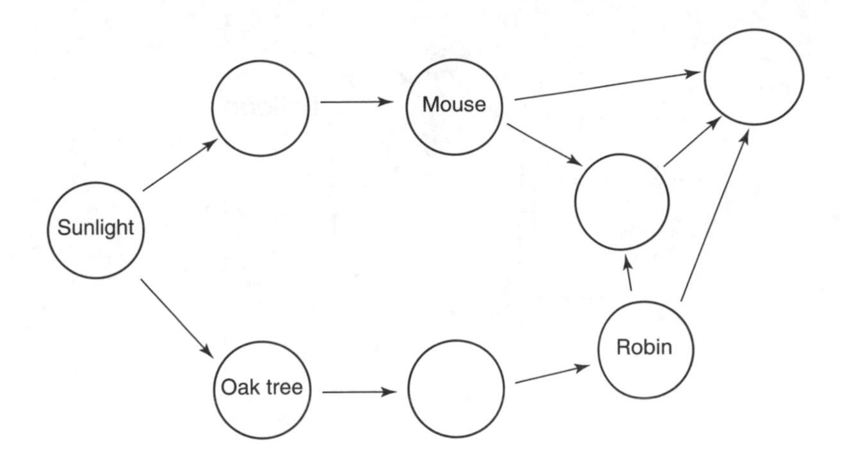

1. Caterpillar
2. Corn
3. Hawk
4. Snake

52.* Complete the food web by filling in each of the empty circles with the number of the correct animal or plant from the list. Remember that the arrows represent energy flow and go from the provider to the user. [2]

53. Give an example of a carnivore, a herbivore, and a producer in this food web. [1]

Carnivore: ______________________________

Herbivore: ______________________________

Producer: ______________________________

54.* If the corn crop failed one year, what would most likely happen to the robin population? [1]

____

____

* Reproduced from IEA TIMSS 2003 Science Released Items for Grade 8, TIMSS International Study Center, Boston College, MA

55.* The water in the tube is heated, as shown in the diagram. As the water is heated, the balloon increases in size. Explain why. [1]

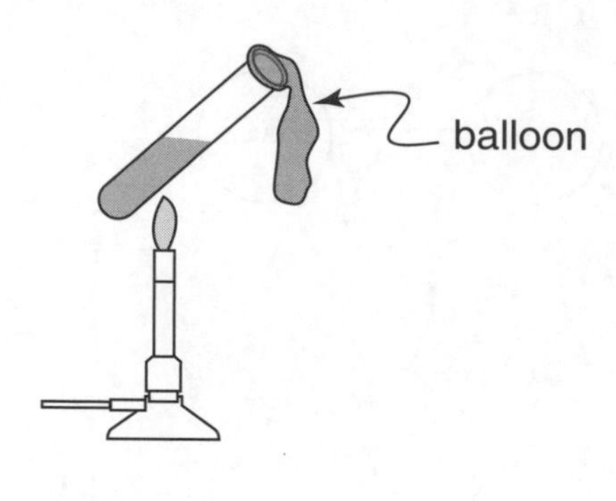

____

____

Base your answers to questions 56–59 on the diagram and information below.

The Punnett square below represents the cross between two black guinea pigs.

Black Female
Guinea Pig

	B	*b*
B	*BB*	*Bb*
B	*BB*	*Bb*

Black Male
Guinea Pig

Key:
B = black fur gene (dominant)
b = white fur gene (recessive)

56.* Identify two offspring from the Punnett square that could produce white guinea pigs if they were crossed. Explain your answer. [2]

Offspring: __________ × __________

Explanation:

____

____

* Reproduced from IEA TIMSS 2003 Science Released Items for Grade 8, TIMSS International Study Center, Boston College, MA

57. What percentage of the offspring in the Punnett square will grow black fur? [1]

____

58. According to the Punnet square, what is the probability of an offspring inheriting two genes for black fur? [1]

(Express your answer as a fraction or a percentage.) ______________

59. Explain why both parents have black fur even though their genes for fur color are not exactly the same. [1]

____

____

____

60. A student on vacation has gone fishing from a dock on a lake several days. The student noticed that she caught more fish at certain times of the day than at others. She hypothesized that time of day had something to do with the number of fish she caught. The student wants to design an experiment to test this hypothesis.

a. Identify the independent (manipulated) variable in the experiment. [1]

____

b. Identify the dependent (responding) variable in the experiment. [1]

____

c. Identify two factors that should be held constant in the experiment. [1]

(1) ______________________________

(2) ______________________________

Base your answers to questions 61–64 on the diagram and information below.

A student put a spring that was pushed together a lot and clamped between two toy cars on a track. It looked similar to the diagram below.

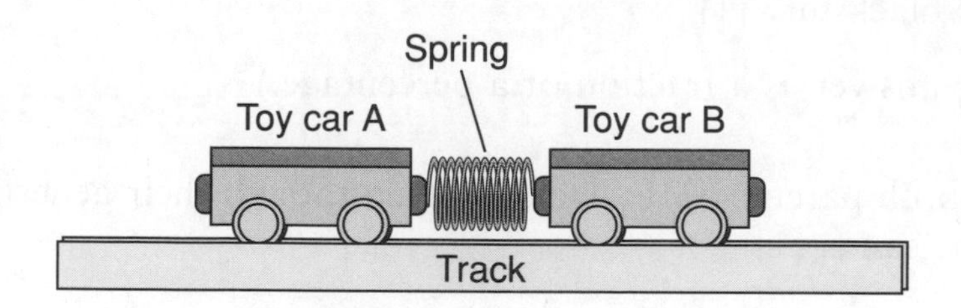

61. On the diagram above, draw and label arrows to show the motion of toy car *A* and toy car *B* when the spring is released. [1]

62. If the student wanted to measure the average speed of toy car A after the spring was released, why would it be important to mark the starting position of the car? [1]

____

____

____

____

63. How would adding a mass to toy car *A* affect the overall motion of the car? [1]

____

____

____

____

64. If the track were very long, what would eventually happen to the motion of toy car *A*? Explain why. [1]

Base your answers to questions 65–68 on the information and chart below.

A teacher asked a class to perform an experiment to compare the heating rate of soil with that of water. To do this, the students were given the following materials:

two heat lamps	two thermometers	one sample of water
two bins	one sample of soil	one timer

They were instructed to heat a sample of soil and a sample of water with heat lamps, measuring the temperature of each sample once a minute for 8 minutes.

65. There are many experimental variables that must be controlled in order to perform this experiment accurately. State three variables that should be controlled in this experiment. [1]

1. ______

2. ______

3. ______

66. The chart below shows the results of this experiment recorded by one group of students.

Time (min)	0	1	2	3	4	5	6	7	8
Soil temp. (°C)	20	21.5	22.5	24	26	28.5	30.5	32	33.5
Water temp. (°C)	20	21	22	22.5	24	24.5	25	25.5	26

Construct a line graph on the grid provided by following the steps below.

a. Use a ■ to plot the temperature of the soil, and a ▲ to plot the temperature of the water. [2]

b. Connect the ■ with a solid line. Then connect the ▲ with a solid line. [1]

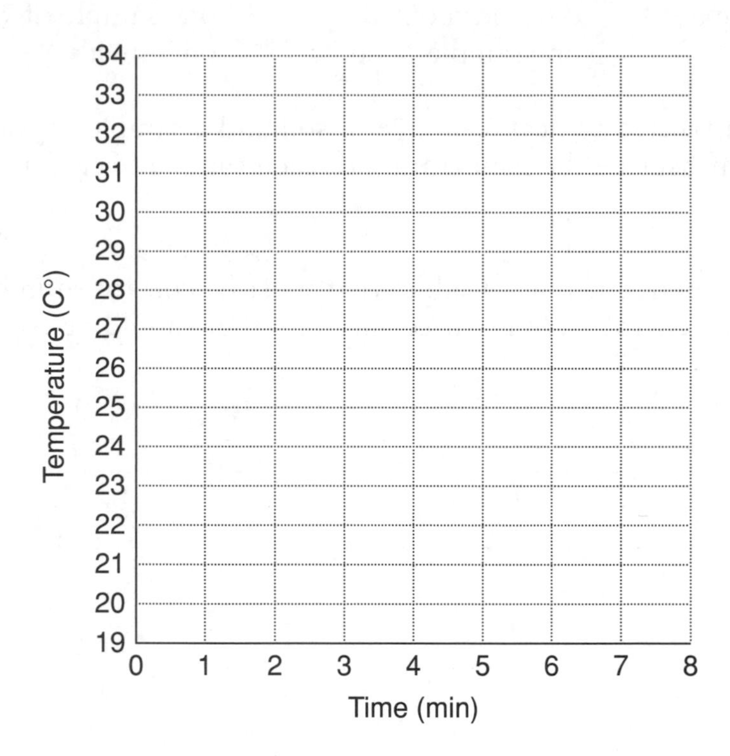

67. Calculate the rate of heating of the soil in the first 4 minutes. Use the equation below. [1]

$$\text{Rate of heating} = \frac{\text{difference in temperature of soil (°C)}}{\text{time of heating (min)}}$$

$$\_\_\_\_\_\_\_\_\_\_ = \frac{°C}{min}$$

68. According to the graph, what will the temperature of the water be after 10 minutes if the temperature increase continues at the same rate as in minutes 6 through 8? [1]

________________ °C

Base your answers to questions 69 and 70 on the map below, which shows air temperatures, in degrees Fahrenheit, recorded at the same time at weather stations across North America. *(The air temperature at location A has been deliberately left blank.)*

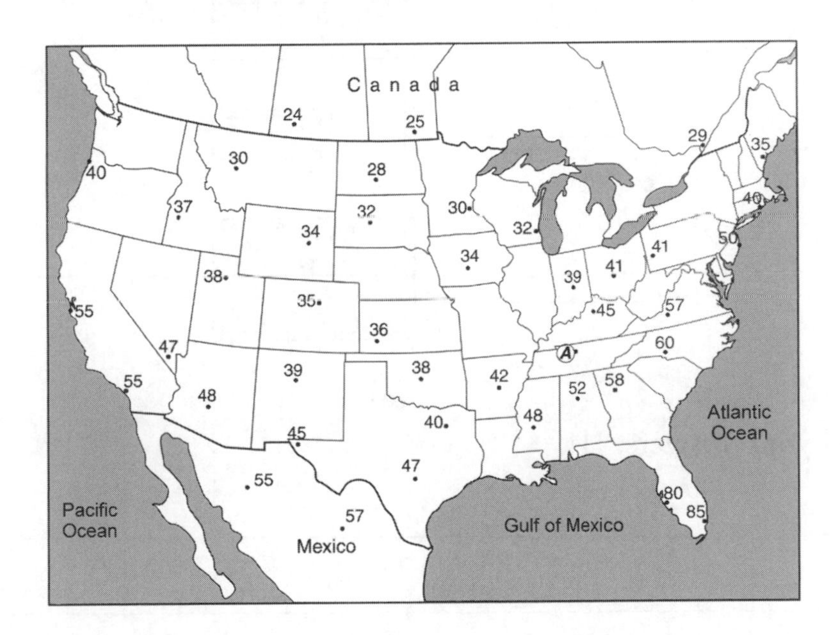

69. An isoline connects points of equal value. On the map above, use smooth, curved solid lines to draw the 30°F, 40°F, and 50°F isolines. [2]

70. What is the most probable air temperature at location *A*? [1]

________________ °F

71.* Figure A below shows a box that contains a material that could be a solid, a liquid, or a gas. The material is then put into a box four times as large, shown in Figure B.

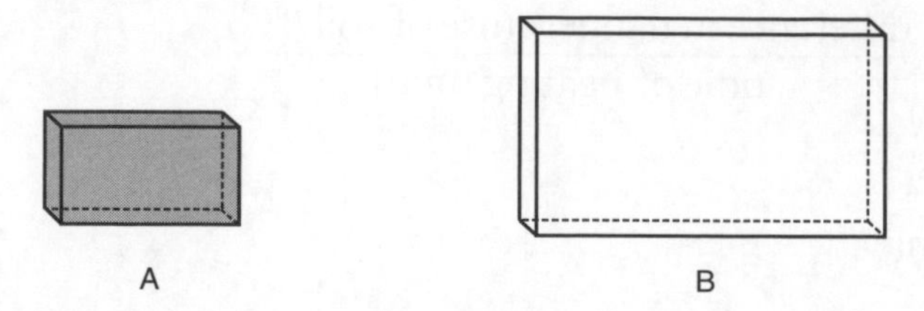

a. Look at the figures below. They show how the different types of material will look when put into the larger box. Identify which figure shows a solid, which shows a liquid, and which shows a gas. (Write the word *solid*, *liquid*, or *gas* on the line next to each figure below. Use each word only once.) [1]

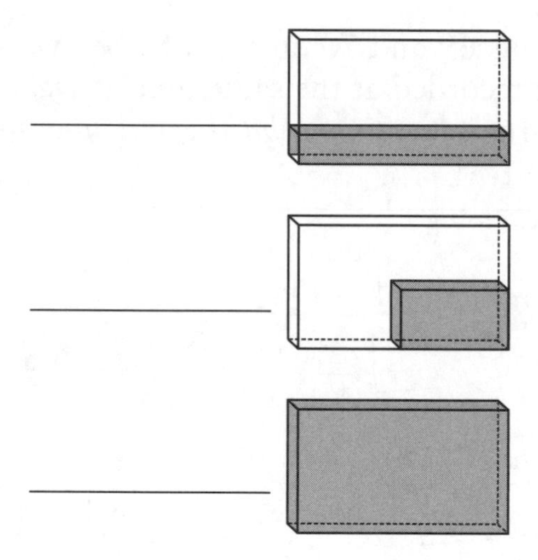

b. Explain your answers. [1]

____

____

____

____

____

* Reproduced from IEA TIMSS 2003 Science Released Items for Grade 8, TIMSS International Study Center, Boston College, MA

Base your answers to questions 72 and 73 on the diagram below, which shows the Sun's rays striking Earth at a position in its orbit around the Sun.

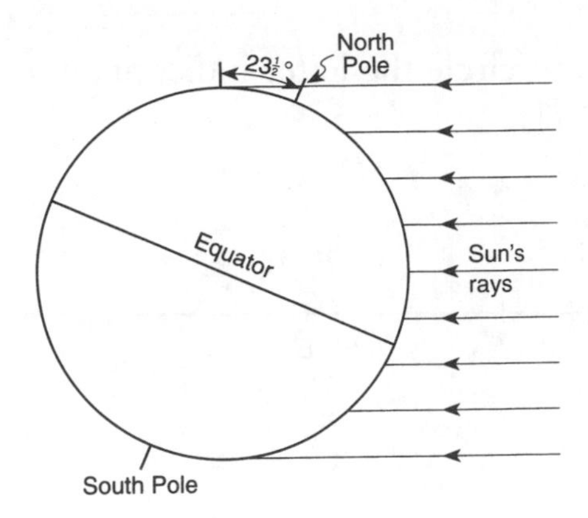

72. On the diagram, neatly and accurately shade the area of Earth that is in darkness. [1]

73. What season is beginning in the Northern Hemisphere? [1]

____

74. On the United States time zone map provided below, indicate the standard time in each time zone when it is 9 A.M. in the Central Time Zone. The dashed lines represent the standard-time meridians for each time zone. (Be sure to indicate the time for all three zones.) [1]

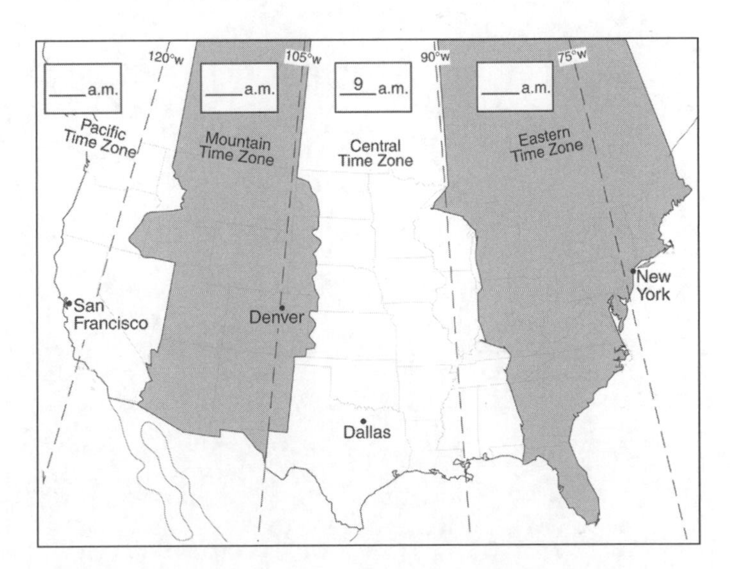

75.§§ A space station is to be located between Earth and the Moon at the place where Earth's gravitational pull is equal to the Moon's gravitational pull.

a. On the diagram below, circle the letter indicating the approximate location of the space station. [1]

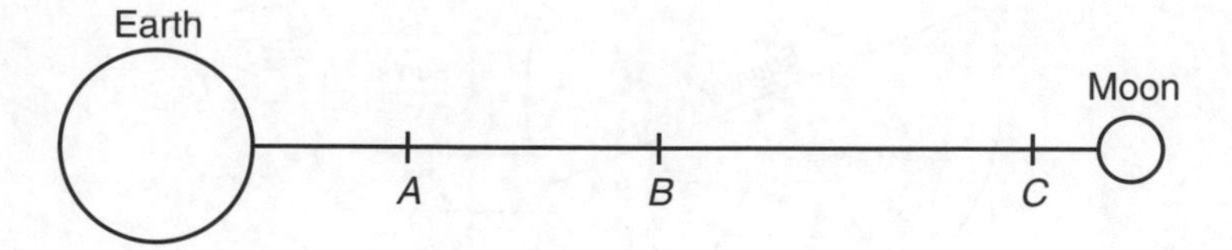

b. Explain your answer. [1]

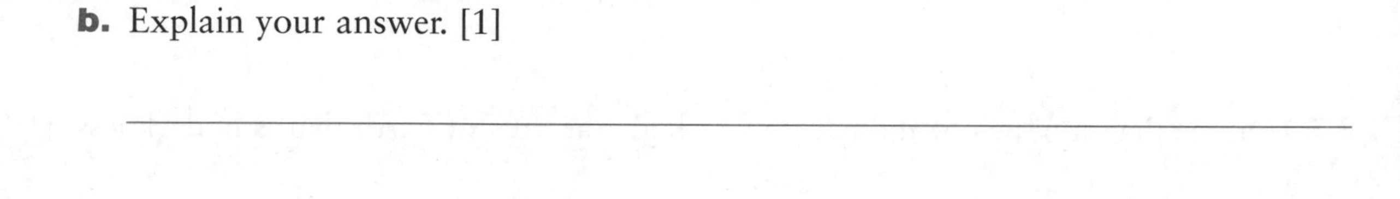

NEW YORK STATE INTERMEDIATE-LEVEL SCIENCE TEST SCORING KEY FOR MULTIPLE-CHOICE QUESTIONS

1. 3	**6.** 4	**11.** 4	**16.** 3	**21.** 2	**26.** 2	**31.** 1	**36.** 3	**41.** 4
2. 1	**7.** 3	**12.** 3	**17.** 1	**22.** 2	**27.** 1	**32.** 4	**37.** 3	**42.** 2
3. 2	**8.** 3	**13.** 2	**18.** 2	**23.** 2	**28.** 3	**33.** 1	**38.** 1	**43.** 3
4. 3	**9.** 2	**14.** 3	**19.** 1	**24.** 4	**29.** 2	**34.** 2	**39.** 2	**44.** 1
5. 2	**10.** 2	**15.** 2	**20.** 2	**25.** 4	**30.** 3	**35.** 2	**40.** 1	**45.** 1

ANSWERS EXPLAINED

Part I

1. 3 Traits are carried on chromosomes. All humans grow from a single-celled zygote with 46 chromosomes that forms when an egg cell from the mother containing 23 chromosomes joins with a sperm cell from the father containing 23 chromosomes. Thus, a son's chromosomes come from both mother and father, and his traits will be inherited from both his mother and his father.

2. 1 All the organisms in group 1 have legs, those in group 2 do not.

Wrong Choices Explained

2 Organisms in both groups have eyes; e.g., humans in group 1 have eyes, and snakes in group 2 have eyes.

3 Organisms in both groups have a nervous system.

4 All the organisms in both groups have skin except flies.

3. 2 By definition, fossils are the remains of living things and fossil fuels are fuels formed from the remains of living things.

4. 3 The acid in acid rain typically forms from gaseous nitrogen and sulfur oxides that combine with water in the atmosphere to form nitric and sulfuric acid. The main source of nitrogen and sulfur oxides is the burning of fossil fuels.

Wrong Choices Explained

1 Wastes from nuclear power plants consist of radioactive materials. These are stored and not released into the environment (except by accident) and do not form the acids in acid rain.

2 Spills from chemical plants would be in liquid form and probably seep into the ground, not enter the atmosphere and cause acid rain.

4 Aerosol spray cans do not contain the gaseous nitrogen and sulfur oxides that cause acid rain.

5. 2 The word *reacts* indicates that a chemical change is occurring. When chlorine gas (an element) reacts with sodium metal (an element), the compound sodium chloride (or salt) is formed.

6. 4 To test the idea that amount of sand in soil affects growth, the other variables need to be held constant (e.g., sunlight, water). Only choice 4 shows sunlight and water being held constant while the sand/soil ratio is changed.

Wrong Choices Explained

1 Placing the plant in a dark cupboard while keeping the other variables the same would test the effect of sunlight on plant growth, not the need for sand in the soil.

2 Placing the plant in a dark cupboard and removing sand from the soil would change two variables; therefore, you would not know if your experimental result was due to the change in light or the change in amount of sand.

3 Taking away water while keeping the other variables the same would test the effect of water on growth, not the need for sand in the soil.

7. 3 In a mixture, the components keep their individual properties. Because there are black and white specks, there are two different substances, each with its own properties. Thus, the powder is a mixture.

Wrong Choices Explained

1 A solution is a mixture that is a uniform blend of two substances, having the same composition and properties throughout. Most solutions are liquids. Because the specks in the powder have different properties and the powder is a solid, the powder is not a solution.

2 Although a compound is a form of matter in which two elements are joined together in a fixed ratio, the compound has its own properties. If the compound is pure, those properties will be uniform throughout the sample. Since the specks in the powder have different properties, the powder is not a pure compound.

4 An element is a single substance and has a single set of properties. Because the specks in the powder have more than one color, the powder is not an element.

8. 3 Gills allow oxygen dissolved in water to pass into a fish's blood and carbon dioxide in the fish's blood to pass out into the water. Thus, the function of gills is gas exchange. In humans, this function is carried out in the lungs.

9. 2 Most water plants carry out photosynthesis. The presence of the light is a further clue that photosynthesis is involved in the production of gas *X*. Because oxygen is a product of photosynthesis, gas *X* is most likely oxygen.

10. 2 By definition, fertilization is the joining of an egg cell and a sperm cell.

11. 4 In humans, digestion of food begins in the mouth and continues in the stomach and small intestines. But, it is in the villi of the small intestines that food molecules pass from the digestive system into the bloodstream.

12. 3 Elements are matter, and matter cannot be created or destroyed by normal physical or chemical changes. Thus, when an organism dies, the molecules of which it is composed are broken down by decomposers and recycled into the environment.

13. 2 The central principle of cell theory is that all living things are composed of cells. Only photosynthetic organisms make their own food and produce oxygen, and only some, not all, living things cause human diseases.

14. 3 The process described in the questions maintains body temperature at a constant level. This process is called homeostasis.

15. 2 By convention, the scientific name of an organism consists of its genus and species.

16. 3 The lenses of a microscope cause the image seen to be upside down and reversed, i.e., up and down are reversed, and left and right are reversed.

17. 1 The pictures show an amoeba engulfing another organism, a process called phagocytosis. In simple terms, the amoeba is eating the organism, or obtaining food.

Wrong Choices Explained

2 Getting rid of wastes involves substances inside the an amoeba passing out of the amoeba. The picture shows an object being taken into the amoeba.

3 Reproduction would result in formation of a second organism. The pictures show only one amoeba.

4 Inheritance involves the passing on of traits via reproduction, not the taking in of materials from outside the amoeba.

18. 2 The graph shows an increase in oxygen over time; thus oxygen is being produced. Because oxygen is a product of photosynthesis, the organism in the test tube is probably carrying out photosynthesis. Of the choices, only the green plant carries out photosynthesis.

19. 1 The diagram shows approximately two cells spanning the 1-mm field of view. One millimeter divided by two cells equals ½ mm per cell.

20. 2 In order for snow to occur, air temperatures must be at or below the freezing point of water (0°C). Although both town *B* and town *D* had temperatures below freezing, only town *B* recorded precipitation. Therefore, the snow occurred in town *B*.

Wrong Choices Explained

1 In order for snow to form, temperatures must drop below the freezing point of water, i.e., below 0°C. The temperatures in town *A* remained above 0°C; therefore, it could not have snowed.

3 In order for snow to form, temperatures must drop below the freezing point of water, i.e., below 0°C. The temperatures in town C remained above 0°C; therefore, it could not have snowed.

4 Although town *D* had temperatures below freezing, town *D* recorded no precipitation. Therefore, it did not snow in town *D*.

21. 2 Butterflies undergo complete metamorphosis, developing from egg to larva (caterpillar) to pupa to adult. Eggs are shown in 1, larva in 3, pupa in 2, and adult in 4.

22. 2 Reducing sperm production would interfere with insect reproduction. The fewer insects that reproduce, the fewer insects in the total population.

Wrong Choices Explained

1 If sperm production is prevented, no reproduction would take place and no offspring, male or female, would be produced. Therefore, the number of females would decrease, not increase.

3 Production of a new species would involve changing the genetic material of the insects. Preventing sperm production does not change the genetic material; it prevents sperm cells that contain it from forming.

4 Presence or absence of sperm does not prevent the act of mating; it only prevents the production of a fertilized egg as a result of the act.

23. 2 Cats are mammals. Therefore, cats are most closely related to other mammals. Because whales are also mammals, cats are most closely related to whales.

Wrong Choices Explained

1 Crocodiles are reptiles.

3 Frogs are amphibians.

4 Penguins are birds.

24. 4 Note that in photosynthesis, energy is a reactant, whereas in respiration, energy is a product. Therefore, photosynthesis uses energy and respiration produces energy.

25. 4 During the process of mitosis, chromatin first condenses into chromosomes (*B*), then the chromosomes reproduce and line up in pairs at the center of the cell (*A*), then the chromosomes separate, with one of each pair moving into each half of the cell (*D*), and finally a new cell membrane forms between the chromosomes (*C*).

26. 2 Two structures found in plant cells but not in animal cells are chloroplasts and a cell wall. In the diagram, *B* points to a chloroplast and *C* points to the cell wall.

Wrong Choices Explained

1 In order to be correct, both letters must point to structures that are found only in plant cells. Letter *A* points to the cell nucleus and letter *B* points to a chloroplast. Because both plant and animal cells have a nucleus, observing a nucleus would not enable one to identify a cell as a plant cell.

3 Letter *A* points to the cell nucleus and letter *D* points to cytoplasm. Both of these are found in both plant and animal cells.

4 Letter *D* points to cytoplasm and letter *B* points to a chloroplast. Cytoplasm is found in both plant and animal cells.

27. 1 The cardiovascular system consists of blood and the organs that circulate blood, i.e., the heart, the blood vessels, and the blood.

Wrong Choices Explained

2 The atria, ventricles, and valves are parts of the heart, which is only one organ in the cardiovascular system.

3 Lungs are part of the respiratory system, body cells are present in most organ systems, and digestive organs are part of the digestive system, not the cardiovascular system.

4 Plasma, red blood cells, and white blood cells are part of the blood, which is only one part of the cardiovascular system.

28. 3 Refer to the picture of the human digestive system on page 67 and note the path food follows: esophagus, stomach, small intestine, large intestine.

29. 2 Weathering and erosion break down and carry away rocks exposed on mountains. Over time, the mountain becomes more rounded and smaller. Thus, the mountains in picture *B* have been exposed to weathering and erosion for a longer period of time than those in picture *A*; i.e., the mountains in picture *B* are older.

Wrong Choices Explained

1 Over time, a mountain becomes smaller as rock material is broken down and carried away by weathering and erosion. Because the mountain in picture *A* is a larger mountain than the one in picture *B*, it has been exposed to weathering and erosion for less time. Therefore, the mountains in picture *A* are younger, not older.

3 There is no indication of structure in the diagrams beyond height and shape. Therefore, no conclusions can be drawn about how they were formed or their age.

4 Being located in a different hemisphere (e.g., eastern vs. western) would not necessarily affect weathering and erosion. Therefore, it could not explain the differences in the appearance of the mountains in the two pictures, nor their age.

30. 3 Only the illuminated portion of the Moon is visible from Earth. As the Moon orbits Earth, different portions of the half that is illuminated by sunlight are visible from Earth.

31. 1 Average yearly temperatures are closely related to the intensity of the sunlight received at a location. Latitude affects the angle at which sunlight strikes Earth's surface, and therefore its intensity. Sunlight is most intense near the Equator and becomes less intense as you approach the poles. Therefore, a location is likely to have an average yearly temperature similar to that of location *A* if it is located a similar distance from the Equator. Location *X* has a latitude of 30°N; location *A* with a latitude of 30°S is a similar distance from the Equator and will therefore have a similar average yearly temperature.

32. 4 To test the effect of ramp height on speed, wheel size and mass should be held constant. The three trials in which wheel size and mass are constant but ramp height varies are *H*, *V*, and *Y*.

Wrong Choices Explained

1 Since *G*, *H*, and *I* have the same ramp height, no conclusions could be drawn about the effect of changing ramp height.

2 Although *I*, *W*, and *Z* have different ramp heights, trial *W* also has different wheels. Therefore, you would not know whether any difference in speed was due to the ramp height or the wheels.

3 Although the ramp height in *X* is different from that in *U* and *W*, the masses carried in the carts in these three trials are all different. Therefore, you would not know whether any difference in speed was due to the ramp height or the mass.

33. 1 Earthquakes tend to occur where forces acting on the crust cause it to break and move. Volcanoes tend to occur where Earth's crust is broken and magma is able to rise to the surface through the breaks. Both of these conditions exist along tectonic plate boundaries.

Wrong Choices Explained

2 Boundaries between deep and shallow water also occur in places where there is little or no earthquake or volcanic activity.

3 Earthquakes and volcanoes are not caused by water meeting water.

4 Because many volcanoes and earthquakes occur on land, they are not related to high ocean temperatures.

34. 2 Although Mercury is close to the Sun, it has almost no atmosphere, and any heat absorbed by its surface is lost again as it is radiated back into space. The chart shows that Venus has an atmosphere composed mainly of carbon dioxide. Carbon dioxide is a greenhouse gas that allows shortwave light radiation to penetrate to Venus' surface but blocks longwave heat radiation from escaping, resulting in a higher surface temperature.

35. 2 A galaxy is a group composed of billions of stars. A solar system consists of a single star orbited by planets. The planets in a solar system are much smaller than the star. Thus, the correct order from smallest to largest is planet, star, solar system, galaxy.

36. 3 As a river moves from a mountainside to a plain, the slope of the land decreases. The less the land slopes, the slower the water flows. Because the slower-moving water does not erode the land as quickly, the channel is less deep and the water tends to spread out over the surrounding land, making the river wider. Thus, compared to a fast-moving river on a mountain slope, a river on a plain will be slower and wider.

37. 3 Liquids expand when heated. When a thermometer is placed in hot water, heat is transferred through the glass by conduction and heats the liquid inside, causing it to expand. Because the liquid is constrained by the surrounding glass, it rises upward inside the tube of the thermometer.

38. 1 According to the diagrams of the balances, box *Z* has a mass greater than 10 grams, box *Y* has a mass equal to 10 grams, and box *X* has a mass less than 10 grams. Therefore, box X has the least mass.

39. 2 The angle of incidence is equal to the angle of reflection. Therefore, the angle between the dotted line and the incoming ray should be the same as the angle between the dotted line and the reflected ray as shown in choice 2.

40. 1 Less dense materials float in more dense materials. The rising (floating) of the helium-filled balloon in the surrounding air indicates that it is less dense than the air.

Wrong Choices Explained

2 Air resistance is a force that opposes motion, and would act on the balloon regardless of its direction of motion.
3 Gravity acts on all objects that have mass. Because the helium balloon has mass, gravity will act on it.
4 Wind is a horizontal movement of air, not a vertical movement. Therefore, wind would not move the balloon upward, but sideways.

41. 4 A wet towel left in the Sun becomes dry because the liquid water in the towel changes into water vapor. The change from the liquid phase to the gaseous phase is called evaporation.

42. 2 Concentration is the ratio of solute (salt) to solvent (water). To make a solution half as concentrated, you must double the amount of solvent, i.e., add 100 ml of water. ($10 \div 100 = 10\%$; $10 \div 200 = 5\%$)

43. 3 According to the chart, there was a last-quarter phase on August 13 and a new moon phase on August 20. Thus, on August 17, the Moon would be in the phase between last quarter and new moon, or the old crescent phase.

44. 1 In the Northern Hemisphere, the summer season occurs when Earth's North Pole is tilted farthest toward the Sun. This situation occurs in position *A*.

45. 1 During a lunar eclipse, the Moon moves into Earth's shadow. Earth's shadow is cast on the side of Earth facing away from the Sun. Therefore, during a lunar eclipse the Moon is on the side of Earth directly opposite the Sun as shown in choice 1.

Part II

46. Consumers eat producers to obtain the energy in the food molecules they contain. Producers make food molecules out of molecules of carbon dioxide and water using the energy in sunlight. Thus, the food from which both producers and consumers obtain energy can be traced back to the Sun.

47. A primary consumer is an organism that obtains its energy by eating producers (plants). Acceptable responses include: insect *or* small fish.

48. Secondary consumers obtain their energy by eating primary consumers. Acceptable responses include: heron *or* large fish *or* frog.

49. Chlorophyll is a green pigment contained in plants cells. Acceptable responses include: green algae *or* rushes *or* water lily.

50. *a* loose sugar

b In order to dissolve, sugar has to come in contact with water. In general, the more surface area exposed to water, the faster the rate at which the sugar dissolves. In sugar cubes, more of the sugar is on the inside of the cube where it does not come in contact with the water. The smaller particles of loose sugar have more surface area per unit of mass that comes into contact with the water. Thus, the loose sugar dissolves faster than the sugar cubes because the loose crystals have more surface area than the sugar cubes.

51. *Z W Y X*

52.

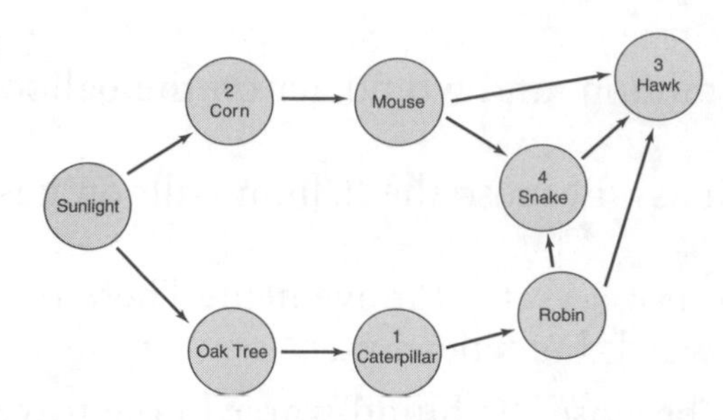

53. Carnivore: hawk *or* snake *or* robin
Herbivore: mouse *or* caterpillar
Producer: corn *or* oak tree

54. The robin population will decrease. An acceptable explanation is based on predators (snakes, hawks) eating more robins if mice die because the corn crop failed.

55. A correct explanation will state that the balloon expands because of increased *pressure* of air/gas/water vapor when the tube is heated. A response that states that water evaporates, increasing pressure inside the balloon, is also considered correct. Simply stating that air rises, or that it expands when heated, is not enough of an explanation because in order for the balloon to expand, the air must exert enough pressure on it to cause it to stretch.

56. Offspring: *Bb* × *Bb*
Explanation: Since white fur is a recessive gene, two *b* genes must be present in a guinea pig for it to have white fur (*bb*). The only offspring that can contribute a *b* gene are the two *Bb* guinea pigs.

57. 100%. The gene for black fur (*B*) is dominant and will mask any recessive white gene present in a guinea pig. Because all the offspring in the Punnett square have a dominant black gene, all will grow black fur.

58. According to the Punnett square, two out of four offspring will have two genes for black fur (*BB*). Thus, ½ or 50% of the offspring will have two genes for black fur.

59. Each guinea pig has *two* genes for fur color, one from its mother and one from its father. According to the key, there are two types of fur color genes: a dominant gene for black fur (*B*) and a recessive gene for white fur (*b*). Dominant genes block recessive genes from showing up in an organism. Thus, a guinea pig with even one dominant black gene will have black fur because the recessive white gene is blocked. Because the male has two dominant black genes, it has black fur. Although the female carries one recessive white gene, it is masked by the other dominant black gene and the female is also black.

60. *a* independent variable: time of day
b number of fish caught
c correct responses include, but are not limited to: fishing location, type of bait (or lure), type of fishing line, type of tackle (float, no float; sinker; and so on).

61.

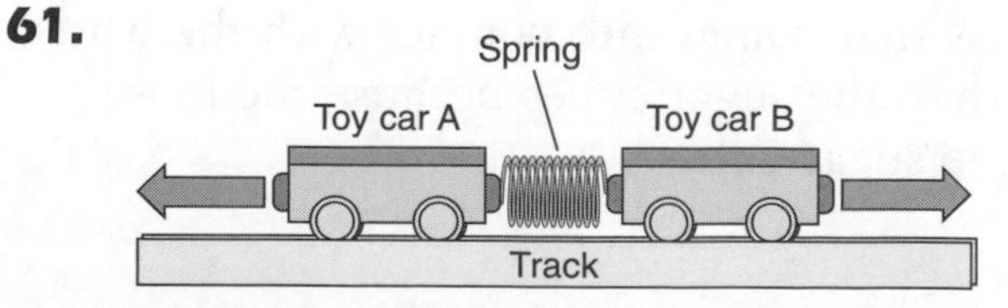

62. Speed = distance ÷ time. In order to measure the distance traveled by toy car *A* in a given time, you need two points—a starting position and a position a given time later.

63. If mass is added to toy car *A*, it will not move as far or as fast. The force exerted by the spring would be the same, but the mass would be greater, so the acceleration of the toy car would be less.

64. Toy car *A* would eventually stop. Once the spring expanded to its normal length, it would no longer exert a force on the toy car. Then, friction with the track would act against the motion of the car unopposed, causing the car to decelerate until it came to rest.

65. Experimental variables to be controlled include, but are not limited to: volume of soil, volume of water, wattage of heat lamp, distance of heat lamp from sample of soil and water, placement of thermometer, container for soil and water. (Any three would be acceptable.)

66.

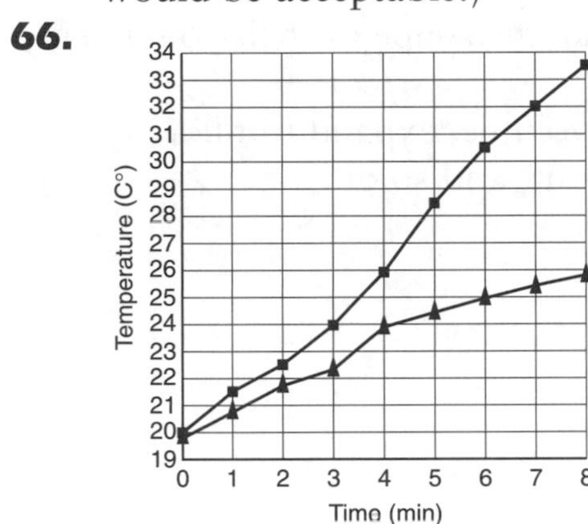

67. $$\text{Rate of heating} = \frac{\text{difference in temperature of soil (°C)}}{\text{time of heating (min)}}$$

$$\text{Rate of heating} = \frac{6°\text{C}}{4\text{ min}} = 1.5\,\frac{°\text{C}}{\text{min}}$$

68. 36.5°C

69.

Canada
PACIFIC OCEAN
Mexico
GULF OF MEXICO
ATLANTIC OCEAN
A

70. 46°F (±2)

71. *a*

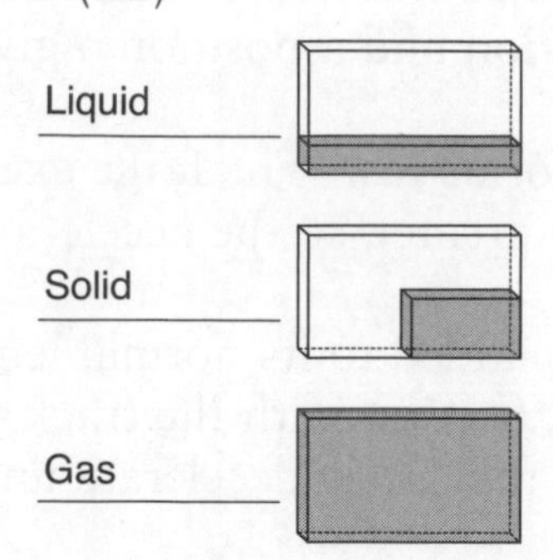

b Your explanation should refer to at least one property of each of the three phases that can be used to tell them apart:
Liquids: flow, or take the shape of their container, seek the lowest level, have a definite or fixed volume, and so on.
Solids: have a definite or fixed shape, do not take the shape of their container, are hard objects, and so on.
Gases: do not have a definite or fixed shape or volume, expand to fill a container of different size or shape, can spread out, and so on.

72.

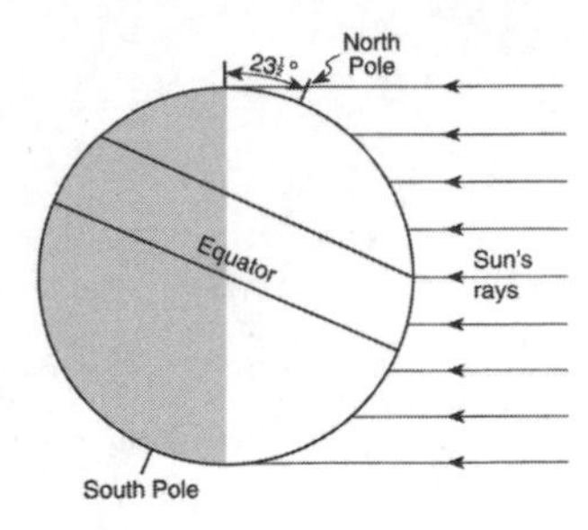

73. Summer

74. Eastern Time Zone: 10 A.M.; Mountain Time Zone: 8 A.M.; Pacific Time Zone: 7 A.M.

75. *a* Circle point *C*.

b Your explanation should state that gravitational pull depends on mass and distance; thus, the station must be closer to the Moon because the Moon's mass is less than that of Earth.